Series: Large Language Models for Critical Applications

Series Editor: Amit Kumar Tyagi

Volume

LLMs

Introduction, Background, Applications, Challenges, Limitations and Future Scope

Books in the Series

- LLMs: Introduction, Background, Applications, Challenges, Limitations and Future Scope
  Amit Kumar Tyagi (ed.), 2027

Series: Large Language Models for Critical Applications

Series Editor: Amit Kumar Tyagi

Volume

LLMs

Introduction, Background, Applications, Challenges, Limitations and Future Scope

Editor

Amit Kumar Tyagi

National Forensic Sciences University

Gandhinagar, Gujarat, India

CRC Press

Taylor & Francis Group

Boca Raton London New York

CRC Press is an imprint of the

Taylor & Francis Group, an **informa** business

A SCIENCE PUBLISHERS BOOK

First edition published 2027
by CRC Press
2385 NW Executive Center Drive, Suite 320, Boca Raton FL 33431

and by CRC Press
4 Park Square, Milton Park, Abingdon, Oxon, OX14 4RN

CRC Press is an imprint of Taylor & Francis Group, LLC

Library of Congress Cataloging-in-Publication Data (applied for)

ISBN: 978-1-041-29850-2 (hbk)
ISBN: 978-1-041-29855-7 (pbk)
ISBN: 978-1-003-77165-4 (ebk)

DOI: 10.1201/9781003771654

Typeset in Times New Roman
by Prime Publishing Services

Preface to the Series

Today the rapid advancement of Large Language Models (LLMs) is transforming the landscape of artificial intelligence, redefining how machines understand, generate, and interact with human language. From applications like agriculture, healthcare, transportation, finance to law, cybersecurity, and public governance, today LLMs are increasingly being used in various domains where accuracy, reliability, and ethical responsibility are essential.

This book series titled "*Large Language Models for Critical Applications*" aims to discuss the opportunities, challenges, and implications of integrating LLMs in critical LLMs are increasingly highlighting the need for advancements and innovations that ensure the responsible use of critical applications in today's smart, AI-driven era.

This book series titled "*Large Language Models for Critical Applications*" explores the opportunities, challenges, and implications of deploying LLMs in critical environments. As LLMs become more widespread, there is a growing need for innovations that ensure their safe and responsible use in today's AI-driven world. While LLMs demonstrate remarkable capabilities in natural language processing and generation, their application in critical domains requires careful consideration. Various issues such as model hallucination, bias, explainability, data privacy, and security vulnerabilities pose significant risks if not properly addressed. So, this series emphasizes not only on the technical foundations of LLMs but also on the frameworks, methodologies, and safeguards necessary to ensure trustworthy and responsible deployment.

Each volume in this series brings together contributions from researchers, practitioners, and domain experts, offering interdisciplinary views on designing, evaluating, and implementing LLM-driven systems. Topics covered include robust model architectures, domain adaptation, human-in-the-loop systems, regulatory compliance, and ethical AI practices. This book series also includes real-world case studies and practical insights to bridge the gap between theory and application.

Hence, the objective of this series is to provide a comprehensive resource for academicians, industry professionals, policymakers, and students seeking to understand the potential of LLMs and their convergence with critical application domains.

We hope this series will serve as a valuable platform for advancing research and guiding the responsible use of LLMs in shaping the future of critical applications.

Preface

Large Language Models (LLMs) have emerged as a cornerstone of artificial intelligence, revolutionizing how machines process, generate, and understand human language at scale. Introduced in the early 2010s, their foundational architecture draws from the Transformer model proposed by Vaswani et al. in 2017, which enabled parallel processing of sequential data through self-attention mechanisms. This breakthrough addressed limitations of recurrent neural networks (RNNs) and long short-term memory (LSTM) units, allowing LLMs to handle vast contexts—up to millions of tokens in modern iterations like GPT-4o and beyond. Trained on internet-scale corpora via unsupervised pretraining on next-token prediction, followed by supervised fine-tuning and reinforcement learning from human feedback (RLHF), LLMs exhibit emergent abilities such as few-shot learning, reasoning, and creativity, mimicking human cognition without explicit programming.

The background of LLMs traces a trajectory from early prototypes like GPT-1 (2018) to behemoths like PaLM 2 and Llama 3, fueled by exponential compute scaling under laws akin to Moore's. These models, often parameterized in trillions, leverage distributed training across GPU clusters, incorporating techniques like mixture-of-experts (MoE) for efficiency. Their rise coincides with democratization via open-source releases, enabling widespread adoption despite initial proprietary dominance by labs like OpenAI and Google DeepMind.

Several applications span in several domain like: in natural language processing, LLMs power chatbots like ChatGPT for customer service, code generation via tools like GitHub Copilot, and summarization for legal or medical documents. Multimodal extensions handle vision-language tasks, such as image captioning or drug discovery through protein sequence analysis. In education, they personalize tutoring; in creative industries, they assist writing and art generation. Enterprises deploy them for sentiment analysis, translation, and automated reporting, slashing labor costs while enhancing accuracy.

Also, some challenges faced with this like: Hallucinations—fabricating plausible but false information—undermine reliability, particularly in high-stakes fields like healthcare or finance. Bias amplification from skewed training data perpetuates stereotypes, demanding rigorous debiasing. Privacy risks arise from memorization of personal data, clashing with regulations like GDPR. Compute intensity poses environmental concerns, with training emitting CO_2 equivalent to hundreds of flights, alongside escalating costs prohibitive for smaller entities.

Further some limitations include poor long-term reasoning (e.g., failing complex math without chain-of-thought prompting), and context window constraints, though expanding. They lack true understanding, relying on pattern matching rather than grounded world models, leading to inconsistencies.

The future scope comes with promise: agentic architectures will integrate LLMs with tools for autonomous task execution, as in multi-agent systems for research or planning. Smaller, efficient models via distillation and quantization will edge toward on-device deployment. Advances in multimodal fusion promise unified AI for robotics and AR/VR. Ethical alignment through scalable oversight and constitutional AI will mitigate risks, while federated learning enhances privacy. By 2030, LLMs could underpin general intelligence, transforming society through personalized medicine, climate modeling, and global problem-solving, provided governance evolves apace.

Lastly, we hope that this book will be loved by research community and any corrections are heartly welcome for improving the quality of this book.

Happy learning.

Acknowledgement

First of all, we would to extend our gratitude to our Family Members, Friends, and Supervisors, which stood with us as an advisor in completing this book. Also, we would like to thanks our almighty "God" who makes us to write this book. We also thank Science Publishers, CRC Press (who has provided their continuous support during this important/their valuable time) and our colleagues with whom we have work together inside the college/university and others outside of the college/university who have provided their continuous support towards completing this book on areas like Large Language Models, Gen AI, etc.

Also, we would like to thank our Respected Madam, Prof. G Aghila, Prof. Siva Sathya, our Respected Sir Prof. Ajith Abraham, Prof. Aswani Kumar Cherukuri, Prof. (Dr.) Naveen Chaudhary, Prof. (Dr.) Purvi Pokhariyal for giving their valuable inputs and helping us in completing this book.

Once again, thanks to all.

Contents

1

Large Language Models (LLMs)

Foundation and Architectures

Sanjana Sundar,[1] *S. Vinila Jinny,*[1,*] *Jabanjalin Hilda*[1] and *Shekaina Justin*[2]

1. Introduction

The introduction of LLMs have revolutionized the field of NLP, making machines to understand, generate and reason with human language more accurately and fluently than earlier. Initial system depends on fixed simple rules and statistical approaches, which restricted their flexibility. After more research, it paved path in using neural networks which could easily capture meaning and the context. This progress from statistical approaches to word embeddings, recurrent networks and transformer-based models made LLMs to solve core problems such as handling high-dimensional data, remembering long-term information and processing text perfectly. The evolvement of Transformer in the year 2017 was an important turning point which removed the need of recurrence and made large scale parallel training easier through self-attention mechanism (Vaswani et al., 2017).

Now LLMs are not only greater but also more specialized. Various types of architectures are introduced, which includes encoder-only, decoder-only and encoder-decoder model. Each of this model is designed for specific tasks like text classification, conversation or translation. Further models are made more efficient

[1] Vellore Institute of Technology, Vellore, India.
Emails: sanjana.sundar2024@vitstudent.ac.in, https://orcid.org/0009-0007-4615-1997; jabanjalin.hilda@vit.ac.in, https://orcid.org/0000-0002-5593-5976
[2] Princess Nourah Bint Abdulrahman University, Kingdom of Saudi Arabia.
Email: SJThami@pnu.edu.sa, https://orcid.org/0009-0009-4966-7096
* Corresponding author: vinilajinny.s@vit.ac.in, https://orcid.org/0000-0003-2756-6736

DOI: 10.1201/9781003771654-1

by the introduction of new methods like MoE and SSMs, which helps in managing high computation costs (Mu et al., 2025). This chapter concentrates on how LLMs are built and how they perform, explains its history, important turning points and the recent trends. Also examining their design and real-world use cases, its objective is to explain the working of the model, need of different types of models and where the model is excelling.

2. The Path to Transformer-Based Models: An Evolutionary Perspective

The introduction of LLMs is an important history of persistence and creativity. Today models like OpenAI's GPT series, Google's Gemini and Anthropic's Cluade are excelling like a new magic of intelligence, able to talk, create and reason that was once felt impossible. This evolvement doesn't happen overnight. Multiple years of research and many attempts in solving the impossible challenges in computing resulted in transformer architecture which can understand and use human Language (BL, 2025).

This innovation was not from randomised research work, it was a logical progression in giving solution to a problem. Each novel architecture was introduced by overcoming the shortcomings of the previous, like a series of sophisticated tools that was being forged to tackle complex tasks. To truly understand why today's LLMs work the way they do, we must trace this lineage of ideas. We must see how the community grappled with the ghost of infinite possibilities in language, how they taught a machine to have a memory, and how they finally broke the shackles of processing the world one word at a time (Pióro et al., 2024).

2.1 Neural Representations

The first attempts, back in the pioneering days of the 1950s and 60s, were exercises in pure logic, dominated by symbolic, rule-based systems. Researchers tried to hand-craft intelligence, building systems based on rigid rules.

The Georgetown-IBM experiment of 1954 was a bold attempt to translate Russian into English by programming a computer with a dictionary and a set of grammatical laws (Weizenbaum, 1966). A decade later, ELIZA, one of the first

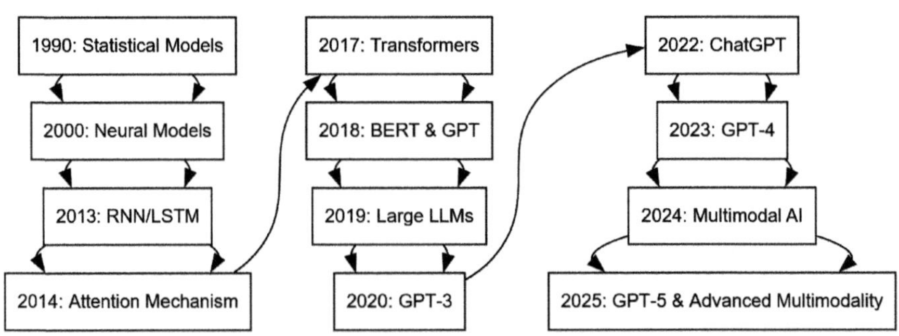

Figure 1: Evolution of NLP models to LLMs.

chatbots, mimicked a psychotherapist by using clever tricks to rephrase a user's statements back at them as questions. These systems were fascinating, but they were brittle. They were like automatons that could perform a perfect waltz, but would fall to pieces if you asked them to dance to a different rhythm (McCarthy et al., 1955).

The first novel idea came out when researchers started to think solution from other end. Instead of instructing the computer with rules in understanding text, researchers made computer to learn from the text. This paved path for introducing Statistical Language Models (SLMs) in the years 1980 and 1990s. These models learned the probabilities of words appearing together. One of the famous models was n-gram model. The working model is that it looks at the last few words (For example, previous 2 words in a trigram) and tries to guess the next word. This approach was simple but very useful at the time, especially for speech recognition and translation (Jelinek, 1997; Manning and Schütze, 1999).

But the n-gram model had a big weakness, called the curse of dimensionality. Human language has almost endless combinations of words. For example, if a dictionary has 100,000 words, then the number of possible 10-word sentences is unimaginably huge. Since a model can only be trained on a small portion of all possible sentences, most new sentences it sees are ones it never encountered before. The model would then assign them a probability of zero, which made it unreliable. This was expressed as a failure to generalize. Furthermore, in n-gram models, words were treated as isolated symbols; the model had no understanding that words with similar meanings, such as "car" and "vehicle," or "happy" and "joyful," were related. For example, the sentence "The car is parked outside" provided no insight for predicting "A vehicle is waiting in the driveway." The turning point was a new idea that reoriented the research for the next two decades. Yoshua Bengio and collaborators, in 2003, in "A Neural Probabilistic Language Model," came up with a new approach to word representation. The model achieved two firsts:

2.1.1 Distributed Word Representation

Instead of assigning a unique identifier for each word, the system found a method to represent words as high-dimensional, rich vectors, known as embeddings (quadeer15sh, 2023). They acted as coordinates, situating words within a high-dimensional map of meaning.

$$P(w_1,w_2,w_3,\ldots\ldots..w_t) = \Pi(t = 1 \text{ to } T)\ P(w_t\,|w_1,w_2,w_3,\ldots\ldots..w_{t-1}) \quad (1)$$

$$P\ (S) = P\ (How) \times P\ (are \mid How) \times P\ (you \mid How, are) \times P\ (? \mid How, are, you)$$

2.1.2 Simultaneous Learning

The system learned both the structure of a meaning-map and sentence probabilities simultaneously within a single neural network. The model itself was a simple feed-forward neural network. It took the coordinates of the last few (n−1) words, concatenated them, passed them through its hidden layers, and used a final softmax layer to produce a probability distribution over the entire vocabulary for predicting the next word. The breakthrough was its ability to generalize. Because similar words

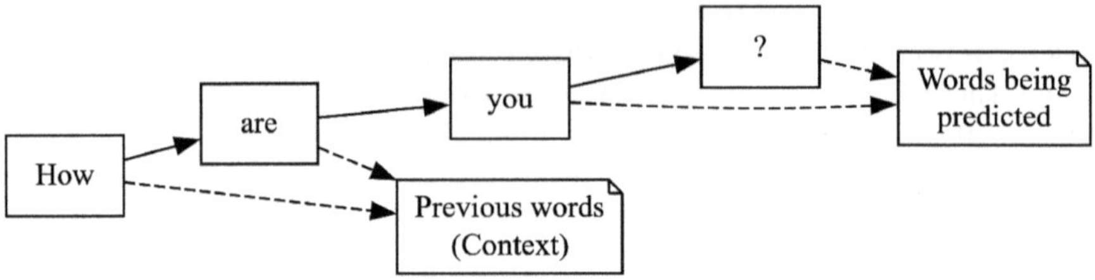

Figure 2: Generation using an N-gram language model.

such as "car" and "vehicle" or "happy" and "joyful" tended to appear in similar contexts, the model learned to place their embedding vectors close together in the vector space. As the neural network is a smooth function of these embeddings, a small change in the input vectors for instance, replacing "car" with "vehicle" would result in only a small change in the output probabilities. Consequently, a training sentence now informed the model about an exponential number of semantically related sentences, effectively mitigating the curse of dimensionality. Even though Bengio's model was a big step forward, it still had one major drawback it worked like an n-gram model with a fixed window of words.

This meant it could only look at a limited number of words at a time. If an important word was outside that window, the model couldn't connect it. To solve this problem, researchers needed new types of neural networks that could handle long-range word relationships.

2.2 *Modeling Sequences: The Rise and Limitations of Recurrent Architectures*

The fixed-size window was too limiting because language often depends on connections that go far back in a text. To solve this, researchers created the Recurrent Neural Network (RNN). Unlike a regular feed-forward network (which only looks at one word at a time without remembering the past), an RNN has a loop that gives it a kind of memory. As it processes each word in a sentence, it updates a hidden state and a summary of everything it has read so far. The hidden state gets passed on to the next step, and information gets able to flow along the sequence. Theoretically, that provided the model with an unlimited memory, the freedom to tie the penultimate word of a chapter to the first.

But simple RNNs were forgetful and suffered from the vanishing gradient problem. Learning in a neural network happens when an error signal is sent backward through the network. In an RNN, this signal has to travel back in time, through every single step. With each step, it gets multiplied, and if the network's weights are small, this signal can shrink exponentially, becoming so faint that the model is unable to learn from its distant past. Essentially making it a machine with amnesia.

The answer to that brain teaser was a piece of neural engineering: the Long Short-Term Memory (LSTM) network, which was created by Hochreiter and

Schmidhuber in 1997 and later refined. LSTMs included a high-end management system for memory (Mohsen et al., 2021). The key element inside an LSTM is a memory cell, and that is basically an information conveyor belt, wrapped in a set of gates acting as intelligent valves.

As seen in Figure 2,

- The Forget Gate learns to look at the incoming information and decide which old memories are no longer needed and should be discarded.

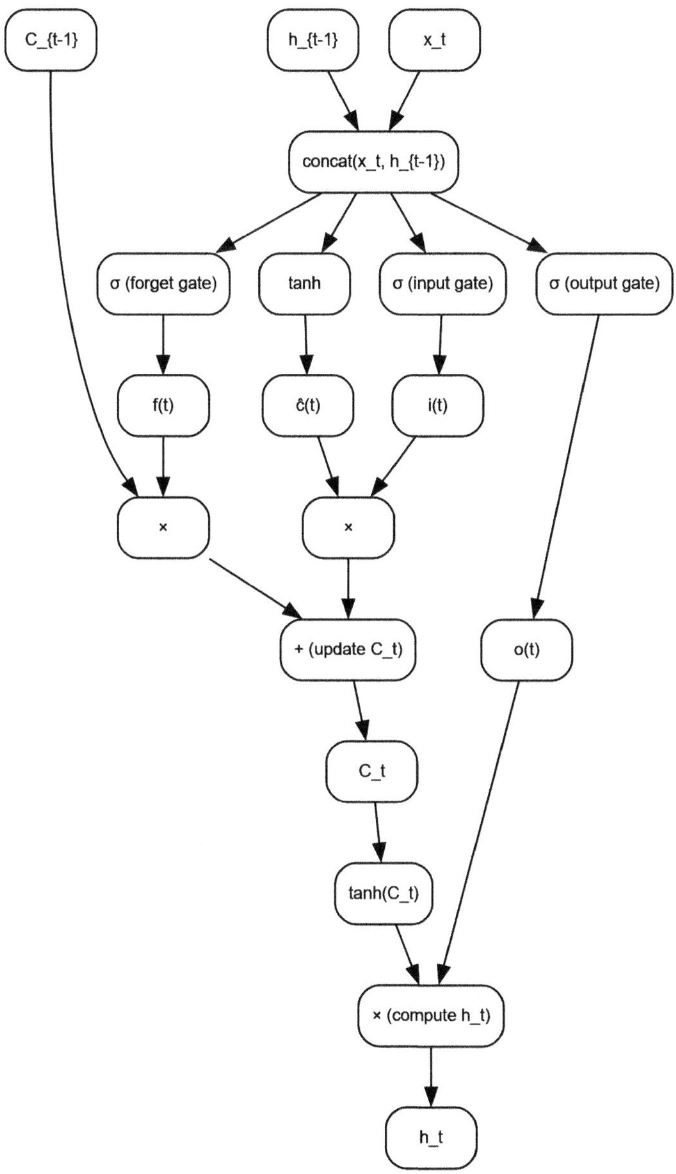

Figure 3: Long short-term memory (LSTM) unit architecture (Mohsen et al., 2021).

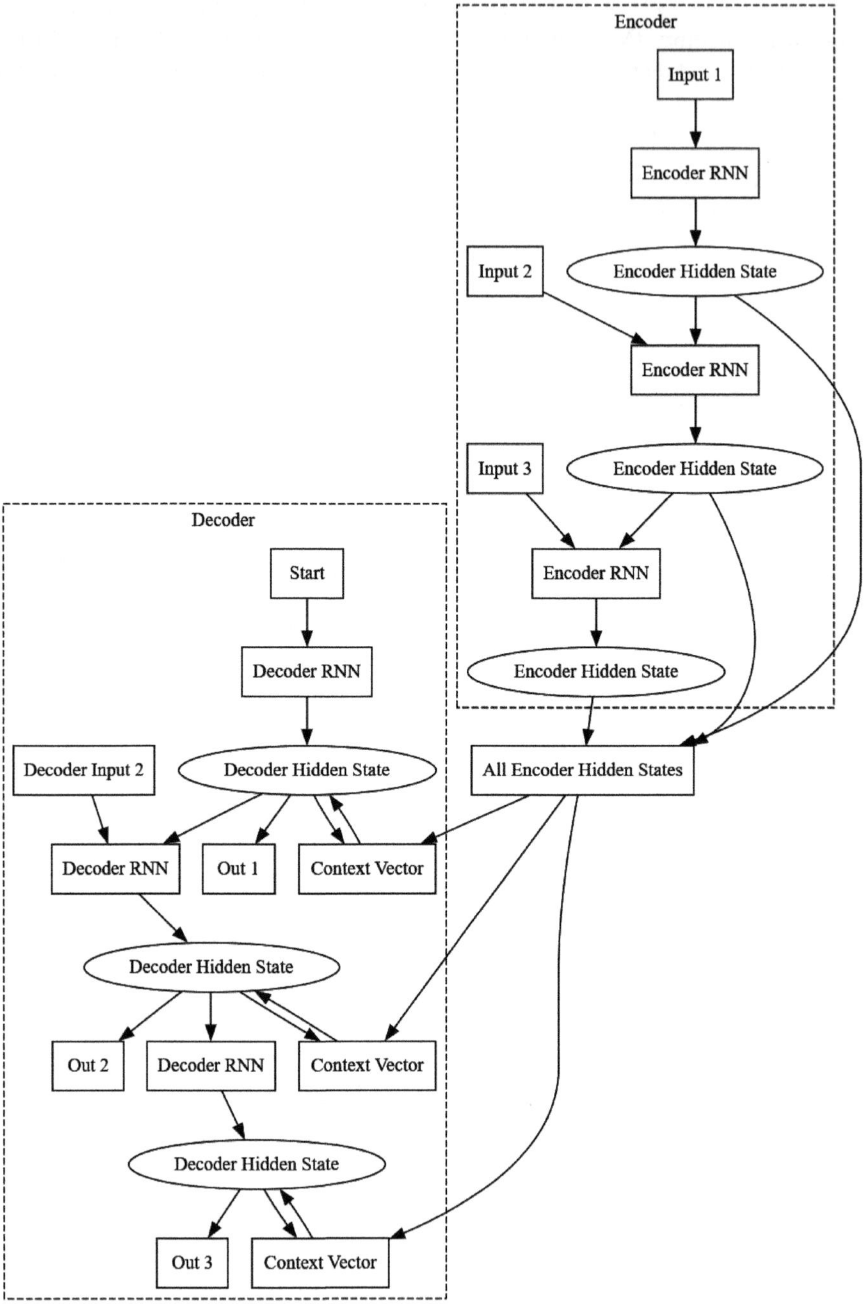

Figure 4: Attention mechanism flowchart (Takyar, 2023).

- The Input Gate decides which new information is important enough to be written to the memory cell.

- The Output Gate controls what part of the memory is relevant to the current moment and should be used to influence the model's output.

These gates gave the LSTM the ability to be incredibly selective about its memory. It could keep a crucial piece of information from page one for hundreds of steps, yet pass over irrelevant information. This ability for carrying out long-term dependencies led the LSTMs and their relatives, i.e., the Gated Recurrent Unit (GRU), to supremacy in NLP for a decade and fuelled exponential gains in translation and speech recognition.

But whereas LSTMs addressed the long-term memory issue, they passed on a fundamental, architectural flaw with all recurrent networks: sequential computation. RNNs and LSTMs process text one word at a time because each step depends on the result of the previous step. This step-by-step process makes it impossible to run computations in parallel, which slows things down (Sit et al., 2020). As sentences and datasets grew larger, training became too slow and inefficient. This problem led researchers to look for a new type of model that could handle all words in a sequence at once.

2.3 The Attention Mechanism: A New Approach to Context

The final conceptual breakthrough that would unlock the Transformer came from an effort to fix a specific problem in machine translation. The standard encoder-decoder models (Tangbasky, 2025) of the time worked by having an "encoder" LSTM read the source sentence and compress its entire meaning into a single, fixed-length "context vector." A "decoder" LSTM would then take this vector and try to generate the translated sentence from it.

The problem was that single vector. It was an information bottleneck. It was like asking a translator to read an entire novel and then summarize it in a single sentence, from which another person would have to reconstruct the entire book. For long sentences, crucial details were inevitably lost in this compression, leading to poor translations. In 2014, Bahdanau et al. (Bahdanau et al., 2015) introduced the attention mechanism as a way to overcome his bottleneck. Instead of relying on a single context vector, the attention mechanism allowed the decoder, at each step of its generation process, to have a direct "look" back at the hidden states from every token in the source sentence. It would then compute a set of attention weights, which represented the relevance of each source token to the current translation step, and use these weights to create a dynamic, weighted context vector tailored for generating the current target token. This effectively created a shortcut, allowing the model to focus on the most pertinent parts of the input sequence at any given time, dramatically improving performance, especially on long sentences.

This technique was an important turning point in AI. Because it showed processing important parts using weighted connections rather than remembering

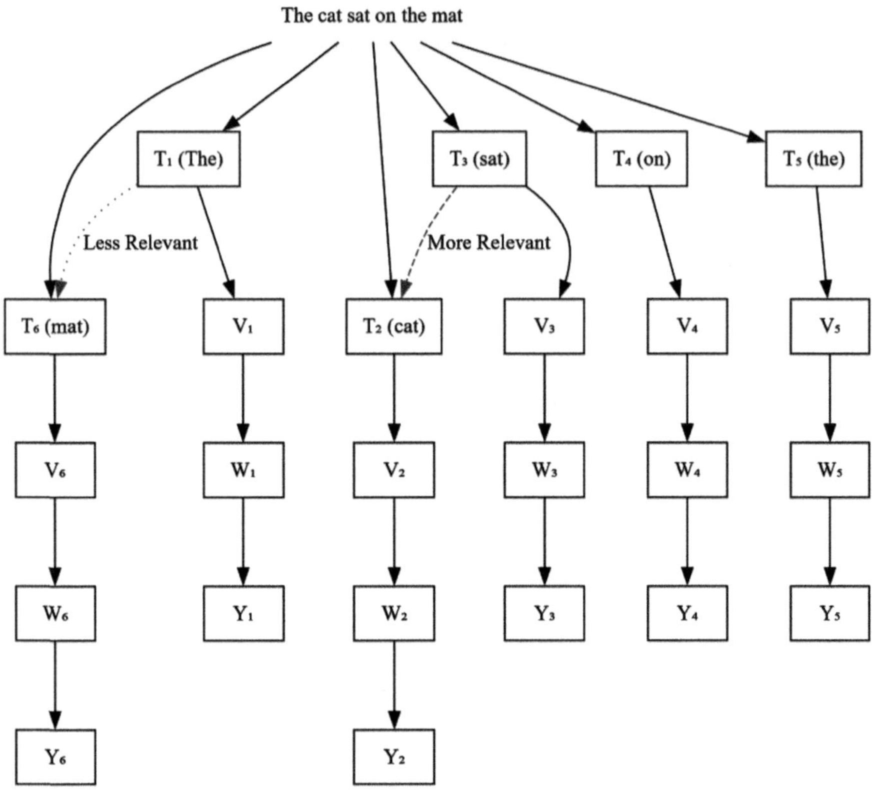

Figure 5: Attention mechanism working illustration.

information step by step. Researchers identified that if this attention works for connecting input and output sequence, it can also be used to connect the words in the same sentence. This technique is self-attention which helps in checking each word with all other words at the same time decide the best connected. Not like RNNs or LSTMs, transformers are processing words in parallel. This reduced computation time and made large, powerful LLMs possible.

3. Deconstructing the Transformer: The Foundational Architecture

The transformer introduced by the research article, "Attention is All you Need" (Vaswani et al., 2017), introduced parallel self-attention mechanism. Word order is preserved using positional encoding.

3.1 Preparing the Input: From Text to Tensors

Initial step prior to transformer processing is converting raw text to numerical tensors which the model can understand. This is performed using tokenization, embedding and positional encoding. Now the transformer can interpret sequence effectively.

3.1.1 Subword Tokenization Strategies

LLMs can understand tokens instead of words or letters. Building unique token is impossible. Using individual characters is also inefficient, as it creates very long sequences. The solution is subword tokenization, a clever compromise that breaks words down into their common constituent parts. This allows the model to have a fixed, manageable vocabulary while still being able to represent any word by piecing together these subword units. The most common algorithms for this are:

- **Byte Pair Encoding (BPE):** Initially a data compression algorithm, BPE was adapted for NLP. The process begins with a base vocabulary of all individual characters (or bytes, in the case of byte-level BPE). It then iteratively scans the training corpus, finds the most frequent pair of adjacent tokens, and merges them into a new, single token that is added to the vocabulary. This process is repeated for a predetermined number of merges, resulting in a vocabulary of common subwords and full words. For example, the pair "e" and "r" might merge to "er", which might later merge with "##" to form "##er".
- **WordPiece:** Utilized by models like BERT, WordPiece is similar to BPE but uses a different merge criterion. Instead of merging the most frequent pair, it merges the pair that maximizes the likelihood of the training data given the current vocabulary. This is calculated by a score: score = (frequency of pair)/(frequency of first × frequency of second). This method prioritizes merges that are more "surprising" or informative over those that simply involve very common individual tokens.
- **SentencePiece:** Developed by Google, SentencePiece is a language-agnostic tokenizer that treats the input text as a raw stream of Unicode characters. Its key innovation is to treat whitespace as a normal character, which it encodes using a special meta-symbol (e.g.,U+2581). This ensures that the tokenization process is fully reversible and lossless, as the original spacing can be perfectly reconstructed. This makes it particularly effective for languages that do not use spaces to delimit words, such as Japanese or Chinese.

3.1.2 Input Embeddings

Once tokenized, each token is assigned a unique ID. This ID is then used to look up a corresponding vector in a massive embedding matrix. Each row in this matrix is a high-dimensional vector (e.g., 768 or 4096 dimensions) that represents a token. The size of this vector, often denoted as d_{model}, is a core hyperparameter of the model. These embedding vectors are initialized randomly and are learned during the training process to capture the semantic meaning of the tokens. This matrix is the model's internal dictionary, where each token is represented by a high-dimensional vector (the d_{model}) that captures its learned meaning. These vectors start out as random numbers but are gradually refined during training until they form a complex "meaning space" where similar words are located close to one another.

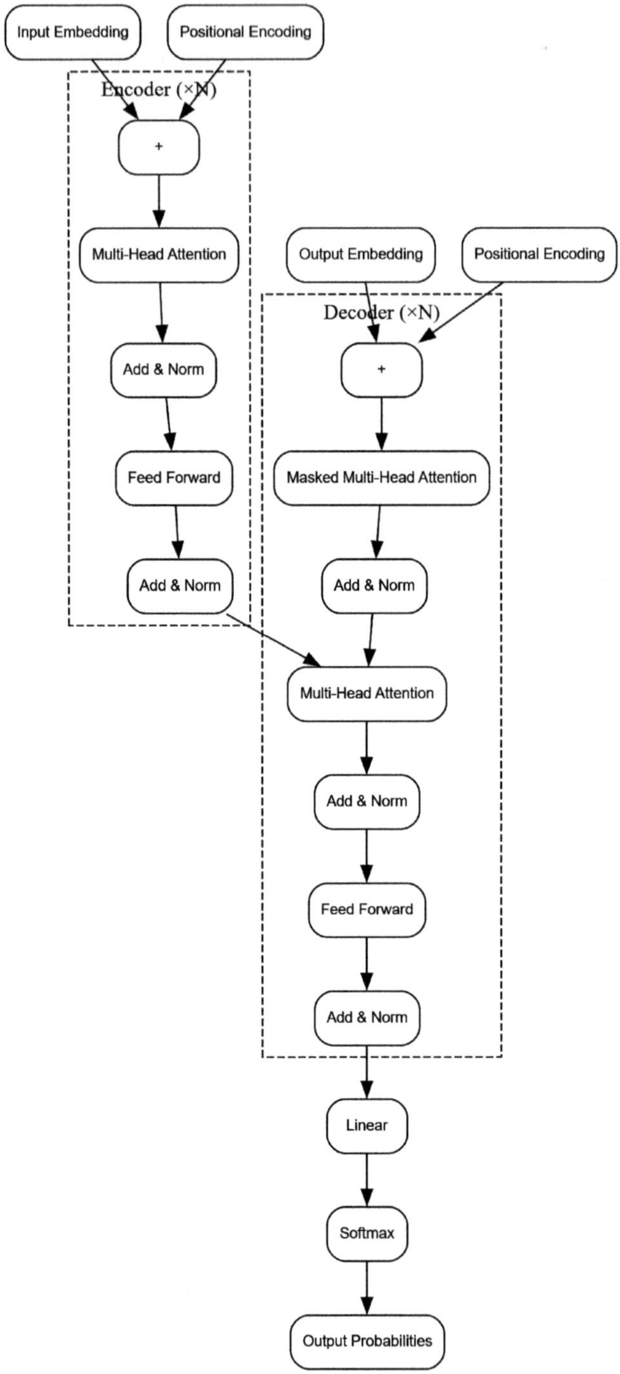

Figure 6: The Transformer - model architecture (Vaswani et al., 2017).

3.1.3 Positional Encoding

This is the Transformer's ingenious solution to its lack of sequential awareness. Since self-attention mechanism looks at all words at once, it has no inherent sense of their order. To fix this, a positional encoding (Lee, 2025) vector is added to each token's embedding. In the original paper, these are not learned but are calculated using a beautiful formula of sine and cosine waves of different frequencies:

$$PE_{(pos,2i)} = \sin\left(\frac{pos}{10000^{\frac{2i}{d_{model}}}}\right) \tag{2}$$

$$PE_{(pos,2i+1)} = \cos\left(\frac{pos}{10000^{\frac{2i}{d_{model}}}}\right) \tag{3}$$

Here, pos is the position of the token in the sequence, d_{model} is the embedding dimension, and i is the index of the dimension within the encoding vector. Each dimension of the positional encoding corresponds to a sinusoid of a different frequency. This scheme was chosen for several reasons: it provides a unique encoding for each position, its values are bounded, and most importantly, it allows the model to easily learn relative positions, since for any fixed offset k, PE_{pos+k} can be represented as a linear function of PE_{pos}. The final input representation for each token is the element-wise sum of its token embedding and its positional encoding. This combined vector carries both semantic information (what the token is) and positional information (where it is in the sequence).

3.2 The Core Engine: Scaled Dot-Product Self-Attention

At the very centre of the Transformer is the self-attention mechanism (Vaswani et al., 2017). Think of it as a dynamic, context-aware routing system. For each word, the model asks a simple question: *"Given who I am, which other words in this sentence matter most for understanding my role right now?"*. This is performed by query,key,value technique, where each word compares itself with other. This measure relevance and thus produces context-aware representation.

- **Query (Q):** A vector representing the current word's question to the others.
- **Key (K):** A vector representing another word's "label," which the Query can match against.
- **Value (V):** A vector representing the actual information that other word contains.

The Q,K,V are generated by multiplying learned weight matrices (W_Q, W_K, W_V) on token's input vector. Attention score are computed by multiplying queries and keys, scaling the scores and applying SoftMax to get weight that decide how much each token attend to others. Applying the attention weights to the value vectors to produce a context-rich representation which shows the relevant words in the sequence.

This process can be expressed concisely in matrix form for all tokens simultaneously:

$$Attention\,(Q,K,V) = \left(\frac{softmax\,(QK^T)}{sqrt(d_k)}\right)V \tag{4}$$

3.3 Enhancing Representational Power: Multi-Head Attention

A single attention mechanism (Vaswani et al., 2017) might learn to focus on one specific type of relationship. To give the model a more holistic view, the Transformer uses multi-head attention. This is like having multiple, parallel attention mechanisms, or "heads," each learning to focus on a different aspect of the language.

Instead of having just one set of weight matrices (W_Q, W_K, W_V), multi-head attention has h sets (where h is the number of "heads," typically 8 or 12). The input vector for each token is linearly projected h times into different, lower-dimensional Q, K, and V representations. The scaled dot-product attention mechanism is then applied independently and in parallel to each of these h projected versions.

This yields h separate output vectors for each token, each capturing different contextual relationships. These h output vectors are then concatenated and passed through a final learned linear projection matrix (W_O) to produce the final output of the multi-head attention layer. This allows the model to jointly attend to information from different "representation subspaces," for instance, one head might learn to track syntactic dependencies while another tracks semantic similarity.

3.4 The Transformer Block: Integrating Components

Multi-head attention is one of two main sub-layers within a standard Transformer block, the fundamental building unit of the model.

3.4.1 Position-wise Feed-Forward Networks

The second sub-layer is a simple position-wise feed-forward network (FFN). After the attention mechanism has updated a word's representation with context, this FFN is applied to each word's vector independently. It's a standard two-layer neural network, with a linear transformation to a higher dimension (typically 4×dmodel), a non-linear activation (like ReLU or GELU), and another linear transformation back to the original dmodel dimension. This gives added computational depth, helping the model to perform complex transformations on the contextualized representations.

3.4.2 Residual Connections and Layer Normalization

To train deep Transformer models, each block, i.e., multi-head attention and feed-forward network, is wrapped with an Add and Norm step.

- **Residual Connections:** The input is summed back to the output of the sub-layer, then normalized.

$$output = LayerNor(x + Sublayer(x)) \tag{5}$$

This idea, borrowed from ResNet in computer vision, gives the gradient a clear path during training. It solves the vanishing gradient problem and makes it possible to train networks with many layers (dozens or even hundreds).

- **Layer Normalization:** Applying after the residual connection, layer normalization stabilizes the training. The vector representation's features for each token are normalized for a mean and a standard deviation value of one. The effect is that the distribution of activations becomes invariant across layers and at training and convergence is speeded up and stability for the whole model is gained.

Working together, these components that are: input processing, multi-head self-attention, feed-forward networks, and stabilizing layer, form a robust and very parallelizable block that may be stacked to build deep and expressive language understanding and language generation models.

4. A Taxonomy of Transformer Architectures

The Transformer was initially a complete system, with an encoder to decode and a decoder to produce. But soon researchers found they could employ both parts on their own, and thus three families of architecture were born. The question of whether to use one is one of purpose. You want to comprehend at a fundamental level, you use an encoder. You want to produce naturally, you use a decoder. And you want to transform, you use both. This section is a field guide to the three species of model, and includes a discussion of their anatomy, their learning philosophies, and their environments.

4.1 Encoder-Only Architectures: For Deep Contextual Understanding

Encoder-only models are the large thinkers within the world of LLM. They are supposed to be experts at comprehending. They are able to look at the entire piece of

Table 1: Types of Transformer Architectures.

Architecture	Attention Mechanism(s)	Information Flow	Pre-training Objective	Example Models	Primary Use Cases
Encoder-Only	Bidirectional Self-Attention	Global, Static Analysis	Masked Language Modeling (MLM)	BERT, RoBERTa	Text Classification, NER, Extractive QA
Decoder-Only	Causal (Masked) Self-Attention	Local, Dynamic Generation	Causal Language Modeling (CLM)	GPT Series, Llama, Claude	Text Generation, Dialogue, Summarization
Encoder-Decoder	Bidirectional Self-Attention (Encoder), Causal Self-Attention & Cross-Attention (Decoder)	Hybrid (Global Analysis -> Local Generation)	Denoising, Text-to-Text	T5, BART	Machine Translation, Abstractive Summarization

text at once, not to produce a response, but to develop a rich, context-aware numerical encoding for each individual word.

- **Architecture:** The architecture consists of a stack of Transformer encoder blocks. The characteristic feature is bidirectional self-attention (IBM, 2024). Every word gets to see each and every other word, both earlier and later. This gives the model a complete, holistic view of the text and allows the model to build a sense of a word based on its context as a whole.
- **Pre-training Task:** The model is pretrained on Masked Language Modeling (MLM). It is a fill-in-the-blank task in which some words in a sentence are masked and the model is required to predict them from the surrounding context. This forces the model to learn the internal structure of the language in both directions.
- **Case Studies:**

BERT (Bidirectional Encoder Representations from Transformers) (Devlin et al., 2018): Introduced by Google in 2018, it was a seminal encoder-only model that revolutionized the field. It demonstrated the power of deep bidirectional pre-training, achieving state-of-the-art results on a wide range of NLU tasks after fine-tuning. BERT was also trained with a Next Sentence Prediction (NSP) objective, where the model had to predict if two sentences were consecutive, though this was later found to be of limited value.

RoBERTa (A Robustly Optimized BERT Pretraining Approach) (Wikimedia Commons, 2025; Liu et al., 2019): Researchers from Facebook AI found that the original BERT was significantly undertrained. RoBERTa kept the same architecture but optimized the pre-training process by:

(1) removing the NSP task

(2) training with much larger batch sizes on a significantly larger text corpus (160 GB vs. 16 GB)

(3) training for a longer duration

(4) using dynamic masking, where a new masking pattern is generated for each training sequence, unlike BERT's static, one-time masking. These changes led to substantial performance improvements over BERT.

Primary Applications:

Encoder-only models are best for discriminative tasks because they analyze the whole text at once. Examples include:

1. Text Classification performs labeling text (e.g., positive/negative sentiment).
2. Named Entity Recognition (NER) helps in finding and tagging names of people, places, organizations, etc.
3. Extractive Question Answering helps in locating the exact answer span inside a passage.

4.2 Decoder-Only Architectures: For Autoregressive Generation

Decoder-only models (Suresh and Puducherry, 2024) are the workhorses of modern generative AI. They are designed to produce fluent, coherent, and contextually relevant text by predicting one token at a time, building upon what has already been generated.

- **Architecture:** The architecture is built as a stack of Transformer decoder layers. The core part is causal self-attention. There is a masking procedure that is utilized to ensure that no word has seen later words. That is, each prediction for the following word is done on the basis of previous tokens, keeping the usual, autoregressive character of language.
- **Pre-training task:** Decoder-only models are fine-tuned using Causal Language Modeling (CLM), a plain next-token prediction task. Given a token sequence, we are looking to predict the next token. The system learns a probability distribution across the vocabulary for the likely next token, and at generation time, it samples from the distribution, inserts the token into the sequence, and repeats.
- **Case Study:** The GPT Lineage (Generative Pre-trained Transformer): The GPT lineage was created by OpenAI. The most striking embodiment of the decoder-only architecture comes from the GPT lineage. From GPT-1 onwards up to GPT-4 and beyond, these systems have proved that scaling the decoder-only architecture in unison with colossal training datasets result in emergent behaviours, i.e., few-shot and zero-shot learning, where tasks are possible without direct training, yet guided only by prompts and demonstrations.
- **Major Uses:** The localized, dynamic generation process characteristic of decoder-only models gives them a strange ability for generative tasks, i.e.:

Open-ended Text Generation: The production of stories, newspaper/magazine articles, and fiction.

Chat and Dialogue Systems: Enabling conversational agents that keep track of context and provide natural responses.

Summary and Question Answers (Generative): Producing abstractive summaries or giving question answers in free-form, in contrast to extracting direct text spans.

4.3 Encoder-Decoder Architectures: For Sequence-to-Sequence Transformations

Encoder-decoder models, or sequence-to-sequence (seq2seq) models, take the best from both architectures. They are designed for those tasks that are composed of converting one information sequence into another, for instance, translations of languages.

- **Architecture:** This architecture uses both an encoder stack and a decoder stack. The encoder processes the entire input sequence bidirectionally, creating a complete, context-rich representation. The decoder then generates the output sequence autoregressively. The key interaction happens in the cross-attention layer within each decoder block. This layer allows the decoder to attend to the output representations from the encoder at every generation step, enabling it to ground its output in the information provided by the input sequence.

- **Pre-training Objectives:** These models are often pre-trained on denoising or text-to-text objectives that mimic transformation tasks.

T5 (Text-to-Text Transfer Transformer): T5, from Google AI, unifies all NLP tasks into a single text-to-text framework. A task-specific prefix is added to the input (e.g., summarize:...), and the model is trained to generate the target text string. Its pre-training objective is a fill-in-the-blank style of denoising where random spans of tokens are masked and the model must reconstruct them.

BART (Denoising Sequence-to-Sequence Pre-training): Developed by Facebook AI, BART (Lewis et al., 2019) is pre-trained as a denoising autoencoder. It takes a corrupted document as input to its bidirectional encoder and is trained to reconstruct the original, uncorrupted document with its autoregressive decoder. The corruption process can manifest in token masking, token deletion, infilling of text, and reordering of sentences and hence is a flexible pre-training approach.

- **Primary Applications:** The hybrid approach of global analysis makes these models a natural fit for sequence-to-sequence tasks:

Machine Translation: Translating a sentence from a source language to a target language.

Abstractive Summarization: Generating a concise summary that captures the key information of a longer document, potentially using words not present in the original text.

Generative Question Answering: Answering a question based on a provided context by generating a new answer sentence.

5. Scaling and Efficiency: The Modern Frontier

The rise of the Transformer is mainly about scaling. Because it processes text in parallel, researchers could build models far larger than ever before. The growth, however, came at a cost. The self-attention mechanism at the core of the Transformer scales computationally quadratically with sequence length. As model size rose from large to colossal, the quadratic bottleneck thus became the core issue in the field. The tradeoff between scaling benefits and the expensive cost of realizing them has characterised today's era of LLM research, bringing with it an "era of efficiency" in which the focus is on clever rather than bigger and bigger models.

5.1 The Challenge of Quadratic Complexity and Scaling Laws

The core computational bottleneck of the Transformer architecture lies in its self-attention mechanism. To compute the attention scores, the model must perform a matrix multiplication between the Query matrix ($n \times d_k$) and the transpose of the Key matrix ($d_k \times n$), where n is the sequence length and d_k is the dimension of the keys. This procedure has time and memory complexity of $O(n^2 \cdot d_k)$. As the sequence length n increases, the quadratic factor overwhelms and processing lengthy input, such as whole books or long codes, becomes infeasible.

Regardless of such a limitation, research has continually demonstrated that scaling improves performance nevertheless. This relationship is described by neural scaling laws, which empirically model how a model's performance (typically measured by cross-entropy loss) improves as three key factors are increased: the number of model parameters (N), the size of the training dataset (D), and the amount of compute used for training (C).

- **Early Scaling Laws (Kaplan et al., 2020):** Initial studies demonstrated that model loss decreases as a predictable power-law function of N, D, and C, as long as the other two factors are not bottlenecks. This work suggested that for a fixed compute budget, the best performance was achieved by training very large models, even if they did not fully converge.
- **Chinchilla's Law (Hoffmann et al., 2022):** This landmark study from DeepMind refined these laws. By training over 400 models of varying sizes on different amounts of data, they discovered that for optimal compute efficiency, the model size and the number of training tokens should be scaled in roughly equal proportion. This implied that many previous large models, such as GPT-3, were significantly undertrained, they were too large for the amount of data they were trained on and would have performed better if they had been smaller but trained on more data.

These scaling laws create a clear but challenging path: better models require more parameters and more data. But the quadratic cost of attention makes this an ever-more-expensive proposition, creating a powerful incentive to find more efficient architectures.

5.2 Mixture-of-Experts (MoE): Scaling Parameters, Not Compute

The Mixture-of-Experts (MoE) design helps models grow bigger without slowing them down.

- **Core Idea:** Instead of one feed-forward network (FFN), an MoE layer has many experts (parallel FFNs). A small gating network chooses just a few experts (like 2 out of 64) for each token.
- **Efficiency (Sparse Activation):** Even though the model has a huge number of parameters, only a few are used at a time. This means we can build trillion-parameter models that run as fast as much smaller ones.
- **Implications:** MoE has become a dominant architecture for state-of-the-art open-source models, such as Mixtral 8x7B, Grok-1, and DeepSeek-V3. These models demonstrate that it is possible to achieve the performance of extremely large dense models while maintaining the inference speed and computational

footprint of a much smaller model, making high-performance LLMs more accessible.

5.3 State Space Models (SSMs): A Return to Recurrence with Linear Complexity

While Mixture of Experts (MoE) improves efficiency by optimizing the feed-forward layers, State Space Models (SSMs) (Somvanshi et al., 2025) represent a more fundamental shift by aiming to replace the quadratic attention mechanism itself. SSMs combine the advantages of RNNs and CNNs, enabling them to model long-range dependencies while still supporting parallel training with linear or near-linear computational cost. A notable example is the Mamba framework, a modern SSM architecture that competes with Transformers in tasks such as language modeling. Mamba introduces Selective State Space (S6), which dynamically adapts its parameters based on the input, allowing the model to selectively remember or forget information in a way similar to attention but far more efficiently. By supporting both recurrent-style processing and parallel convolutional computation, Mamba achieves linear-time complexity, making it well suited for very long sequences where Transformers become slow. Together, innovations like MoE and Mamba signal a new era of efficient large language models, where architectural design is as important as model size.

6. A Comparative Analysis of Architectural Performance

A model's effectiveness is judged by benchmarks like GLUE, SuperGLUE, and SQuAD, which show that different architectures excel at different tasks and should be chosen based on the application.

6.1 Performance on Natural Language Understanding Benchmarks (GLUE & SuperGLUE)

GLUE (General Language Understanding Evaluation) and its more challenging successor, SuperGLUE, are benchmark suites designed to evaluate a model's general natural language understanding capabilities. They include tasks such as sentiment analysis (SST-2), textual entailment (RTE, MNLI), and paraphrase detection (MRPC), all of which require a strong understanding of the full context within one or two sentences. During the late 2010s and early 2020s, research placed heavy emphasis on these benchmarks, which strongly influenced model development and favored architectures with bidirectional encoders capable of capturing holistic sentence-level meaning.

- **Encoder-Only Dominance:** Encoder-only models like BERT-Large and RoBERTa excelled on GLUE and similar benchmarks. DeBERTa even surpassed human performance on SuperGLUE (89.9 vs. 89.8) thanks to its bidirectional attention, which builds rich contextual understanding.

Table 2: Comparative Analysis of Architectural Performance (Nangia and Bowman, 2018; Hagen, 2021; Scilit, 2024).

Model	GLUE Score (Avg)	SuperGLUE Score (Avg)	SQuAD v1.1 F1	SQuAD v2.0 F1
Human Baseline	87.1	89.8	89.5	92.2
BERT-Large	82.1	~84.5	90.9	83.1
RoBERTa-Large	88.1	~88.0	94.6	89.4
BART-Large	88.8	88.8	94.6	88.8
T5-11B	90.3	88.9	92.2	93.7
DeBERTa (1.5B)	90.8	89.9	91.9	90.7

Note: Benchmark scores evolve rapidly. The values presented are representative scores from the provided sources to illustrate relative performance. Some scores are from base models while others are from large or ensemble models, as indicated in the sources. SuperGLUE scores for older models are often harder to find as the leaderboard is dominated by newer entries.

- **Encoder-Decoder Success:** Encoder-decoder models like T5 and BART also performed very well. For example, T5's 11B parameter model nearly matched human performance on SuperGLUE with a score of 88.9. BART was designed to be exceptionally great at both understanding and generating and, indeed, equaled RoBERTa's robust GLUE performance. The robust bidirectional encoders in their frameworks are able to perform the robust deep thinking required, and their decoders are designed to produce the classification labels.
- **Decoder-Only Lag in Fine-Tuning:** When fine-tuned in the traditional way, decoder-only models like the GPT (Sharkey and Treleaven, 2024) series generally underperform on GLUE and SuperGLUE. Their unidirectional, causal attention is fundamentally less suited for tasks that benefit from seeing both left and right context simultaneously. However, as these models grew in scale, they developed powerful zero-shot and few-shot learning capabilities. When prompted with task instructions, very large decoder-only models can achieve competitive, and sometimes superior, performance without any fine-tuning, though this is a different evaluation paradigm.

6.2 Performance on Extractive Tasks (SQuAD)

The Stanford Question Answering Dataset (SQuAD) is a reading comprehension test. The task is to retrieve the exact substring within a passage that is the answer to a question. SQuAD 2.0 added unanswerable questions to make the task harder, requiring deep, bidirectional understanding similar to GLUE.

- BERT was a landmark, with BERT-Large among the first models to surpass human-level performance on SQuAD 1.1 F1.
- Later encoder-only and encoder-decoder models continued to set new records on both SQuAD 1.1 and 2.0, showing that bidirectional encoders are very effective for these tasks.

Table 3: Summary of Transformer Architectures.

Architectural Family	Core Analogy	Key Attention Mechanisms	Primary Pre-training Objective	Core Strength	Ideal Use Cases	Example Models
Encoder-Only	The Analyst	Bidirectional Self-Attention	Masked Language Modeling (MLM)	Deep Contextual Understanding	Text Classification, Named Entity Recognition, Extractive Question Answering	BERT, RoBERTa, DeBERTa
Decoder-Only	The Creator	Causal (Masked) Self-Attention	Causal Language Modeling (CLM)	Fluent and Coherent Generation	Open-Ended Text Generation, Dialogue Systems, Creative Writing	GPT Series, Llama, Claude
Encoder-Decoder	The Translator	Bidirectional (Encoder), Causal & Cross-Attention (Decoder)	Denoising/ Text-to-Text	Sequence-to-Sequence Transformation	Machine Translation, Abstractive Summarization, Generative Question Answering	T5, BART

6.3 Synthesis: Matching Architecture to Task for Optimal Performance

Experiments show a clear pattern: there's no single best architecture that it depends on the task.

- Bidirectional encoders (Encoder-Only and Encoder-Decoder) are best for tasks requiring deep contextual understanding, because they can see the whole text at once, making them ideal for discriminative tasks.
- For tasks that require fluent and creative text generation, the autoregressive nature of Decoder-Only models is unparalleled. Their next-token prediction objective is perfectly aligned with the generative process.
- Encoder-Decoder models are the versatile hybrids, excelling at sequence-to-sequence transformations that require both a deep understanding of an input and the controlled generation of an output, making them ideal for machine translation and abstractive summarization.

7. Conclusion: An Ongoing Evolution

The history of large language models (LLMs) is a gradual scientific evolution, not a single breakthrough.

- o Early challenges includes N-gram models were limited by the curse of dimensionality and Neural embeddings solved this by representing words in dense, distributed spaces.

- Long Short-Term Memory (LSTM) networks Captured long-range dependencies and Sequential processing made them less efficient.
- Attention mechanisms Solved the sequential bottleneck of LSTMs.
- Transformers Eliminated recurrence entirely and Provided a scalable foundation for modern LLMs.

Model architecture families like Encoder-only (e.g., BERT): Deep analyzers for tasks needing bidirectional context, Decoder-only (e.g., GPT): Text generators producing words autoregressively and Encoder–decoder (e.g., T5): Sequence transformers, converting one sequence into another.

Efficiency-focused innovations like Quadratic cost of attention and scaling challenges led to MoE and SSMs. These improve performance without proportional increases in computation.

Future outlook is Identifying bottlenecks and designing task-specific architectures will make LLMs more powerful, efficient, and accessible.

References

Bahdanau, D., Cho, K. and Bengio, Y. (2015). Neural machine translation by jointly learning to align and translate. In Proceedings of the 3rd International Conference on Learning Representations (ICLR), San Diego, CA, USA. https://arxiv.org/abs/1409.0473

BL, P. (2025, July 1). The history of SLMs: From rule-based to transformer-based. Medium. https://medium.com/@pbl/the-history-of-slms

Devlin, J., Chang, M.-W., Lee, K. and Toutanova, K. (2018). BERT: Pre-training of deep bidirectional transformers for language understanding. arXiv preprint arXiv:1810.04805. https://arxiv.org/abs/1810.04805

Hagen, A. (2021, January 6). Microsoft DeBERTa surpasses human performance on SuperGLUE benchmark. Microsoft Research. https://www.microsoft.com/en-us/research/blog/microsoft-deberta-surpasses-human-performance-on-the-superglue-benchmark/

IBM. (2024, December 5). Attention mechanism. https://www.ibm.com/think/topics/attention-mechanism

Jelinek, F. (1997). Statistical methods for speech recognition. MIT Press.

Kaplan, J., McCandlish, S., Henighan, T., Brown, T. B., Chess, B. and Amodei, D. (2020). Scaling laws for neural language models. arXiv preprint arXiv:2001.08361. https://arxiv.org/abs/2001.08361

Le, J. (2019, May 23). Recurrent neural networks: The powerhouse of language modeling. *Built In*. https://builtin.com/data-science/recurrent-neural-networks-powerhouse-language-modeling.

Lee, F. (2025, May 6). What is positional encoding? IBM. https://www.ibm.com/think/topics/positional-encoding.

Lewis, M., Liu, Y., Goyal, N., Ghazvininejad, M., Mohamed, A. and Zettlemoyer, L. (2019). BART: Denoising sequence-to-sequence pre-training for natural language generation, translation, and comprehension. arXiv preprint arXiv:1910.13461. https://arxiv.org/pdf/1910.13461

Liu, Y., Ott, M., Goyal, N., Du, J., Joshi, M. and Stoyanov, V. (2019). RoBERTa: A robustly optimized BERT pretraining approach. arXiv, 1. https://doi.org/10.48550/arxiv.1907.11692

Manning, C. D. and Schütze, H. (1999). Foundations of statistical natural language processing. MIT Press.

McCarthy, J., Minsky, M. L., Rochester, N. and Shannon, C. E. (2006). A Proposal for the Dartmouth Summer Research Project on Artificial Intelligence, August 31, 1955. AI Magazine, 27(4): 12. https://doi.org/10.1609/aimag.v27i4.1904

Mohsen, S., Elkaseer, A., and Scholz, S. G. (2021). Industry 4.0-oriented deep learning models for human activity recognition. IEEE Access, 9: 150508–150521. https://doi.org/10.1109/access.2021.3125733

Mu, S., Zhou, J., He, X. and Sun, J. (2025). A comprehensive survey of mixture-of-experts: Algorithms, applications, and challenges. arXiv preprint arXiv:2503.07137. https://arxiv.org/abs/2503.07137

Nangia, N. and Bowman, S. (2018). Human vs. Muppet: A conservative estimate of human performance on the GLUE benchmark. https://woollysocks.github.io/assets/GLUE_Human_Baseline.pdf

Pióro, M., Baranowski, R. and Piętak, Ł. (2024). MoE-Mamba: Efficient selective state space models with mixture of experts. arXiv preprint arXiv:2401.04081. https://arxiv.org/abs/2401.04081

quadeer15sh. (2023, June 25). Introduction to language modelling using RNNs. Kaggle. https://www.kaggle.com/code/quadeer15sh/introduction-to-language-modelling-using-rnns

Scilit. (2024). Comparative analysis of state-of-the-art Q A models: BERT, RoBERTa, DistilBERT, and ALBERT on SQuAD v2 dataset. https://www.scilit.com/publications/ea5580a03d6b243a67c9ffa74dcd6431

Sharkey, E. and Treleaven, P. (2024). Comparison of BERT vs GPT. arXiv. https://arxiv.org/pdf/2405.12990

Sit, M., Demiray, B. Z., Xiang, Z., Ewing, G. J., Sermet, Y. and Demir, I. (2020). A comprehensive review of deep learning applications in hydrology and water resources. Water Science and Technology 82(12): 2635–2670. https://doi.org/10.2166/wst.2020.369

Somvanshi, S., Islam, M. M., Sultana, M. M., Bin, S., Chhetri, G. and Das, S. (2025). From S4 to Mamba: A comprehensive survey on structured state space models. arXiv. https://arxiv.org/abs/2503.18970

Suresh, S. and Puducherry, S. (2024). Towards smaller, faster decoder-only transformers: Architectural variants and their implications. arXiv. https://arxiv.org/pdf/2404.14462

Takyar, A. (2023, November 9). Exploring the power of attention mechanism in deep learning. LeewayHertz. https://www.leewayhertz.com/attention-mechanism/

tangbasky. (2025, May 14). The comparison between the encoder and the decoder. Medium. https://pub.towardsai.net/the-comparison-between-the-encoder-and-the-decoder-d567b0417d9e

Vaswani, A., Shazeer, N., Parmar, N., Uszkoreit, J., Jones, L., Gomez, A. N. et al. (2017). Attention is all you need. arXiv preprint arXiv:1706.03762. https://arxiv.org/pdf/1706.03762

Weizenbaum, J. (1966). ELIZA—a computer program for the study of natural language communication between man and machine. Communications of the ACM, 9(1): 36–45. https://doi.org/10.1145/365153.365168

Wikimedia Commons. (2025). BERT on sentence classification [Diagram]. https://upload.wikimedia.org/wikipedia/commons/1/1a/BERT_on_sentence_classification.svg.

2

Core Components and Architectures of LLMs

Nisha Rathore,[1] *Manjushree Nayak,*[1,*] *Shikha Tiwari*[1] and *Syed Danish Hassan*[2]

1. Introduction

Large Language Models (LLMs) belong to one of the major breakthroughs in artificial intelligence in recent years, and they were trained on large datasets. These models have demonstrated competent skills in generating and understanding natural language as well as reasoning (Vaswani et al., 2017). LLMs are now an integral part of modern AI systems that run everything from conversational agents like ChatGPT and Bard to code generation support, content creation, scientific research, or business intelligence. Their development is considered a critical shift towards general-purpose models for specific-task machine learning models which need very minimal task-specific supervision and can be used across large domain ranges and different applications (Brown et al., 2020).

Large Language Models are fundamentally based on the transformer architecture. In the seminal paper by Vaswani et al., they introduced it. Self-attention, a major novelty in transformers, enabled models to relate contextual relationships and long dependencies within a sequence of data. Earlier architectures, such as convolutional neural networks and recurrent neural networks suffered from scalability issues as well as problems with context retention; this development addressed those shortcomings. The transformer has become the main base for most state-of-the-art NLP systems

[1] Amity School of Engineering & Technology, AMITY UNIVERSITY, Raipur, Chhattisgarh, Manth (Kharora), State Highway 9, Baloda Bazar Road, 493225.
Emails: nisha.rathor271@gmail.com, https://orcid.org/0009-0009-6471-4540;
stiwari@rpr.amity.edu, orcid id-0000-0002-4877-2777
[2] Data Science University of Europe for Applied Sciences, Dubai.
Email: syeddanish.hassan@ue-germany.de, https://orcid.org/0009-0000-3534-1786
* Corresponding author: manjushreesai44@gmail.com, https://orcid.org/0000-0001-6383-780X

DOI: 10.1201/9781003771654-2

since it supports models (such as encoder-only BERT1 to decoder-only generative models like GPT-4) that cover a wide range of model types.

The scale of these LLMs is unprecedented. These models are usually trained on a set of billions to trillions of tokens and trained on a combination of diverse datasets which includes texts from books and websites, scholarly works and even code repositories. This allows LLM to create a rich internal representation of the world's knowledge, which is the LLMS scale, these models are able to complete a number of tasks even without clear training for a specific task. Emergent Behaviour describes how LLM shows the problem, the zero-shot region and even some shooting learning, which was considered unique to human intelligence. LLM is at a distance from many types of complex building blocks, which are the customized heads for self -meditation teams and output from many heads, ranging from symbols and building -in. In addition, training of the model is an important engineering problem in itself, loaded with the need for sophisticated adaptation methods, distributed computer systems and labor-intensive data sets that ensure balanced and high-quality data. A large part of the behavior of these models is treated by using training processes such as direct preference adaptation, which improves models with RLHF and improves the adjustment for human values.

This chapter seeks to unpack the basic components and LLM architectures to give the reader a solid understanding of the processes involved in the construction, training and optimization of LLM. It examines a wide range of architectural paradigms, from mixture of mixing to expert to codes and even only decoder models, while exploring the advantages and disadvantages, as well as the use cases of each. We also record the retraining-augmented generation (RAG) and advanced methods for parameter-efficient fine tuning (PEFT) and inference optimizations such as KV-Cache (Shazeer et al., 2019) and flash attention (Dao et al., 2022). Social, research and industrial effects of LLM-fortify to increase as they continue to move on. Understanding how LLMs works internally enables stakeholders to address efficiency, safety, interpretability and management challenges in a more efficient way. This chapter offers both a conceptual and technical road map to understand and navigate the rapidly growing landscape in large language models.

1.1 What is an LLM?

We look beyond Arabs a few billions from Arabs to a few billions from Arabs to a few billions, which most parts are kept through autoregressive (only decoder) training (Brown et al., 2020), which also belongs to the sequence of sequence (encoder) type (Raffel et al., 2020). In addition, it is important to note that the definition can be changed. They include the extent of what we can think of as a regular-manual text formation, and they also work well when very long signs are given as inputs, even then they sometimes add components (such as codic interpreters, calculators and information recovery).

1.2 Overview of the Architecture

The architecture of large language models is a complex synthesis of many basic ingredients that eventually work together to enable models to understand, treat and generate human equal languages. In short, LLM layered calculation blocks are conducted in a way that is capable of participating in prolonged addiction and sophisticated linguistic structures on large-scale text data. Although each LLM family applies its uniqueness and specialties, they are attracted to the transformer architecture that has developed to become a major paradigm for NLP functions.

A workflow of an LLM that includes processing through tokens, manufactured—in, transformer blocks and decoding. Each of these stages affects the performance, scalability and efficiency in models. Start by changing the raw text of a series of discrete elements that are usually referred to as symbols. Modern large language model (LLM) uses either subward or byte pair coding (BPE) strategies, which allows for a broader and diverse lexicon at the traditional order level-level. This method reduces the vocabulary by creating small components while maintaining adequate adaptability to clarify rare or complex words. Tokens, code, notation and polyglot can be handled by model through this form of materials; Out-off-volume complexity is non-existent.

After the token of the text, a manufactured—each of these words maps in a high-dimensional room. The model can learn reference and meaning through these manufactured—in related to syntax between various symbols and semantic conditions. Status code is added to build—in sequence order. While modern versions such as rope—rotating position setting or Alibi—attention with linear bias are improved in a better way and supports prolonged references, standard transformers use sinus -shaped position coding.

Multi-Head Self-Addion (MHSA) and Feed Association Neural Network (FFN) are the constituent components of the main level transformer blocks, which are arranged in a stack. In self -confidence, every symbol of all other symbols of follow-up; Therefore, there is a potential for long stay or global addiction between elements. Multi-head attention lets different heads focus on different types of linguistic conditions at the same time and adds expressive power to retrieve fine linguistic details. Training includes normalization layers and remaining connections to ensure a stable gradient current and fast convergence.

After a linear output layer, a softmax function can be used to map back the hidden treated states through the process to the vocabulary spaces, which generates probabilities over multiple next-door options. It produces text that lays out meaningfully and logically coherence. Different levels of richness, creativity and alternatives of uncertainty could be introduced in inference time using the rear beam search, core sampling, and top-k sampling.

An architectural overview of LLMers offers an architectural construction of components independently and together to deal with larger and more diverse data sets. The interesting aspects of LLMS, is the human expert-like task selections generated from the importance of token recordings (fast defaults), built- in transformer blocks (fast and computational) and organized through highly sophisticated decoding apparatus of artificial intelligence. Many more opportunities are available for innovating offer improvements and developments of language generation and understanding through AI.

2. Mathematical and Computational Foundations

2.1 *Subword Units and Tokenization*

Strong tokenization strikes a balance between sequence length and vocabulary size. Frequent pairs are combined to create a condensed subword vocabulary using Byte-Pair Encoding (BPE). To maximize corpus likelihood, Unigram language models (Sentence Piece) use a probabilistic token selection process (Hazeer et al., 2018). By avoiding out-of-vocabulary edge cases, byte-level BPE streamlines code handling and multilingualization. Context length, throughput, and downstream alignment are all significantly impacted by the tokenizer selection.

Design Factors to Take Into Account

Average sequence length and embedding table size are traded off in vocabulary size (e.g., 32 k–100 k). Errors on noisy text are decreased by byte-fallback and normalization (NFKC, case-folding). Maintaining punctuation granularity and whitespace can enhance syntax fidelity for code models.

2.2 *Positional Encodings and Embeddings*

Dense vectors are projected with token ids. Absolute (sinusoidal) or relative schemes are used to add positional information (Sannrich et al., 2015). Extrapolation and effective long-context interpolation are made possible by Rotary Position Embeddings (RoPE), which rotate query/key vectors by a complex phase proportional to position. With varying latency and length generalization trade-offs, alternatives like learned relative positions and ALiBi (attention with linear biases) are popular.

2.3 *The Transformer Block*

Every Block Calculates

1. LayerNorm (pre-norm) on the hidden state h input
2. Multi-Head Self-Attention (MHSA): softmax $(QK^{\tau}/\sqrt{dk}+b)$ for attention weights; queries Q, keys K, and values V.
3. Dropout and residual add
4. MLP: a gated activation (GELU, SwiGLU) separating two linear transforms
5. Dropout and residual add

Autoregression is enforced by causative masking. Grouped-Query (GQA) and Multi-Query Attention (MQA) minimize perplexity loss at inference by sharing keys and values across heads.

2.4 Laws of Scaling and Optimization

Adafactor and Adam/AdamW are standard. Training is stabilized by weight decay and learning-rate schedules (cosine decay with warmup). Loss is related to compute, data, and parameters by empirical scaling laws; compute-optimal training strikes a balance between parameters and tokens ("train on more data before making the model larger") (Kudo et al., 2018). At scale, distributed training, mixed precision (FP16/BF16), and gradient checkpointing are needed.

3. Core Architectural Families

Over the past few years, the architecture of large language models (LLMs) has rapidly evolved, leading to the emergence of multiple distinct model families, each tailored to particular objectives, tasks, and deployment constraints. All these designs are based on the transformer architecture, but there are differences in the way the attention mechanism is used, the information that moves through the network, and the distribution of computational resources (Press et al., 2021). To choose the design that best fits their use cases, researchers and practitioners must have a thorough understanding of these architectural families.

Decoder-only causal transformers, encoder-decoder (sequence-to-sequence) models, encoder-only models, mixture-of-experts (MoE) architectures, and efficient attention variants intended for long-context modeling are, in general, the main architectural families of LLMs. We go into more detail below about the benefits and drawbacks that each family has to offer.

3.1 Transformers that are Decoder-Only (Causal)

Popular models like OpenAI's GPT series, Meta's LLaMA, and Anthropic's Claude are powered by the decoder-only transformer, which has emerged as the predominant design paradigm for contemporary LLMs (Dai et al., 2019). Each token in these models can only focus on itself and the tokens that come before it in the sequence because they are trained using a causal self-attention mechanism. The model is generative by nature because of this structure, which enforces autoregressive next-token prediction.

Given all of the previous tokens, this architecture model takes a sequence of tokens and predicts a probability distribution over the next token. The objective in training is to minimize cross-entropy loss between predicted next token and actual next token. At inference time, the model samples one token at-a-time. The generated token is fed back into the sequence.

Key Advantages

Simplicity of design: The objective for training is plain, involving next-token prediction only; complex mapping between input and output sequences does not have to be performed.

Efficient inference. One major plus for these models is that the use of Key-Value (KV) caching retains previously computed attention keys and values, so they do not have to be recomputed during generation. This brings down computational complexity from $O(T^2)$ per step to O(T) where T is the length of the sequence.

Unified modeling: In the same setup, only-decoder models take care of different types or kinds, like words, codes, and even pictures (when joined with see tokens).

Alignment readiness: They make room for detailed chat helpers by adding in a smooth way with after-teaching methods.

Limitations

Prefix conditioning only: It limits conditioning to a prefix of tokens. Therefore, it might look like there is some disadvantage on tasks that place a very clear distinction between input and output, such as translation.

Implicit bidirectionality: It does not have the explicit bidirectional context that encoder models do have, even though it learns to capture bidirectional relationships through large amounts of training data.

Lack sufficient controllability: In the absence of additional techniques such as special tokens or adapters, specific aspects of generation may not be easily controlled.

3.2 Sequence-to-Sequence (Encoder–Decoder) Models

Encoder-decoder architectures are otherwise referred to as sequence-to-sequence, or simply seq2seq models. They were designed so that the input and output sequences can be managed independently. It was neural machine translation architectures such as Google's T5 (Text-to-Text Transfer Transformer) and BART that first brought this design to popular attention (Press et al., 2021).

In this configuration:

By processing the input sequence in both directions, the encoder creates contextual embeddings that fully convey the input's meaning.

After that, the decoder creates the output sequence while using cross-attention mechanisms to attend to the encoder's representations.

Because of this explicit separation, the model can perform tasks like these where the input and output clearly differ from one another:

Translation by machine

Summarization of documents

Retrieving information while reranking

Answering questions

Training Goals

Most encoder-decoder models use span corruption or denoising objectives, while some use next-token prediction (Dettmers et al., 2022). For example, T5 encourages a richer understanding of structure and semantics by masking arbitrary spans of text and then training the model to predict the missing tokens.

Advantages

Flexibility: It is easy to fit encoder-decoder-based systems into complex pipelines, such as in RAG where the response generated by the decoder uses the retrieved documents as input to the encoder.

Structured conditioning: Clear separation of input and output improves performance on tasks where the two differ significantly.

Trade-offs

Increased latency: Even higher latency than just a decoder model because cross-attention layers slow down all stages of training and inference with more computation.

Higher memory usage: Even higher memory usage because more memory is required after storing the encoder representation for each token.

More complex implementation: A more intricate implementation: For optimum results, encoder and decoder sizes must be perfectly matched.

Despite these trade-offs, encoder-decoder models remain the dominant architecture for structured NLP tasks and systems where fine control over input-output relationships is required.

3.3 Encoder-only Models

Models that only use encoders BERT-based type models have no generative component. Since BERT cannot be used to generate text, it does not fall in the classic large language model category, although it is an integral part of many LLM architectures (Frantar et al., 2022). However, typical use cases where they outperform include read and fetch tasks such as semantic search, text classification, document clustering, and similarity ranking.

Retriever or reranker Encoder-only models have mostly turned popular as retrievers or rerankers in the RAG pipelines of modern environments of LLMs. As an example, a model like BERT could be used to create embeddings for documents inside a vector database such that it enables retrieval that is both very fast and highly accurate for the most relevant information to be fed into a generative LLM.

Advantages

Context flows both ways. By looking at words that come before and after, these models get a better read by taking in the whole feel of a string. They tend to be less large than fully creative LLMs and learn more quickly. Good for later controlled sharpening of skills, making field-specific embedding. View of Previous and Next Tokens. By taking both previous and next tokens into account, these models improve perception by looking at the entire context of a sequence. They are usually much smaller than fully generative LLMs and are much faster to train. Great for downstream supervised fine-tuning, domain-specific embedding generation.

Cons

Cannot generate text directly. Needs to be used together with generative models for end-to-end systems.

3.4 Mixture of Experts (MoE) Architectures

With Large Language Models (LLMs) scaling to hundreds of billions or trillions of parameters (Frantar et al., 2022), such as CoPilot and GPT-3, computation efficiency becomes paramount. Mixture-of-Experts (MoE) architectures are a useful approach of conditional computation, where only a designated subset of the model's parameters, known as experts, are made active for each token. Typically, at the token level, a router network activates the experts, maintaining processing time low although the actual parameters are very high because of the sparsity in computation from only a portion of the experts being activated. Current leading implementations include Mixtral, Switch Transformers, and GShard.

Advantage

Scalability: Allows for the development of very large models without the cost of computation scaling linearly.

Specialization: Have specialists working on different token patterns or different domains.

Challenges

The largest challenge is trimming the load balance to ensure that all specialists are leveraged equally so that no specialists is underutilized or oversaturated. Other challenges are possibilities for communication overhead—since expert computation is always split further in hybrids with hardware accelerators you will inject latency—and stabilization of training since optimization might be even more complexity with sparse updates which would probably require much more auxiliary learning loss and hyperparameter tuning.

3.5 Long-Context Variants and Effective Attention

For standard transformer attention, the calculation of extremely long contexts becomes impossible due to its quadratic nature in terms of sequence length. Specialized methods of achieving effective long-context variants are required as LLMs take on tasks with large documents, codebases, or conversations.

Key Approaches Include

1. Local or Sparse Attention: limits attention to certain patterns (for example, dilated or block-sparse patterns) or windows of fixed size eliminates a few global dependencies to bring down computation from being quadratic to almost linear.
2. Kernel/Linear Attention: uses kernel feature maps in approximating the SoftMax attention function offers sub quadratic scaling but with a potential drop in performance when compared to exact attention.

Table 1: Comparative Overview of Core LLM Architectures.

Architecture	Key Features	Strengths	Limitations	Example Models
Decoder-only (Causal)	Uses causal self-attention for autoregressive next-token prediction. Simplest design for generation.	- Efficient generation with KV caching. - Unified modeling for text/code. - Simple objective. - Easy to align with RLHF/DPO.	- Only prefix conditioning. - Bidirectional context is implicit. - Limited control over output structure.	GPT series (OpenAI), LLaMA (Meta), Claude (Anthropic)
Encoder–Decoder (Seq2Seq)	Separate encoder for inputs and decoder for outputs with cross-attention bridges.	- Explicit input-output separation. - Ideal for translation, summarization, RAG. - Flexible conditioning and controllability.	- Higher latency due to cross-attention. - Increased memory requirements. - More complex architecture.	T5 (Google), BART (Meta), mT5
Encoder-only	Bidirectional encoding for semantic understanding without generation capability.	- Strong context comprehension. - Ideal for retrieval, ranking, embeddings. - Lightweight and efficient.	- Cannot generate text directly. - Needs integration with generative models for end-to-end tasks.	BERT, RoBERTa, DistilBERT
Mixture-of-Experts (MoE)	Sparse activation of a subset of experts per token using learned routers.	- Scales to trillions of parameters efficiently. - Specialized experts improve performance. - Lower compute per step.	- Complex training with load balancing. - Communication overhead between experts. - Stability issues.	GShard, Switch Transformer, Mixtral
Efficient Attention/ Long-Context	Optimized attention mechanisms like sparse, linear, or recurrent for handling long sequences.	- Sub-quadratic attention cost. - Enables >100 k token contexts. - Essential for long documents and dialogues.	- Approximation may reduce quality. - More complex implementation and tuning required.	Transformer-XL, Long former, ALiBi, RoPE-based models

3. Recurrence and Memory Tokens: Segment-level recurrence, as in models like Transformer-XL that pass memory representations over segments. Tokens for external memory allow the model to explicitly store and fetch old data.
4. Position Encoding Strategies: Interpolation, ALiBi (Attention with Linear Biases), and RoPE (Rotary Position Embeddings) scaling are techniques that make possible context windows of 100,000 tokens or more. Stability and coherence are maintained through ultra-long-sequence training.

In domains where context length can become extremely large, for example, genome sequencing, legal document analysis, or multi-turn dialogue systems, attention mechanisms have to be highly efficient.

4. The Training Stack

4.1 Data Filtering and Curation

Web pages, books, code, math, scientific texts, and multilingual sources are examples of high-quality, varied corpora that are essential. Heuristic/learned classifiers and quality filters (language id, perplexity, toxicity heuristics) eliminate boilerplate and spam, while deduplication stops memorization. Domain balance is controlled by temperature mixing.

4.2 Goals

The autoregressive (next token) loss: maximizes $\sum_t \log p\,(x_t \mid x<t)$.

Masked/denoising losses: include bidirectional context for encoders and span corruption for seq2seq. Auxiliary losses include tool-use grounding, contrastive (for instruction embeddings), and routing balance (MoE).

4.3 Distributed Training

Data parallelism: Gradient aggregation across replicas is a form of data parallelism. Tensor/model parallelism requires fused kernels and careful layout; splits matrices across devices. Pipeline parallelism: carefully schedule bubbles and micro batch across layers.

Optimizer state partitioning: ZeRO partitions optimizer states, gradients, and shard parameters.

Table 2: Typical pretraining corpus composition (illustrative).

Source	Share	Notes
Web crawl (cleaned)	40–60%	language-balanced, de-duplicated
Books & long-form	10–20%	improves coherence and world knowledge
Code repositories	10–20%	boosts reasoning and tool-use
Academic papers	5–10%	math and technical reasoning
Multilingual web	5–15%	coverage across languages

4.4 Stabilization and regularization

Instability is decreased by dropout, label smoothing (seq2seq), adaptive gradient clipping, RMSNorm/LayerNorm selection, and meticulous initialization (1/√d). In transformers for LLMs, activation functions (SwiGLU) frequently perform better than ReLU/GELU.

4.5 Post-Training Alignment

Supervised fine-tuning (SFT): Curated instruction-response pairs are used in supervised fine-tuning (SFT).Preference learning: Direct Preference Optimization (DPO), which fits a Bradley–Terry preference model without on-policy rollouts, or Reinforcement Learning from Human Feedback (RLHF) with reward models are two methods for preference learning.

Constitutional strategies: self-criticism and revision based on rules lessens the need for human labels.

Safety filters: include red-teaming datasets, refusal policies, and prompt and output classifiers.

It starts with Figure 1. To ensure quality, it begins by gathering data followed by cleaning and filtering that data. Unprocessed information is turned into organized

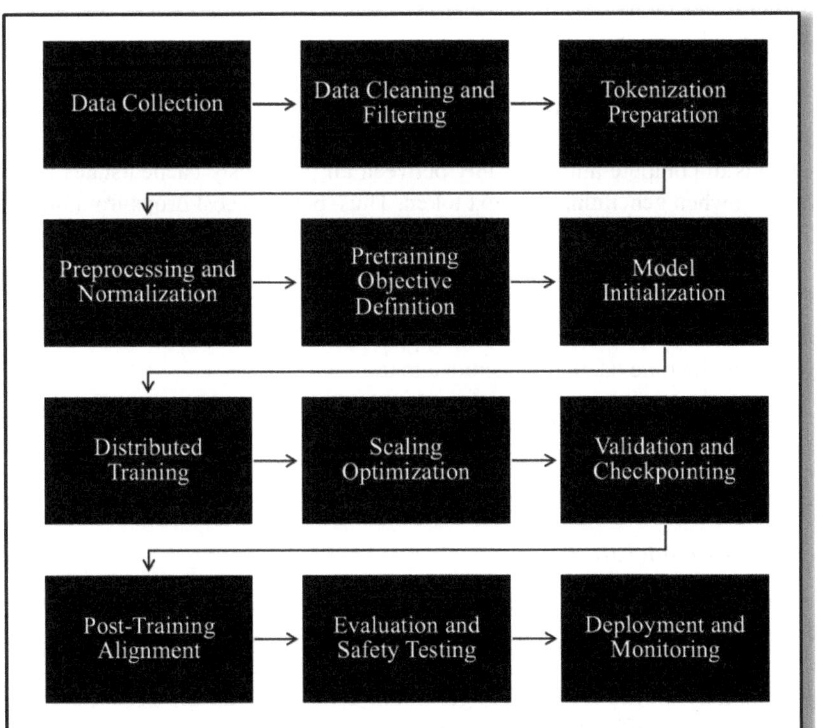

Figure 1: The pipeline for training and after-training.

forms suitable for model training through steps of tokenization preparation and preprocessing/normalization.

5. Inference and Serving Architecture

Equally important to the training procedure is the inference and serving architectures of Large Language Models. If training builds and tunes the model, then inference determines how well and quickly it can be realized in actual practice. The end goal for inference is real-time or near real-time output generation with low latency and high throughput while maintaining memory and computational costs, plus any other controllable overheads. Huge, long context windows and highly computationally complex make large-scale deployment of LLMs extremely difficult. This section discusses some of the main ingredients and techniques that account for effectiveness in inference. They comprise serving topologies, KV caching, quantization, decoding techniques, and FlashAttention optimizations.

5.1 KV Throughput and Caching

Self-attention is one of the main performance snags during LLM inference. The computational cost of naïve implementation is $O(T^2)$ per sequence where T denotes the length of the sequence. This has to be so because with every new token being generated, attention has to be recalculated over all previously processed tokens. Particularly for models with long contexts with 32k or 100k tokens, this quadratic scaling soon becomes unaffordable.

Key-Value (KV) caching is what solves it. What happens is that the model stores in cache the attention keys and values as it generates, rather than recalculating them at every step of generation, which is what would happen if not using KVs. Now the model has to compute attention only between all previously cached states and one new token when generating the next token. Thus, per-step cost drops from quadratic O(T2) to linear O(T).

Benefits of KV Caching

Reduced latency: It enables real-time applications e.g. code assistants and chatbots by reducing the time taken to generate a token.

Increased throughput: It supports the concurrent generation of multiple users or sessions.

Scalability: It avoids redundant calculation hence supporting long-context generation.

Practical Considerations

The systems ought to be resource efficient and they should understand about cache retention. There are cache eviction policies, like least-recently-used (LRU) that will clear some memory if large numbers of sessions are considered at once, while still holding users who are actively generating text. Some installations share caches across GPUs in an effort to manage very large batch sizes.

Impact on Performance

All of the optimizations mentioned allow LLMs to handle thousands of requests at the same time on significantly less hardware considerations. This is essential to making large commercial deployment.

5.2 Techniques for Decoding

Decoding policy is, at its core, describing how the model decides tokens during generation. KV cache optimization only improves the speed of the computational graph. Quality, diversity, and speed, hinge on the decoding policy.

Greedy Search

At every step in greedy decoding the token with the highest probability is picked.

Pros: Deterministic and Fast.

Cons: Can lead to repetitive and generic text, No originality or variation.

Beam Search

At each step, a few possible sequences is remembered and picks the one that seems most likely.

Advantage: By looking at several guesses, the sense is made better.

Disadvantage: The downside is computational cost plus maybe less variety since all beams go to similar outputs.

The Top-K Sampling

Top-k sampling restricts the selection to the top kk highest-probability tokens and selects a random sample from each of them rather than always choosing the most likely token.

Benefits: Enhances creativity and diversity by introducing controlled randomness. Use case: Conversational agents and creative writing.

Top-p Nucleus Sampling

Tokens from the smallest set whose cumulative probability is greater than a threshold p are sampled in top-p sampling.

Benefits: Context-adaptive; keeps uncommon, low-probability tokens from being chosen too frequently. For instance, ChatGPT and other contemporary chat systems make extensive use of it.

Scaling of Temperature

Logits are subjected to a temperature parameter τ prior to SoftMax:

The distribution is sharper and favors high-probability tokens and more deterministic outputs when τ is smaller than 1.0. The distribution is flattened by higher τ (> 1.0), which increases creativity and randomness.

Speculative Interpretation

Speculative decoding is a new method for lowering latency:

A smaller, quicker draft model makes bulk predictions about a series of tokens. These tokens are validated by the larger target LLM. The draft tokens are accepted if they are accurate; if not, changes are made immediately. For high-traffic deployments, this method greatly increases throughput by lowering the number of forward passes needed from the larger model.

5.3 Memory Footprint and Quantization

Modern LLMs typically have billions of parameters meaning they need hundreds of gigabytes of memory to store both weights and activations. Quantization solves this problem by reducing the memory and bandwidth requirements at the cost of decreasing accuracy not only for model weights but also for activations.

Quantization in 8-bit and 4-bit

8-bit quantization reduces memory usage by up to 75% with practically no performance degradation since the parameters are stored as 8-bit integers rather than as 16- or 32-bit floating-point values. 4-bit quantization reduces memory by yet another half, however, in order to preserve accuracy, careful scaling and zero-point adjustments have to be made.

Quantization After Training

Pre-trained models can also be quantized without retraining. Examples of methods that enable this include AWQ (Activation-aware Weight Quantization) and GPTQ (Generative Pre-trained Transformer Quantization).

Quantization Aware Training (QAT)

By imposing quantization effects during training, QAT develops models that are less sensitive to degradation from reduced precision.

QLoRA for Fine-Tuning

QLoRmA (Quantized Low-Rank Adaptation)

By setting most parameters to be frozen and only training a small number of low-rank adapters, it is possible to effectively fine-tune large, quantized models using QLoRA (Quantized Low-Rank Adaptation). This allows fine-tuning on consumer hardware, e.g., with memory-constrained GPUs.

Benefits of Quantization

Reduces the requirement of memory on GPU hence reduces the cost of deployment. Improves hardware usage, and energy use. Allows large models to run on small groups or edge tools.

5.4 Fused Kernels and FlashAttention

Transformers' attention mechanism is memory-bound as opposed to compute-bound. Conventional implementations create a bottleneck by requiring multiple memory reads and writes for operations such as matrix multiplications, dropout, and softmax.

I/O-aware kernel design is introduced by FlashAttention to address this:

High-bandwidth memory (HBM) traffic is significantly decreased by tiling and combining the attention computation into a single pass. Yields notable speedups, particularly for inference and long-sequence training. Gives very large context windows numerical stability.

Fused kernels are utilized to integrate several lightweight operations into a single kernel call in addition to Flash Attention. Among the examples are:

combined bias-activation-dropout and RMSNorm sequences. optimized calculations for rotary embedding.

Impact on performance:

lowers the overhead of kernel launches.Increases the use of the GPU. Allows sequences with 100,000 tokens or more to be handled steadily and effectively. Because of its scalability advantages, Flash Attention is now a common feature in many high-performance LLM implementations.

5.5 Topologies for Serving

The topologies in production—how you place LLM to work, basically depend upon the size of models, latency requirements, and available hardware resources. Some topologies that have found their place in real-world scenarios are shared below:

1. **GPU Inference on a Single Node**

   This is a setup at ground level when one single GPU has enough capacity to host the whole model.

   Advantages: Low latency, easy to implement.

   Cons: Memory of the GPU, not used for very large models. Edge deployments and small to medium models.

2. **Tensor Parallelism**

   On one node, multiple GPUs split the large weight matrices of the model. Each GPU computes partial results that will finally be aggregated.

Pros: Even larger models can be enabled which won't hit memory ceilings of one GPU. Cons: May be bottlenecked by communication overhead between GPUs.

3. **Pipeline Parallelism**

   Spreads the model over GPUs by layers. The pipeline takes care of each micro-batch sequentially.

   Pros: Does good for very deep models.

   Cons: If pipeline bubbles are not scheduled right, they reduce efficiency.

4. **Sharding over multi-nodes**

   Extends pipelines and tensor parallelism across more than one server.

   Pros: Fits very big models (e.g., GPT-4).

   Cons: Needs fault tolerance and advanced networking infrastructure.

5. **Speculative Decoding Integration**

   Mixes a model in the loop with hardware running on less expensive equipment with concurrency. While the main model checks outputs, the draft takes care of most token predictions.

The detailed procedure of Key-Value (KV) caching across transformer layers during autoregressive text generation is shown in Figure 2. A new token is added to the model at the start of the process, known as input token initialization. The embedding layer transforms the token into a numerical representation first. In Layer 1, the model retrieves the Keys (K) and Values (V) for every previously processed token from the cache and calculates the Query (Q) for the current token. After that, redundant calculations are removed by appending the newly generated K and V for the current token to the cache. Each new token adds a row to the KV table of each layer as a result of this process being repeated across several layers, creating a visualization of cache growth. The final layer output, which predicts the next token effectively and with less latency, is then produced by using the updated cache for attention computation.

Table 3: Serving topologies and trade-offs.

Topology	**Strengths**	**Limitations**	**Use cases**
Single-GPU	Low latency, simple	Limited context/params	Edge, prototypes
Tensor-parallel	Utilizes multiple GPUs per node	Intra-node bandwidth bound	Medium models
Pipeline-parallel	Fits deeper models	Bubble overhead	Very deep stacks
Multi-node sharding	Scales parameters/context	Networking/consistency	Enterprise serving
Speculative decoding	Big latency wins	Needs draft model	High-QPS chat

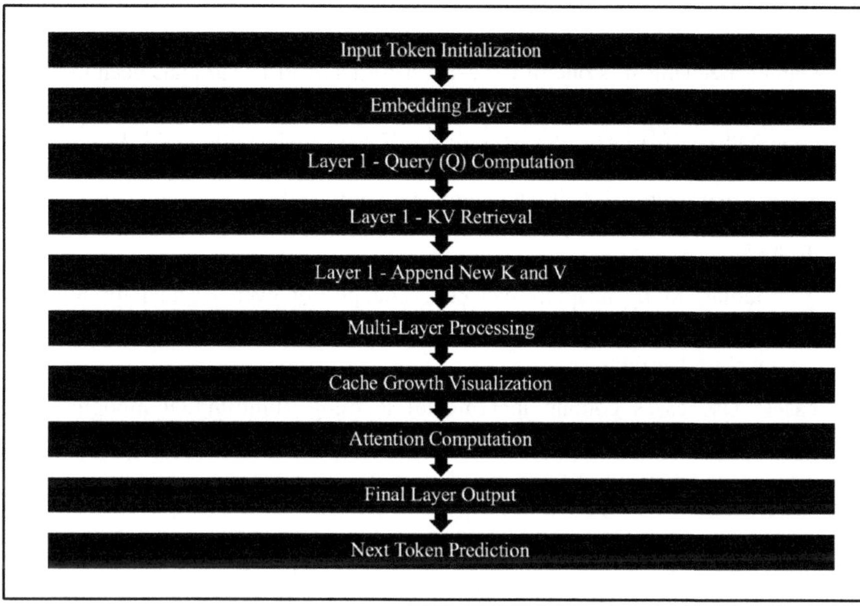

Figure 2: KV-cache layout across layers (schematic).

6. Peft or Parameter-Efficient Fine-Tuning

Large language models (LLMs) are often pretrained with enormous datasets involving billions-or even trillions-of parameters. While this pretraining does yield very powerful general models, significant challenges are presented when trying to tune these models for specific tasks or domains. Traditional fine-tuning requires changing every parameter of the model-which is computationally expensive and requires massive hardware resources. In addition, it is also not practical to store and deploy separate copies of the entire model for each task.

LLMs may be made ready for new tasks without full retraining of the model by using Parameter-Efficient Fine-Tuning PEFT. In place of that, PEFT methods only train a small number of task-specific parameters while freezing most backbone parameters. Although it still has strong capabilities, this means a significant reduction in both storages overhead compared to previous systems and computing needs. For these reasons as well, PEFT has emerged as the default methodology for scientists and organizations with limited capacity or when a scientist has multiple downstream tasks to address.

The primary objective of PEFT is to create lightweight features through both efficiency and modularity that can be attached to relatively frozen model in such a way that it can specialize on new tasks without undergoing catastrophic forgetting and/or overfitting. PEFT produced many mechanisms each one having a different combination of these two principles.

6.1 Adapters

The adapter mechanism is one of the earliest forms of PEFT and has been the most common form used in practice. Adapters work by inserting small bottleneck Multi-Layer Perceptron (MLPs) modules into the transformer architecture, traditionally inserted after feedforward and attention layers. During fine-tuning, The base model parameters are not altered. Only the parameters of the adapters are trained.

Advantages:

Highly modular: Many adapters can be exchanged or even stacked for different tasks. By picking the best adapter for every job without retraining the main part, it allows multi-task learning.

Use cases: Use cases contain incremental learning, multilingual modeling and domain adaptation.

6.2 Low-Rank Adaptation, or LoRA

Since LoRA implements a fine methodology between quality and resource usage, it has naturally become one of the best techniques under PEFT. To remember, low-rank matrices AA and BTBT are included in transformer projection layers, for instance within the feed-forward or attention networks therein inside the original architecture of LoRA. The low-rank update that LoRA trains replaces updating the whole weight matrix WW such that:

$$W' = W + AB^{\tau}$$

Where:
A and B have a much smaller rank r, significantly reducing trainable parameters.

Benefits:

Only 0.1% to 1% of the parameters of the model are required to be trained. There is only a small increase in inference latency. The reintegration of LoRA updates back into the original model for deployment is seamless. Use cases involve fine-tuning large models such as LLaMA or GPT on domain-specific downstream tasks, e.g., financial, legal, and medical text.

6.3 Prompt/Prefix Tuning

Prefix Tuning teaches the model a series of virtual tokens either by inserting them into attention keys and values or just prepending them to the input sequence. These virtual tokens guide the model's behavior towards specific tasks without altering the main weights of the model, very effective for seq2seq models like T5.

Advantages:

Extremely lightweight with < 0.1% trainable parameters. No structural changes to the model.

Limitations:

Achieving optimal performance may require careful prompt design; you might need to craft your prompts thoughtfully.

6.4 Debias Scaling and IA3

IA3 is (Input-Output-Activation Scaling). The scaling factors are learned per channel applied to the activations of the model. This does not introduce extra matrices or layers, as with LoRA or adapters. It has very high parameter efficiency because only scaling factors are trained.

Use Cases: Highly effective domain adaptation and bias reduction in LLMs.

Integration of Low-Rank Adaptation into the transformer model's attention mechanism for effective fine-tuning is depicted in Figure 3. The setup begins with input tokens and subsequently embeddings realized at the embedding layer. In turn, the attention mechanism gets activated by these embeddings passing through the transformer layer such that values (V), keys (K), and queries (Q) are outputted. Normally, these calculations are performed via a fixed projection matrix W. This method is enhanced by LoRA's low-rank adaptation module by computing low-rank updates without altering initial frozen weights through adding two small trainable matrices AA and BB.

Table 4: PEFT methods at a glance.

Method	**Tunable params**	**Placement**	**Pros**	**Cons**
Adapters	1–5%	After sublayers	Modular, composable	Latency overhead
LoRA	0.05–1%	Wq, Wk, Wv, Wo, Wff	Excellent trade-off	Rank selection tuning
Prefix-tuning	< 0.1%	Key/Value prompts	No weight edits	Longer prompt, memory
IA3	< 0.1%	Channel scales	Minimal overhead	Task-specific sensitivity

7. Utilizing Tools and Retrieval-Augmented Generation (RAG)

Large Language Models (LLMs) are designed with massive static datasets that endow them with considerable knowledge of the world. Nevertheless, the knowledge is static at the time of training, and LLMs are unable to retrieve real-time data and act upon the new facts unless they are retrained. Also, LLMs hallucinate more because the generative output depends upon their internal criteria rather than external data, which has the effect of making the output seem plausible and somewhat true.

Retrieval-Augmented Generation (RAG) integrates an external retrieval mechanism into the generation process. It grounds its responses by retrieving relevant documents or data at query time rather than generating answers purely from memory,

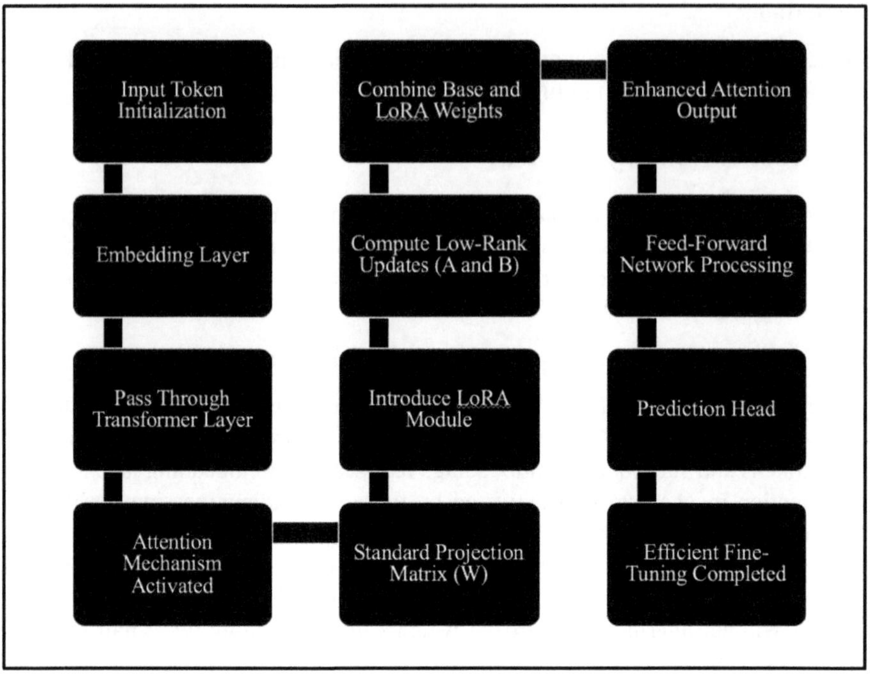

Figure 3: LoRA injection into attention projection.

which significantly reduces hallucinations as well as increases the level of factual accuracy, relevance, and adaptability.

7.1 The RAG Architecture

The RAG framework consists of four core components, each playing a crucial role in grounding the LLM with external knowledge:

1. **Dense Retriever (Bi-Encoder)**
   - A dual-encoder model, often based on BERT or similar transformer architectures, is used to generate embeddings for both user queries and documents.
   - It uses a dual encoder, mostly based on BERT or some related transformer architecture, to generate embeddings for user queries and documents. The mapped embeddings support this objective and send a similarity between documents based on an embedding space in common.
   - As a result, there should be more accurate results, as the similarity is based on semantics, and not just keyword matching.
2. **Vector Store (Approximate Nearest Neighbor Index)**
   - A vector database stores the embeddings of documents and offers an intuitive search/retrieve process.

Table 5: RAG components and options.

Component	**Typical model**	**Notes**
Dense retriever	Dual encoder (e.g., Contriever)	Embedding dimension 768–1024
Vector DB	HNSW/IVF-PQ/DiskANN	Trade precision vs. memory
Reranker	Cross-encoder (monoT5)	Improves relevance, adds latency
Generator	Decoder-only LLM	Must handle long context

- The algorithm should return the most relevant documents in seconds from the Approximate Nearest Neighbor (ANN) process using techniques such as HNSW (Hierarchical Navigable Small World graphs) or IVF-PQ (Inverted File Product Quantization).
- This step will also allow for real-time searching of large corpuses of documents ranging from hundreds of thousands to millions of documents.

3. **Reranker (Cross-Encoder)**
    - After this initial retrieval step, the top scoring candidates will be passed to a cross-encoder reranker for another pass at an even more rigorous and measured rescoring of the top scoring candidates.
    - A cross-encoder will pay more attention in this rescoring to the micro interactions between the query and documents as a means to improve document relevancy.
    - A state-of-the-art model such as monoT5, for instance, can be applied here to achieve state-of-the-art ranking performance.
4. **Generator (LLM)**

The LLM is the generator that takes the snippets that were retrieved, and it is these documents which form the basis of the answer generation. The LLM will synthesize its own internal knowledge with the external, retrieved context to come up with an accurate, supported answer. This enhances the credibility of the answer, and any tendency for hallucination to occur is decreased, because the system can now provide citation.

5. **Benefits of RAG**
    - Knowledge dynamically updated without having to retrain the base model.
    - Source-grounded responses lead to increased transparency as well as factual accuracy.
    - Flexible association with many information sources, such as data sets, news stories, and academic articles.

7.2 Using tools and calling functions

RAG enhances the factual basis of responses, although there remain some tasks that text generation cannot complete alone, such as doing calculations, executing real code, performing some small, fixed logic checks, and querying structured databases to receive information. Recently, state-of-the-art LLMs have also enabled the use of tools, allowing them to be used as orchestrators of external functionality to take controllable actions.

By using RAG for the retrieving capability, along with tools for tasks with actions, LLMs can act as agents; not only generating text, but acting as problem solvers able to reason, act, and acquire and use external knowledge. In other words, it enables and allows for the emergence of intelligent agents, safely and successfully, engaging complex, dynamic environments.

8. Assessment and Diagnosis

Evaluating LLMs is nuanced. Unlike analytic machine learning models that can only be evaluated on fixed datasets, LLMs must be evaluated for many different capabilities—reasoning, factuality, safety, efficiency. There are benchmark tasks in the field including industry standard benchmarks like MMLU, and BIG-bench, and yet there is the addition of retrieval-grounded verification for factuality checking, and self-consistency voting to check self-consistency into account. As models improve these benchmarks can soon saturate, the benchmarks can used against themselves with models leaking performance during the evaluation process (e.g., if the test set overlapped with the pretraining).

Methods should include several approaches for robustness:

- Adversarial splits from held-out suites of tasks evaluate generalization to novel data.
- To reduce hallucination there will be factuality checks integrating retrieval-grounded check, itself-consistency voting.
- Audit safety and abuse bias: to check whether a model produces harmful/abusive content or incorrect refusals for valid requests.
- Compute and environmental costs measured as latency, tokens-per-second, joules-per-token.

Diagnostics are essential for interpretation of model behavior. While tests for representation linearity provide a measure of plausible stabilizations of internal embeddings, other measures—namely slots for the attention patters of models, provide a view into the processing of information within a model's internal system. Perplexity under perturbations can also be used in a controlled manner to display both brittleness and instability under some type of perturbation. Once modes of failure are discovered, some possible forms of remediation can be employed including contrastive decoding, unlearning bad habits or possibly even exploiting modular adapters. All of this leads to an overall assessment of different models to assess the accuracy, safety and efficacy of LLMs within the operational system and practice that is ground in real operational world contexts.

9. Governance, Security and Safety

As Large Language Models (LLMs) are increasingly incorporated in mission-critical applications, adapting their safe, secure, and effective use becomes increasingly important. Safety-oriented solutions primarily focus on defining constraints to lessen unsafe actions, including malicious outputs, violations of privacy, and/or deceitful actions. Red teaming provides the ability to leverage AI for adversarial testing of a system, as in other areas, and refusal training gives the model the capacity to refuse unsafe directives. Further enhancement of safe and responsible practice could also be supported through constitutional AI, with models being provided a set of ethics-based rules.

It is important to note as well, a range of threats from a security association:

With RAG and tool using systems, prompt injections—where bad input changes the LLMs conduct and/or exfiltration of private data—are even more concerning. Jailbreaks are considered an attack on breaking the system level instructions to induce unwanted model behavior. Mitigating the issues for privacy of training data, in part represented as memorization, has to be supported through differential privacy systems and data redaction.

Finally, there are multiple methodologies, statistical or cryptographic, that may be employed to validate picture generated by the system and system to prevent abuse.

Effective governance frameworks establish accountability and transparency between both sides of the deployment; at the very least, the developers should be accountable to the users of the system. Governance frameworks should also create an incident report mechanism to identify failures and should proactively include usage policies that set forth the ethical contingencies that the system will not violate. Model cards should also document capabilities and potential risks around use. Finally, domain-specific governance to support deployment in high-stakes industries- and healthcare, for example, is informed by the data protections outlined by HIPAA and financial regulations on data security.

10. Limitations and Open Challenges

While LLMs can be encouraging, they also have innumerable weaknesses and raise research questions for which much more research should be conducted as a means toward constant and continuous progress.

1. **Power and Computation:** Reliability and Hallucinations: While training frontier-scale LLMs will involve computers running at scale and consuming energy, which creates an emergent concern about its environmental impact. Future possibilities will have better optimizers, low salience training, a smarter curriculum of learning, or waste-based selection options.
2. **Hallucination and Reliability:** Massive language models are notoriously known for generating hallucinations, and nearly all of the content that has lapsed is real, but wrong. There is now a different way to get around it: organizations can choose to use an external verifier model, and retrieval augmented generation (RAG) will greatly increase the likelihood that generated content is more factual. Organizations could never reliably eliminate hallucinations and in many, if not all cases, people cannot unambiguously validate the content of a hallucination.
3. **Long-Horizon Reasoning and Planning:** Conventional transformer models do not have the capability to handle large scale multi-staged reasoning. There are a

few hybrid models that might include reasoning modules such as search-based reasoning, program interpreters, external memory, etc.

4. **Representative and Quality of Data:** Scraped datasets from the internet are noisy and biased, but there will have to be a pretraining process, as well as a data governance process, to get better data (as well as bias mitigation).
5. **Robustness to Distribution Shift:** New domain, new style, drop in performance. It is nice to be able to use PEFT, adapters, etc., but yes with supervision.
6. **Interpretability:** LLMs are black boxes, but we are only now starting to see some of the internal notation of model behavior through the early work in mechanistic interpretability, probing, causal tracing, etc. But a substantial amount of work will need to happen prior to us being anywhere close to the levels of openness and confidence we all would like to see.

References

Brown, T. B., Mann, B., Ryder, N., Subbiah, M., Kaplan, J. D., Dhariwal, P. et al. (2020). Language models are few-shot learners. Advances in Neural Information Processing Systems 33: 1877–1901.

Dai, Z., Yang, Z., Yang, Y., Carbonell, J., Le, Q. V. and Salakhutdinov, R. (2019). Transformer-XL: Attentive language models beyond a fixed-length context. Proceedings of the 57th Annual Meeting of the Association for Computational Linguistics 2978–2988. https://doi.org/10.18653/v1/P19-1285

Dao, T., Fu, D., Ermon, S., Rudra, A. and Ré, C. (2022). FlashAttention: Fast and memory-efficient exact attention with IO-awareness. Advances in Neural Information Processing Systems 35: 16344–16359.

Dettmers, T., Lewis, M., Belkada, Y. and Zettlemoyer, L. (2022). GPT3.int8(): 8-bit matrix multiplication for transformers at scale. Advances in Neural Information Processing Systems 35: 30318–30332.

Devlin, J., Chang, M.-W., Lee, K. and Toutanova, K. (2019). BERT: Pre-training of deep bidirectional transformers for language understanding. Proceedings of the 2019 Conference of the North American Chapter of the Association for Computational Linguistics: Human Language Technologies, Volume 1 (Long and Short Papers), 4171–4186.

Fedus, W., Zoph, B. and Shazeer, N. (2022). Switch transformers: Scaling to trillion parameter models with simple and efficient sparsity. Journal of Machine Learning Research 23(120): 1–39.

Frantar, E., Ashkboos, S., Hoefler, T. and Alistarh, D. (2022). GPTQ: Accurate post-training quantization for generative pre-trained transformers. arXiv.

Hu, E. J., Shen, Y., Wallis, P., Allen-Zhu, Z., Li, Y., Wang, S. et al. (2022). LoRA: Low-rank adaptation of large language models. In International Conference on Learning Representations.

Kingma, D. P. and Ba, J. (2014). Adam: A method for stochastic optimization.

Kingma, D. P. and Ba, J. (2014). Adam: A method for stochastic optimization. arXiv.

Kudo, T. and Richardson, J. (2018). SentencePiece: A simple and language-independent subword tokenizer and detokenizer for neural text processing. arXiv

Lepikhin, D., Lee, H., Xu, Y., Chen, D., Firat, O., Huang, Y. et al. (2020). GShard: Scaling giant models with conditional computation and automatic sharding. arXiv.

Ouyang, L., Wu, J., Jiang, X., Almeida, D., Wainwright, C., Mishkin, P. et al. (2022). Training language models to follow instructions with human feedback. Advances in Neural Information Processing Systems 35: 27730–27744.

Press, O., Smith, N. A. and Lewis, M. (2022). Train short, test long: Attention with linear biases enables input length extrapolation. In International Conference on Learning Representations.

Radford, A., Narasimhan, K., Salimans, T. and Sutskever, I. (2018). Improving language understanding by generative pre-training. OpenAI.

Radford, A., Wu, J., Child, R., Luan, D., Amodei, D. and Sutskever, I. (2019). Language models are unsupervised multitask learners. OpenAI.

Raffel, C., Shazeer, N., Roberts, A., Lee, K., Narang, S., Matena, M. et al. (2020). Exploring the limits of transfer learning with a unified text-to-text transformer. Journal of Machine Learning Research 21(140): 1–67.

Rafailov, R., Sharma, A., Mitchell, E., Manning, C. D., Ermon, S. and Finn, C. (2023). Direct preference optimization: Your language model is secretly a reward model. Advances in Neural Information Processing Systems 36: 53728–53741.

Raffel, C., Shazeer, N., Roberts, A., Lee, K., Narang, S., Matena, M. et al. (2020). Exploring the limits of transfer learning with a unified text-to-text transformer. Journal of Machine Learning Research 21(140): 1–67.

Sennrich, R., Haddow, B. and Birch, A. (2016). Neural machine translation of rare words with subword units. Proceedings of the 54th Annual Meeting of the Association for Computational Linguistics (Volume 1: Long Papers), 1715–172.

Shazeer, N. (2019). Fast transformer decoding: One write-head is all you need.

Shazeer, N. and Stern, M. (2018). Adafactor: Adaptive learning rates with sublinear memory cost. Proceedings of the 35th International Conference on Machine Learning, 4596–4604.

Vaswani, A., Shazeer, N., Parmar, N., Uszkoreit, J., Jones, L., Gomez, A. N. et al. (2017). Attention is all you need. Advances in Neural Information Processing Systems, 30.

3

Large Language Models (LLMs) in Aerospace
Use Cases and Applications

Vineet Bhatia,[1,*] *Sumati Sidharth*[2] and *Bhavesh Bhatia*[3]

1. Introduction

The aerospace sector represents the intersection of human innovation and truly revolutionary technology. Because lives depend on the aerospace industry functioning without error, every aspect of it must operate flawlessly. The amount of information necessary to develop new aircraft or coordinate all of the aircraft flying in the skies is staggering. Engineers have been using careful examination, analysis, and specialized software programs developed to assist with this work for many years; however, things are beginning to change. With advances in the capabilities of LLMs and Generative AI, there are exciting possibilities. The significant characteristic of LLMs has been their ability to learn from vast amounts of written text and programming code. What makes these technologies so exciting is their ability to learn context, generate coherent responses, and solve complex problems. Applications of LLMs have the potential to help the aerospace industry improve operational efficiency, enhance the safety of all aircraft, and develop more complete solutions to difficult problems. The question moving forward is whether or not LLM tools have the capacity to assume the weighty responsibility of performing work associated with the aerospace industry (Azam-Khan et al., 2024; Prasad, 2025). LLMs have the potential to aid in various applications in the aerospace industry; some examples include the

[1] Center For Airborne Systems, DRDO Bangalore.
[2] Defence Institute of Advanced Technology Pune.
Email: sumatisidharth@gmail.com, https://orcid.org/0000-0002-1879-596X
[3] Western University Onterio Canada.
Email: bhavesh.b@protonmail.com, https://orcid.org/0009-0007-9680-3842
* Corresponding author: bhatiavineet.in@gmail.com, https://orcid.org/0009-0006-1133-8914

DOI: 10.1201/9781003771654-3

automation of monotonous documents, the examination of aircraft safety reports, assisting engineers in resolving design issues, and assisting aircraft technicians in quickly obtaining answers to their aircraft troubleshooting issues. Think about this: an LLM could read through thousands of pilot reports , engineers report and spot warning signs of safety issues much faster than a team of people could. Or it could help an engineer by suggesting different design options based on what the aircraft needs to do (Jonk et al., 2022; Nanyonga et al., 2025a).

But here's the problem. Aerospace isn't like other industries where you can afford mistakes. When planes are involved, everything has to be absolutely right. There are multiple problems prohibiting LLMs from fulfilling their potential as operable towards the aviation and aerospace industries. In addition to producing inaccurate or incomplete output due to hallucination, we still do not understand the rationale for how LLMs generate answers to questions. This creates a lack of accountability and/or responsibility in regards to sensitive data. Additionally, LLMs require a large monetary investment, as well as a significant source of computing resources, further prohibiting their use as tools for commercial aviation and aerospace (Farquhar et al., 2024).

Currently, research conducted on Aerospace Application of LLM is inconsistent throughout. Various research studies have explored various potential applications of LLM, however, there is no cohesive compilation of knowledge pertaining to LLM uses to make up a complete overview of LLMs as a potential application for aerospace. This chapter's objective is to lead the effort to systematize the different ways LLM could be applied to aerospace, to examine the evidence to determine which applications are reasonable for future development and to describe the safety requirements that must be established for those applications. The purpose of this chapter, therefore, is also to conduct an in-depth investigation into the functioning of the LLM technology, the problems with the real-world application of that technology, and how to best move forward with the future use of LLM technology.

2. Background & Theoretical Foundations

LLMs fundamentally represent a class of artificial neural networks classified as being based on the transformer architecture. The vast majority of neural networks of this type have billions to trillions of parameters (or variable weight settings), which essentially serve as adjustable weights to enable pattern recognition across massive datasets, through extensive networks of interconnected parameters. With more parameter settings available, models can develop increasingly sophisticated pattern recognition mechanisms by processing text in increasingly subtle ways. Among other things, LLMs have the capability to:

- Natural Language Understanding (NLU): The ability to comprehend the meaning, sentiment, and intent behind text.
- Natural Language Generation (NLG): The ability to produce coherent and contextually relevant text.
- Reasoning: The capacity to infer, deduce, and solve problems based on information provided in text.

- In-Context Learning: The ability to learn new tasks from a few examples provided in the prompt, without requiring explicit fine-tuning.

2.1 Taxonomy of Large Language Models (LLMs) in Aerospace

A structured taxonomy categorizing the applications, use cases, and implementations of LLMs across aerospace domains.

Through this approach, we can examine the application areas based on different data types, safety needs, and performance metrics. Figure 1 provides an overview of the taxonomy as it relates to the various applications of LLMs in the aerospace domain (Table 1 summarises this information).

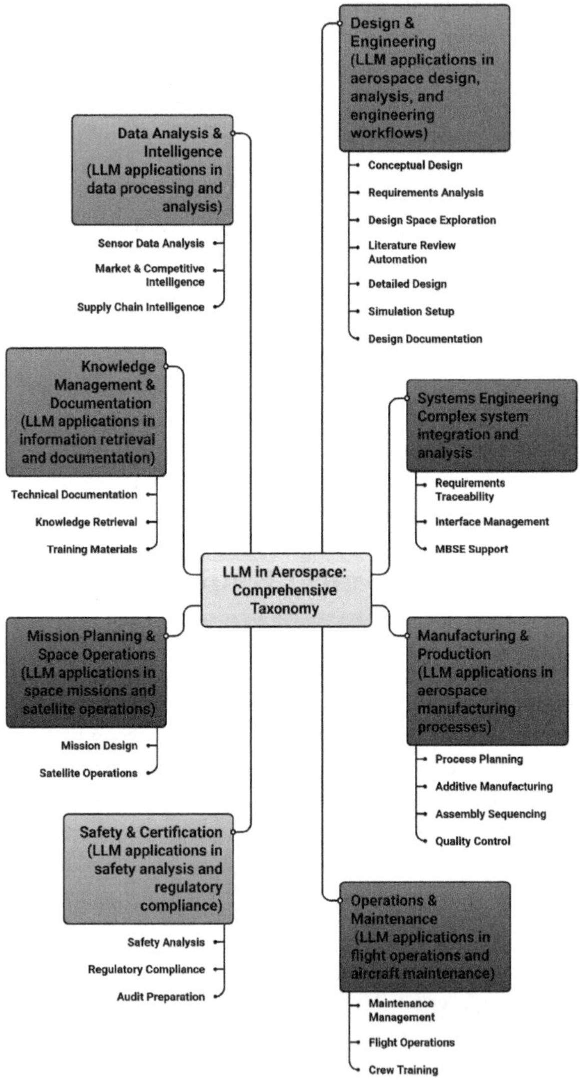

Figure 1: Taxonomy of LLM in Aerospace.

Table 1: Taxonomy of LLMs in Aerospace domain.

Category	**Application Area**	**Description**	**References**
R&D and Engineering Design	Concept Generation	Generating novel design concepts, rapid prototyping, and early-stage feasibility assessments for aircraft and spacecraft	Rane et al., 2024; Mangortey et al., 2025
	Design Optimization	Streamlining Computer-Aided Design (CAD) and Computer-Aided Engineering (CAE) workflows, providing design suggestions, and automating parameter studies	Neural, 2025
	Simulation Support	Assisting in preparing, running, or interpreting results from Computational Fluid Dynamics (CFD) and structural simulations	Avila et al., 2025; Liang et al., 2025
	Technical Documentation Generation	Drafting and updating design reports, technical specifications, and recommendations	
Aerospace Manufacturing	Process Design & Optimization	Automated generation of manufacturing process plans, assembly instructions, and process optimization suggestions	Liu, 2024
	Material Selection	Recommending materials based on requirements and regulations, synthesizing material property databases	Liu et al., 2025
	Tool and Equipment Retrieval	Efficiently searching and explaining technical manuals, tool references, and parts databases	
	Quality Assurance	Assisting in compliance checks and generating assurance arguments to demonstrate system safety and reliability	Graydon and Lehman, 2025a
Operations & Maintenance	Predictive Maintenance	Analyzing sensor data, fault logs, or maintenance reports to predict component failures and recommend interventions	Brian Connolly, 2025
	Troubleshooting Assistance	Stepwise fault diagnosis and aircraft repair guidance based on Aircraft Maintenance manual and historical data	
	Knowledge Extraction	Summarizing maintenance histories, lessons-learned, and distilling best practices from vast knowledge bases	
	Technical Query Handling	Natural language Q&A for engineers and technicians (e.g., "What's the recommended torque for a specific fastener?")	
Flight Operations and Air Traffic Management	Flight Planning	Assisting with flight path optimization, weather risk assessment, and regulatory compliance documentation	

Table 1 contd. ...

...Table 1 contd.

Category	Application Area	Description	References
Regulatory & Assurance	Documentation and Reporting	Automating compliance and event reports for regulatory bodies	Connolly, 2025
	Real-Time Problem Solving	Supporting pilots and controllers with real-time information retrieval and troubleshooting	
	Generation and Assessment of Assurance Cases	Producing or evaluating system safety arguments, hazard analyses, and compliance documentation for certification	
Domain-Specific Adaptations	Audit and Compliance	Assisting regulators and internal auditors in reviewing documentation and verifying adherence to standards	
	Data Governance in Safety-Critical Systems	Ensuring traceability, transparency, and explainability in AI-driven decisions	Lazar, 2024
	Aerospace-Tailored LLMs	Specialized models trained on aerospace engineering corpora, technical manuals, or regulatory texts to improve relevance, reduce hallucinations, and enhance reliability	
Cross-Functional and Emerging Use Cases	Knowledge Graph–Augmented LLMs	Integrating structured domain knowledge to improve factual correctness and support complex reasoning	
	Embedded/Edge Implementation	Deploying compact or quantized LLMs in avionics, maintenance terminals, or onboard systems for real-time support	
	Supply Chain and Manufacturing Optimization	Supporting procurement, supply chain risk assessment, and part traceability	
	Education and Training	Simulating technical situations, generating test questions, and providing on-demand tutoring for aerospace staff	
	Customer and Stakeholder Support	Enhancing customer query resolution, technical sales support, and stakeholder communications for commercial aircraft manufacturers and airlines	
Example Implementations and Trends	Leading Aerospace Companies	Organizations such as Airbus, Boeing, and SpaceX are actively experimenting with LLM-based tools for engineering and manufacturing process optimization	
	Commercial vs. Domain-Specific LLMs	The trend is toward customizing general-purpose LLMs (like GPT-4) with aerospace-specific data and constraints for enhanced accuracy and utility	

Table 2: Comprehensive overview of foundational technologies utilized within LLM's for Aerospace applications.

Concept/ Technology	Description	Key Features	Aerospace Relevance	References
Transformer Architecture	Neural network architecture that processes language through parallel processing rather than sequential processing	• Self-attention mechanisms • Simultaneous evaluation of all words • Superior handling of long-range dependencies	Captures connections between ideas separated by many paragraphs or pages in aerospace documentation (e.g., procedural step on page 40 referencing a warning from page 3)	Vaswani et al., 2023
Self-Attention Mechanisms	Core component of Transformers that determines which parts of a sequence matter most for understanding any given element	• Parallel processing of text • Evaluates relationships between all words simultaneously • Overcomes bottlenecks of sequential processing	Enables understanding of complex technical manuals without the limitations of word-by-word sequential processing	Vaswani et al., 2023
Fine-Tuning	Process of continuing training of a pre-trained LLM on domain-specific datasets	• Uses smaller, specialized datasets • Adapts model to specialized tasks • Improves factual accuracy and relevance • Doesn't require training from scratch	Helps models learn aerospace-specific vocabulary, conventions, and patterns from maintenance manuals, engineering specifications, and regulatory documents	Brown et al., 2020
Retrieval-Augmented Generation (RAG)	Architecture that combines document retrieval with LLM generation to ground responses in verified information	• Searches curated database before generating response • Retrieves actual manual sections • Synthesizes retrieved information into coherent response • References verifiable source material	Addresses hallucination problem by ensuring responses are based on actual approved documentation (e.g., exact torque specifications from maintenance manuals) rather than learned patterns alone	Lewis et al., 2021

2.2 Core Concepts & Models

Table 2 gives a detailed summary of the basic technologies and methodologies that facilitate the effective application of LLMs in aerospace. Use of academic language has been employed in the rewrite as well as the inclusion of sources where appropriate to support specific references made within the text.

Linguistic capabilities of LLM may be applicable in any area of Aerospace by a variety of LLM capabilities provided in Table 3.

The capacity for LLM's transformative effect across multiple areas within the Aerospace sector is further demonstrated in Figure 2 as each opportunity takes advantage of the core abilities of LLM's to solve numerous complex and domain specific issues.

Table 3: LLMs in the Aerospace sector.

Capability	Description	Relevance in Aerospace	Example Application
Natural Language Generation	Creating new, coherent text based on a prompt.	Automating documentation, report generation, and code for simulations.	Generating a flight report summary from raw sensor data.
Natural Language Understanding	Interpreting meaning and intent from text.	Analyzing pilot reports, maintenance logs, and technical specifications.	Identifying and categorizing safety issues from unstructured incident reports.
Reasoning & Inference	Drawing logical conclusions from provided information.	Troubleshooting complex system failures and optimizing complex design parameters.	Diagnosing a system fault based on a detailed text description from a technician.
Retrieval-Augmented Generation (RAG)	Generating text grounded in an external knowledge base.	Ensuring factual accuracy and preventing hallucinations in safety-critical contexts.	Answering a mechanic's question by citing from a verified, up-to-date repair manual.
Code Generation	Writing code for specific tasks.	Automating scripting for simulations and generating flight control logic.	Generating Python code for a computational fluid dynamics (CFD) simulation of an airfoil.

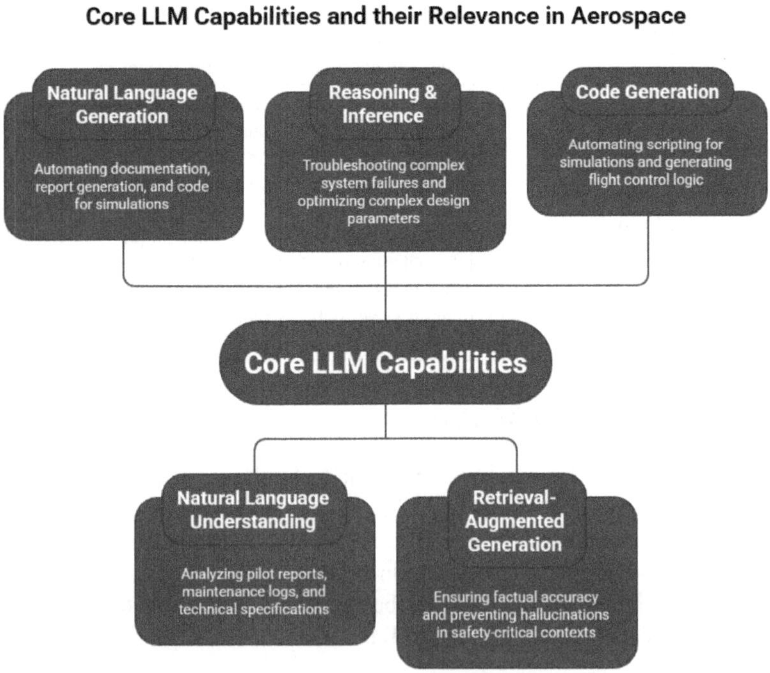

Figure 2: LLMs in Aerospace: A Functional Breakdown.

3. Literature Review

The application of AI in aerospace is not new, with traditional machine learning models long used for tasks like predictive maintenance and trajectory optimization (Abdul Shukor et al. 2025). The aerospace industry has seen significant changes in how it has used Natural Language Processing and Artificial Intelligence since its inception as a rule-based system to today's current use of large text models (see Table 4).

LLMs can act as intelligent assistants, retrieving relevant design principles from vast engineering databases and even proposing modifications to existing models (Allen et al. 2023). A critical debate in this area is the balance between generative freedom and adherence to strict engineering principles. Hallucinated designs, for instance, could lead to catastrophic failures, necessitating robust validation pipelines.

Code generation applications accelerate software development in flight control systems, simulation environments, and test automation(Burhan Sajid 2023), (He et al. 2025). AI-powered tools generate flight control algorithms, create aerodynamic testing simulation models, produce automated test scripts for avionics systems, and generate compliance documentation code that meets stringent aerospace certification standards like DO-178C. This significantly reduces development time while maintaining the high reliability and safety standards required in aerospace applications.

A visual representation of LLM in aerospace technology is provided in Figure 3.

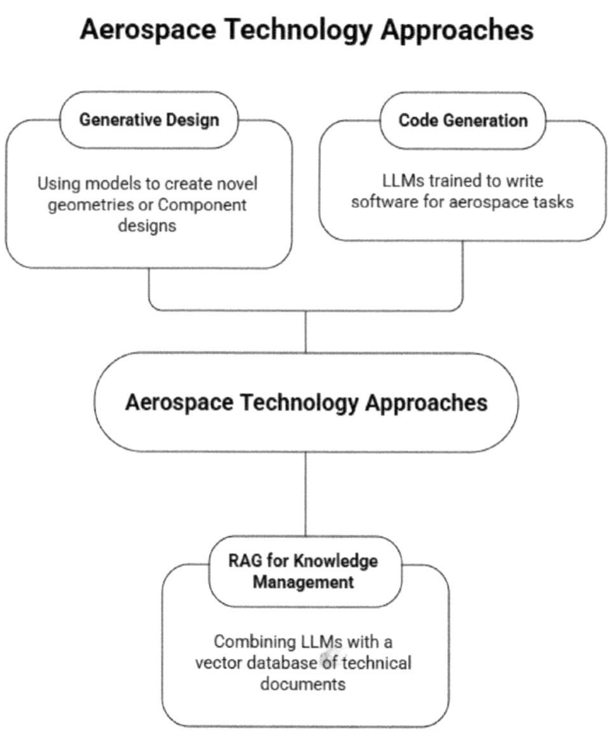

Figure 3: Key Approaches to LLM in Aerospace Domain.

Table 4: Evolution of NLP Technologies in Aerospace.

Era/Period	Technologies Used	Applications	Key Organizations	Capabilities	Limitations	References
Pre-Transformer Era (Early NLP)	• Rule-based systems • Support Vector Machines (SVMs) • LSTMs	• Sentiment analysis of pilot reports • Automated classification of maintenance logs • Parsing and analyzing Aviation Safety Reporting System (ASRS) database • Identifying safety trends	NASA, Federal Aviation Administration (FAA)	• Demonstrated value of NLP for unstructured aerospace data • Automated basic text classification • Pattern identification in safety reports	• Unable to handle semantic nuances • Limited contextual complexity understanding • Could not process natural language variations	Vincent de Vries, n.d.; Carlos Paradis, n.d.
Modern LLM Era	• Transformer architectures • Generative models • Large Language Models	• Generative design • Knowledge-based engineering • Code generation • Complex semantic analysis	Siemens, NASA, Airbus, Boeing	• Deep semantic understanding • Contextual complexity handling • Natural language interaction • Generative capabilities		

Table 5: Summary of LLM Approaches Used In Research with Associated Evaluative Metrics.

Approach	Key Idea	Dataset(s)	Metric(s)	Strengths	Limitations
Generative Design	Using models to create novel geometries or component designs.	Custom datasets of CAD models, simulation data, and engineering specifications.	Aesthetically pleasing output, reduced design time, performance metrics (e.g., strength-to-weight ratio).	Rapid concept generation, exploration of new design spaces.	High risk of generating invalid or non-manufacturable designs, lack of explainability.
RAG for Knowledge Management	Combining LLMs with a vector database of technical documents.	Internal company technical manuals, repair logs, flight manuals.	Factual accuracy (precision, recall), response latency, user satisfaction.	Mitigates hallucination, provides verifiable sources, scalable.	Requires high-quality, up-to-date knowledge base; can fail on complex, multi-step queries.
Code Generation	LLMs trained to write software for aerospace tasks.	Repositories of flight control software, simulation scripts (e.g., C++, Python).	Code correctness, efficiency, security vulnerabilities (static analysis), compliance with coding standards.	Accelerates development, reduces boilerplate code.	High potential for introducing subtle bugs or security flaws, needs rigorous human review.

Table 5 presents an abbreviated summary for the main applications, i.e., key sectors of LLMs in the aerospace industry, as well as an outline of their principal function, specific examples of where they can be utilized, and the advantages provided by these types of models.

4. Key Applications of Large Language Models in the Aerospace Domain: A Comprehensive Analysis and Use Cases

4.1 Aircraft Maintenance

This comparative Table 6 presents a wide-ranging comparison of three different Artificial Intelligence approaches to Maintenance, Troubleshooting/Diagnostic, and Practical Applications of Aeronautical Maintenance Operations.

4.1.1 How RAG Works in Aerospace

If we take an example from the aerospace industry, an engineer will need to know all the specifications of an existing component, or a technician will need a very precise (detailed) step-by-step maintenance procedure applicable to a specific aircraft model.

This diagram at Figure 4 shows the flow of information and various components in RAG in aerospace domain. This Figure visually breaks down the RAG process into these steps:

- User Query:

Table 6: AI approaches for aircraft maintenance.

Approach	Core Technology	Key Features	Data Sources & Methods	Performance Metrics	Capabilities	Advantages	Limitations	References
Hybrid RAG Based Aircraft Troubleshooting	Multi-dimensional retrieval strategy combining Knowledge Graphs (KGs) with LLMs	• Graph-based reasoning • Vector-based retrieval • BM25-based text retrieval • Interactive agent-based assistant	• Structured data (Knowledge Graphs) • Unstructured data (text) • Multi-dimensional extraction from both sources	• F1 score improvement: $\geq 4\%$ • Hallucination reduction: >7% (vs. mainstream RAG baselines)	• AI-guided troubleshooting • Context-aware diagnostics • Comprehensive information extraction • Enhanced diagnostic precision and relevance	• Robust diagnostic results • Combines structured and unstructured data • Better performance than standard RAG	Not specified	Omrani et al. 2024; Xie et al. 2025
Ontology-Integrated Fine-Tuning of LLMs	Aircraft-specific ontology integration with GPT-3.5 fine-tuning	• Hierarchical aircraft component structures • Domain-specific knowledge encoding • Maintenance log enrichment • Physical knowledge integration	• Aircraft-specific ontology • Curated maintenance logs • Encoded hierarchical component structures • Fine-tuned GPT-3.5 model	Significantly outperforms general-purpose GPT-3.5 and GPT-4.0 in component identification and maintenance recommendations	• Defective component identification • Consistent maintenance action recommendations • Advanced maintenance data analysis • Decision-making support	• Enhanced diagnostic accuracy • Better understands physical components • Improved reliability vs. conventional LLMs • Adaptable to other manufacturing domains	Limited to aviation domain initially; requires fine-tuning for updates	Wang et al. 2024

Table 6 contd. ...

RAG Based Aircraft Maintenance	Retrieval-Augmented Generation grounded in verified proprietary data	• Domain-specific knowledge base • Advanced text segmentation • Vectorization (ChromaDB) • Semantic search • Step-by-step troubleshooting	• Aircraft maintenance manuals (Cessna 172, A320, B737) • Policy/regulatory documents • Trusted knowledge base lookup before response generation	• Answer relevance: 0.973 • Content fidelity: 0.807 • Contextual relevance: 0.917	• Accurate troubleshooting support • Automated maintenance form generation • Policy/regulatory update summaries • Minimized hallucination • Reduced manual search time	• Grounded in verified data • High accuracy across metrics • Improved efficiency • Factually sound answers	• Limited aircraft coverage (3 models) • Lacks preventive maintenance features • Users request broader coverage	Hou et al. 2025

- Example Query: "What are the torque specifications for the main rotor blade bolts on the V-22 helicopter, considering the latest service bulletin?"
- Purpose: This is the initial, specific, and often complex question posed by an aerospace professional.

- Aerospace Knowledge Base
  - Content: This represents the repository of all trusted, verified, and specialized aerospace information. It includes:
    - Flight Manuals: Operational instructions for pilots.
    - Maintenance Logs: Historical repair records, inspection data.
    - Sensor Data: Real-time and historical performance data from aircraft.
    - Engineering Blueprints: Detailed design and component specifications.
    - Regulatory Documents: FAA, EASA, military standards, compliance records.
    - Service Bulletins: Updates and modifications from manufacturers.
    - Research Papers: Internal R&D, material science data.
  - Purpose: This is the authoritative source of truth that the LLM needs to reference. These documents are typically pre-processed and indexed (e.g., as vector embeddings) for efficient retrieval.
- Retrieval System
  - Process: When a user submits a query, the Retrieval System's role is to act as an intelligent librarian. It:
    - Encodes the Query: Converts the natural language question into a numerical representation (embedding).
    - Searches the Knowledge Base: Compares the query's embedding against the indexed embeddings of all documents in the Aerospace Knowledge Base.
    - Retrieves Relevant Chunks: Identifies and pulls out the most semantically similar and relevant pieces (chunks or passages) of information from the vast knowledge base.
  - Output: The most pertinent document segments are selected.
- Retrieved Context
  - Visual: A box containing bullet points or brief descriptions of the retrieved information. Arrows point from the "Retrieval System" to this box.
  - Content: These are the specific, verified data snippets pulled from the Aerospace Knowledge Base. For our example query, this might include:
    - "V-22 Helicopter Maintenance Manual - Rotor Assembly, Section 7.2.1"
    - "Service Bulletin SB-An 22-2024-005: Main Rotor Bolt Torque Update"

        - "Engineering Specification ES-Rotor-003: Bolt Material and Stress Limits"
    - Purpose: This forms the factual "evidence" that will guide the LLM's answer.
- Augmented Prompt
    - Process: The original user question is *combined* with the newly retrieved, relevant context. This creates a much richer and more specific instruction for the LLM. The prompt now tells the LLM: "Here is the information you need, now answer this specific question using *only* this information (or heavily prioritizing it)."
    - Example: "Based on 'V-22 Helicopter Maintenance Manual Sec. 7.2.1', 'Service Bulletin SB-A22-2024-005', and 'Engineering Spec ES-Rotor-003': What are the torque specifications for the main rotor blade bolts on the V-22 helicopter, considering the latest service bulletin?"
- Large Language Model (LLM)
    - Process: The LLM receives the Augmented Prompt. Its task is now to synthesize the provided context and the user's question to formulate a coherent, accurate, and relevant response. It's essentially "reading" the provided documents and then answering the question based on what it just read, rather than relying solely on its pre-trained general knowledge.
- Generated Response
    - Content: The LLM delivers a precise, actionable answer, such as:
        - "According to Service Bulletin SB-V22-2024-005 (superseding Maintenance Manual Section 7.2.1), the main rotor blade bolts on the V-22 helicopter require a final torque of 180-185 ft-lbs. Refer to Engineering Specification ES-Rotor-003 for the proper lubrication procedure during installation."
        - Crucially, it can also cite the exact sources from which the information was retrieved.
    - Benefit: This output is highly accurate, contextually relevant, and directly verifiable against the provided aerospace knowledge.

4.1.2 Technical Documentation and Knowledge Management

Technical documentation management represents most significant applications of LLMs in the aerospace domain, addressing the industry's challenge of managing vast repositories of technical knowledge spanning decades of operational history.

- Aerospace companies typically maintain large amount of documentation such as
- maintenance manuals,
- design specifications,
- regulatory compliance documents, and
- historical incident reports.

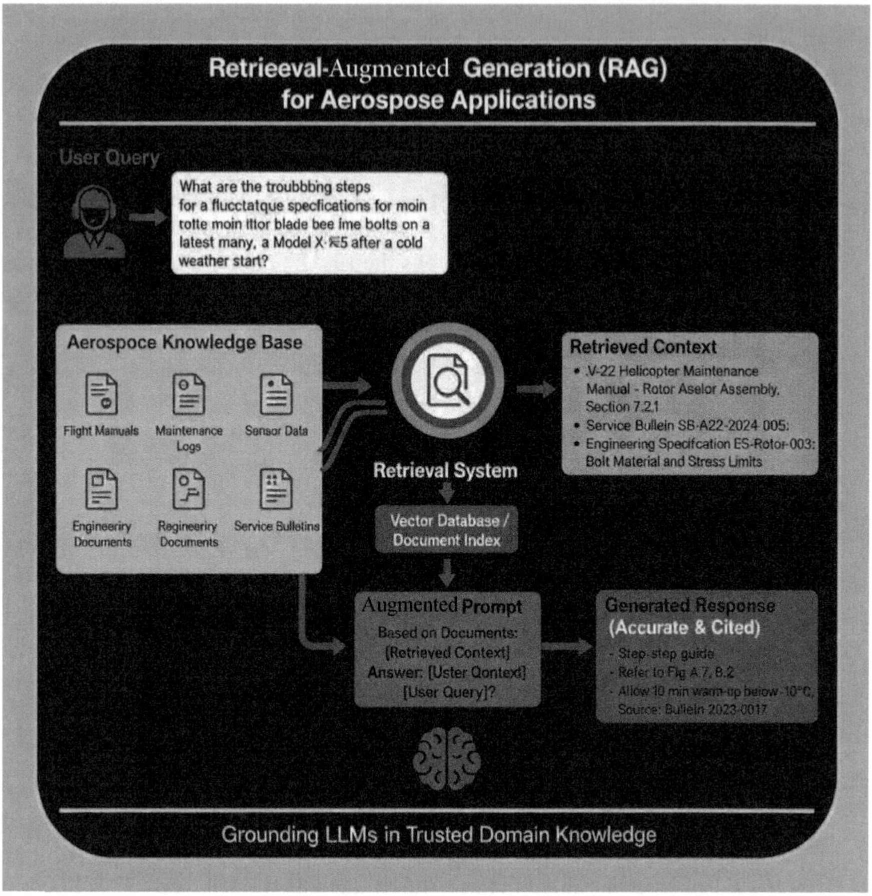

Figure 4: RAG in Aircraft Maintenance.

Traditional document management systems often fall short in providing relevant information, leading to inefficiencies and potential safety risks (Jay Lee and Hanqi Su, 2024).

The Figure 5 also illustrates the process of conversion of manual documentation which is inefficient into standard efficient documentation thus improving interactive knowledge base.

4.1.3 Aircraft Maintenance Knowledge Agent

GenAI is transforming aircraft maintenance document analysis, particularly for service bulletins and airworthiness directives that are critical but challenging for maintenance personnel to interpret due to their technical complexity and volume.

The following chart outlines HCLTech's GenAI solution for Aircraft Maintenance (Anshul Sharma, 2024), breaking down the daily tasks performed by aircraft Maintenance personnel, who use an extensively complex technical library in order to perform workstation-based tasks within the aerospace industry.

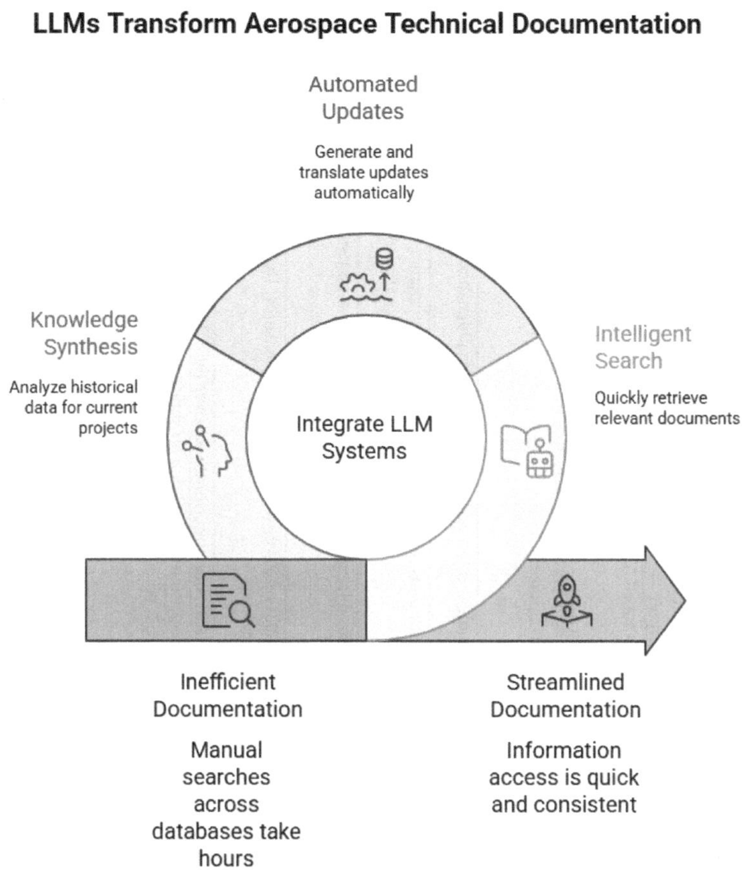

Figure 5: The Role of LLMs in Aerospace Technical Documentation.

The Figure 6 outlines key applications of Large Language Models (LLMs) that are revolutionizing aircraft maintenance by enhancing efficiency, safety, and decision-making for both technicians and engineers.

5. Aerospace Industry Use case

5.1 Boeing Intelligence & Analytics (BI&A) GenAI Framework

Boeing Intelligence & Analytics (BI&A) has established comprehensive expertise in Natural Language Processing and Large Language Models through the development of sophisticated big data analytics systems serving the Intelligence Community, supported by an internal AWS-hosted research laboratory that advances prototype development across AI/ML, NLP, GenAI, DevSecOps, MLOps, and cyber extraction technologies (Boeing Intelligence & Analytics | Strategic Solutions, 2025; Khodabandeh, n.d.).

Table 7: version: System overview for HCL Technologies Aircraft Maintenance Knowledge Agent.

System Component	Description	Details
Solution Name	HCLTech's Aircraft Maintenance Knowledge Agent	Pioneering GenAI solution for aircraft maintenance document analysis
Primary Application	Aircraft maintenance document analysis	Service bulletins and airworthiness directives interpretation
Core Challenge Addressed	Technical complexity and volume of maintenance documents	Documents are critical but challenging for maintenance personnel to interpret
Technology Stack	• Multiple Large Language Models (LLMs) • Retrieval Augmented Generation (RAG) • LLM Agent patterns	Advanced AI architecture combining multiple technologies
Key Capabilities	• Document parsing • Automatic summarization • Translation to plain language • Dynamic querying	Transforms technical documents into accessible information for mechanics and engineers
Integration Platforms	• iMRO/4 • SAP S/4HANA	Seamless integration with existing maintenance management platforms
Automation Features	• Automated updates • Automated alerts • Dynamic query processing	Continuous monitoring and notification system
Query Examples	• List relevant service bulletins for specific aircraft models • Summarize bulletin objectives • Identify required replacement parts	Natural language interaction for common maintenance questions
Operational Benefits	• Reduced manual processing errors • Improved compliance and safety • Enhanced operational efficiency	Transformative shift in MRO (Maintenance, Repair, and Overhaul) operations
Learning Capability	Continuous learning based on user interactions	System refines performance over time through user feedback
Strategic Initiative	Part of HCL Tech's AI & Cloud Native Labs	Broader initiative to deliver industry-specific GenAI applications
Business Impact	• Drive business growth • Operational excellence • Industry-specific solutions	Across aerospace and defense value chain
Target Users	• Aircraft mechanics • Maintenance engineers • MRO personnel	Technical staff requiring document interpretation support
Document Types Processed	• Service bulletins • Airworthiness directives • Technical maintenance documentation	Critical regulatory and technical documents
Reference	Anshul Sharma 2024	Published documentation on the system

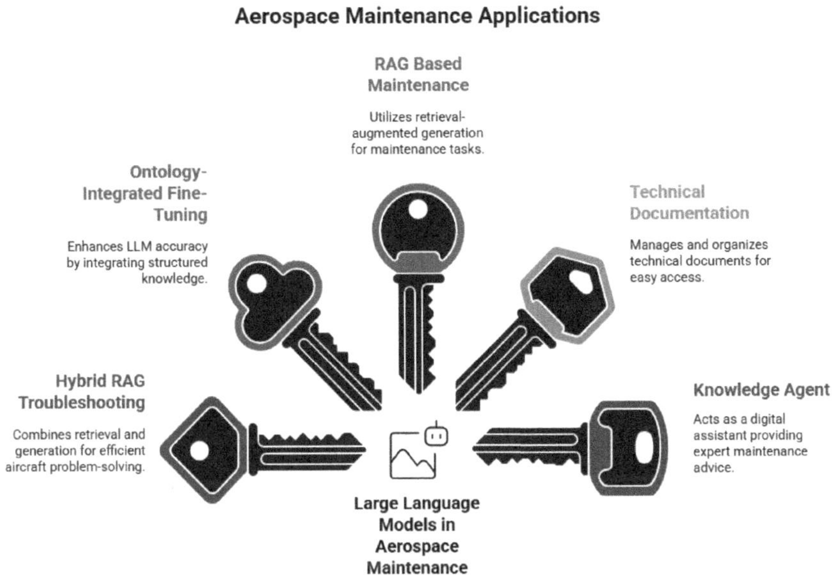

Figure 6: Use cases on of LLM In Aerospace Maintenance.

- Core Expertise: Developed comprehensive NLP and LLM capabilities through big data analytics systems for Intelligence Community, supported by internal AWS-hosted research lab.
- Strategic Partnership: Collaborated with AWS Generative AI Innovation Center (GAIIC) to create GenAI pilot framework addressing real-world intelligence analyst workflow challenges.
- Technical Infrastructure: Built on open-source foundation models, Amazon Bedrock integration, and advanced language models including Claude 2.x series.
- Architecture Design: Expanded framework supports hybrid cloud environments with diverse data sources, open-source LLMs, and open API integrations.
- Tool Chain Integration: Utilizes LangChain and LlamaIndex for sophisticated orchestration of document analysis, summarization, conversational chat, Q&A, and data extraction.
- Multi-Domain Application: Framework adaptable across aerospace engineering, manufacturing optimization, customer service automation, and technical documentation management with minimal customization.
- Security & Compliance: Security and Compliance have been placed at an Enterprise-level and consist of safety measures that limit and control access to sensitive operational environments, along with audit capabilities.

In order to help organizations stay in compliance and remain compliant, the table 8 below contains information on organizations, such as Airbus, NASA, and the Lufthansa Group, that use Large Language Models (LLMs) and Natural Language Processing (NLP) technologies for transforming their technical operations, documentation processes, and maintenance workflows.

Table 8: LLM and NLP Implementations in Aerospace Industry.

Organization	System/ Solution	Primary Application	Technology Used	Key Capabilities	Specific Features	Performance/ Impact	Example Use Cases	Strategic Benefits	References
Airbus	LLM-integrated technical documentation system	Automated manual generation and updates	Large Language Models (LLMs)	• Analyzes engineering change orders • Automatically generates maintenance manual updates • Processes complex engineering drawings • Converts technical data into step-by-step procedures • Multilingual translation capabilities	• Automatic updates to maintenance manuals, installation procedures, and service bulletins • Converts technical specifications into clear procedures • Translates documents into 20+ languages for global operations	• Reduced manual documentation effort by ~ 40% • Improved consistency and accuracy • Ensures consistency in technical communication across different regions	Engineering change orders automatically converted to maintenance manual updates, installation procedures, and service bulletins in multiple languages	• Improved documentation efficiency • Global consistency in technical communication • Clear procedures for maintenance technicians	"How Airbus Uses Generative Artificial Intelligence to Reinvent Itself | Airbus" 2024

NASA	INDUS (NASA-IBM collaborative development)	Scientific domain research and data management	Comprehensive suite of Large Language Models with custom-trained encoders and sentence transformers	• Highly specialized vocabulary and corpus • Domain-specific training for scientific tasks • Enhanced search capabilities • Knowledge extraction • Dataset linkage	• Custom-designed for Earth science, biological and physical sciences, heliophysics, planetary sciences, and astrophysics • Integration with Science Discovery Engine • Integration with Open Science Data Repository • Open distribution on Hugging Face	Significantly outperforms generic LLMs in: • Biomedical QA • Earth science entity recognition • Relevant scientific tasks	Intuitive search, knowledge extraction, and dataset linkage for researchers across multiple scientific domains	• Enhanced research capabilities • Transparency through open distribution • Broader utility for scientific community	Derek Koehl, 2024b; Bhattacharjee et al., 2024

Table 8 contd. ...

...Table 8 contd.

Organization	System/ Solution	Primary Application	Technology Used	Key Capabilities	Specific Features	Performance/ Impact	Example Use Cases	Strategic Benefits	References
Lufthansa Technik	LLM-powered virtual assistants & NLP-based prototype for AMM automation	Aircraft maintenance support and documentation automation	Natural Language Processing (NLP) prototypes and LLM-powered virtual assistants	• Natural language query understanding • Instant access to procedures and documentation • Automated manual references between AMM tasks and defect reports • Document parsing and task identification • Automatic reference generation	• Provides procedural information, safety warnings, technical specifications • Troubleshooting guides • Parses maintenance documentation • Identifies relevant tasks • Generates correct references without manual lookup • Streamlined, sequenced instruction generation	• Reduced average troubleshooting time by 30% • Improved procedural compliance and safety • Drastically reduced time for procedure lookup • Minimized human error in referencing • Faster engine turnarounds • Reduced aircraft ground time	Query Example: "Hydraulic fluid leaking from left main landing gear actuator after hard landing" System provides specific inspection procedures, fluid replacement requirements, operational limitations, and reporting requirements Workflow Example: Engineer diagnoses engine issue Prototype scans defect report Automatically determines all associated AMM tasks Generates streamlined, sequenced instructions	• Speeds up maintenance workflows • Improved accuracy and reduced rework • Technicians spend less time on documentation searches • More efficient engine turnarounds • Automated repetitive, knowledge-intensive tasks • Part of broader AI strategy for operational efficiency	Santiago Erroz Ferrer, 2024

6. Design and Engineering Support

The usage of LLMs in the areas of aerospace design & engineering (which includes studying aerodynamics, materials & structures, propulsion systems, avionics and complete systems integration) is demonstrated throughout the table below (see Table 9 - Krus 2024).

Figure 7 illustrates how Large Language Models (LLMs) can be used throughout the aerospace design lifecycle - from the initial idea to the design being implemented. This will allow designers to improve process efficiency; innovation; and compliance.

7. LLMs in Aerospace: Design and Engineering Support: Industry Use Cases

7.1 Rolls-Royce's

Study 1: University of Cambridge & Rolls Royce - Variational Autoencoders

- Technology: Variational autoencoders for turbomachinery blades (Rotor37)
- Training: CFD simulation-generated blade geometries
- Parameters: Reduced from hundreds to 15 latent parameters
- Benefits: Real-time visualization, efficiency prediction errors reduced > 50%, more accurate and computationally efficient optimization
- Reference: Pongetti et al., 2021

Study 2: Rolls-Royce & Databricks - cGANs for Engineering Design

- Technology: Identify Technology of interest Conditional Generative Adversarial Networks (cGAN) along With the Mosaic Artificial Intellect and Databricks Data Knowledge Base Technology
- Goals: Accelerate design space exploration, reuse legacy simulation data, bypass traditional geometry modeling
- Key Features: Embedding knowledge of failed designs, multi-objective optimization (weight vs. efficiency), robust ML workflows
- Benefits: Reduced cost, faster ML development/deployment, improved accuracy via rapid experimentation, enhanced data governance for aerospace compliance
- Additional: MLflow for transparency/reproducibility, future plans for 3D model generation
- Reference: Jack Kelleher et al. 2024, Databricks 2024, Siemens Software 2025

7.2 Lockheed Martin

Lockheed Martin - LMText Navigator

- System: LMText Navigator - secure, enterprise-wide generative AI platform
- Infrastructure: Powered by NVIDIA DGX SuperPOD

Table 9: The use of Generative Design and AI in Aviation Engineering.

Application Area	Key Technology/ Methodology	Primary Focus	Capabilities & Features	Benefits & Outcomes	Innovation Highlights	References
Topology Optimization	Generative design algorithms	Enhancing structural performance while reducing material utilization	• Computational design approaches • Maintains structural integrity • Achieves weight reduction objectives	• Enhanced structural performance • Reduced material utilization • Weight reduction while maintaining integrity	Establishes foundational principles for applying AI-driven optimization techniques in aerospace structural design	Satya Hanush and Manjaiah 2022
Additive Manufacturing Integration	Integration of generative design with topology optimization	Additive manufacturing applications	• Enhanced spatial exploration capabilities • AI-driven approaches identify novel design solutions • Addresses unique constraints and opportunities of additive manufacturing	• Solutions previously inaccessible through conventional engineering methods • Enhanced spatial exploration compared to traditional topology optimization	Demonstrates how AI-driven approaches can identify novel design solutions; provides critical insights into practical implementation in aerospace manufacturing	Wang et al. 2024
Reinforcement Learning for Lightweight Structures	AI-driven generative design framework utilizing reinforcement learning algorithms	Creating lightweight, manufacturable mechanical structures	• Machine learning agents explore vast design spaces • Identifies optimal structural configurations • Balances weight reduction with manufacturing feasibility	• Measurable improvements in design efficiency • Improved structural performance compared to traditional approaches • Lightweight and manufacturable structures	Groundbreaking framework showing how ML agents can explore design spaces; represents significant advancement in AI application to aerospace structural design	Kumar N et al. 2025 (PMC)

LLM-Based Conceptual Design	Large Language Models (LLMs) for aircraft design	Generating aircraft conceptual design solutions	• Natural language processing capabilities • Rapidly generates and evaluates aircraft design concepts • Intelligent automation • Rapid iteration capabilities	• Significantly accelerates early stages of aircraft development • Rapid generation and evaluation of design concepts	Paradigm shift in conceptual design methodologies; transforms traditional design processes through generative AI	Tong et al. 2025 (Chinese Journal of Aeronautics)
Physics-Based AI Development	Physics-based AI applications; physics-informed AI models	Aerospace design optimization for aerostructures	• Large geometry models trained on numerical simulation data • Aerostructures design acceleration • Maintains accuracy and reliability	• Significantly reduced design cycle times • Maintained accuracy and reliability • Measurable improvements in design efficiency and time-to-market	Demonstrates practical benefits of advanced AI techniques in aerospace design	Graham Warwick 205 AD (Aviation Week)
Generative Design (Comprehensive)	Algorithms linked to physics-based simulation and topology optimization; Integration with AI and LLMs	Product design iteration and optimization across mechanical, electronic, and electrical domains	• Explores dozens or hundreds of design iterations • Balances performance, weight, cost, and manufacturability from the outset • Automates geometric form and documentation processes • Enables automation across multiple engineering domains			

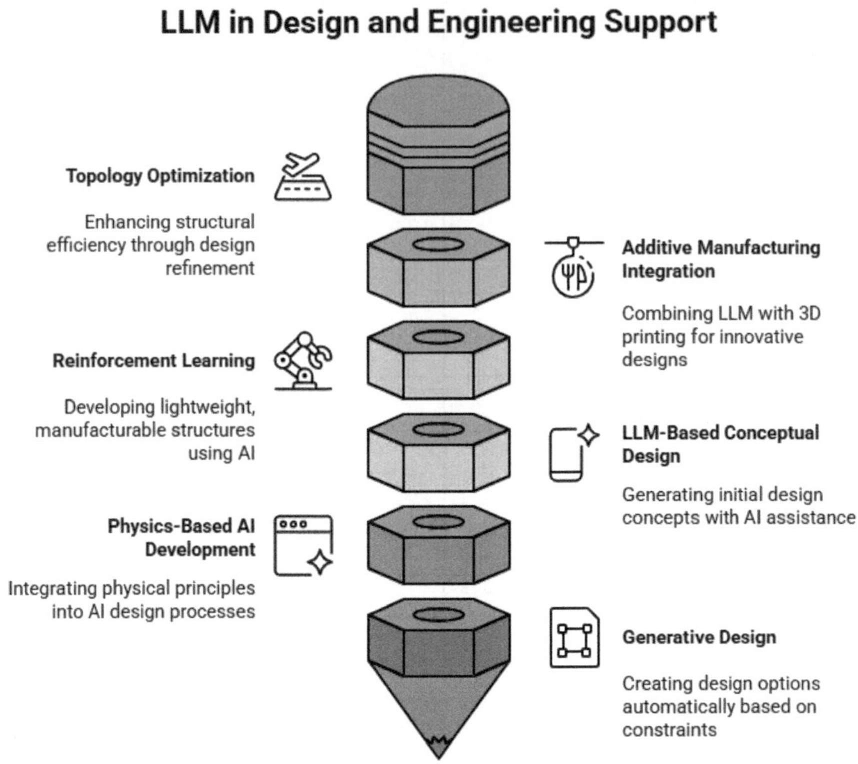

Figure 7: The Role of LLMs in Aerospace Design.

- Development: Internally developed by AI Factory with 1LMX transformation team
- Scope: Enterprise-wide across U.S. operations
- Key Applications:
  - Post-mission analytics
  - Code generation
  - Extracting insights from complex production documentation
- Key Features: Strict data security, supports high-assurance applications
- Integration: Part of Lockheed Martin AI Center (LAIC)
- Approach: Scalable, ethical, and centralized AI implementation
- Scale: Unites over 8,000 engines
- Reference: Mike Baylor 2024

7.3 Space X

LLMs represent an innovative approach to meeting the needs of rocket design by allowing designers to evaluate the range of requirements that define a successful rocket design. Table 10 shows how LLMs have been integrated into every step of the development of a rocket, from the first design to operating the operation, in order to optimize performance, safety, and the cost of operating a rocket (Toby Simonds, 2025).

Key application areas include spacecraft and subsystem design optimization, adaptive mission planning and onboard control, and advanced real-time decision support for critical operations—all of which benefit from LLM-driven insights, automation, and predictive analytics, helping to create smarter and more resilient space missions (Li and Chen, 2025a).

7.4 GE Aviation

GE Aviation deploys LLMs across multiple engineering workflows and detailed engineering specifications. The Table 11 summarize application of LLM in various domain by GE Aviation.

8. Predictive Maintenance and Diagnostics

Large Language Models (LLMs) revolutionize predictive maintenance by processing both structured (sensor data) and unstructured data (logs, notes, manuals). LLMs analyses vast amounts of textual maintenance data, including work orders, inspection

Table 10: Space X application of LLM in various domain.

Application Area	Technology/Method	Specific Use Cases	Benefits
Design Optimization	LLMs with high-fidelity simulations	Multi-objective optimization considering altitude targets, cost constraints, and operational factors	Accelerated design iteration cycles, improved subsystem performance
Flight Path Planning	Reinforcement learning-enhanced LLMs	Motor selection, material choices, mission economic pressure analysis	Outperforms expert engineers and baseline models in trade-off navigation
Real-Time Operations	AI-integrated sensor systems	Landing trajectory adaptation, environmental uncertainty response	Precision landings, improved safety across multiple launches
Mission Planning	Machine learning algorithms	Dynamic task scheduling, trajectory calculations	Enhanced mission scalability and cost reduction
Structural Analysis	AI-powered modeling	Structural integrity checks, material stress analysis	More accurate and efficient engineering assessments
Fuel Optimization	LLM-based modeling	Fuel usage optimization, propellant tank design	Reduced operational costs, improved efficiency
Reusability Enhancement	Continual learning AI models	Stage recovery optimization, landing gear analysis	Cost-effective reuse, minimized launch costs
Manufacturing	Digital twin integration	Production line modeling, resource allocation	Faster build times, improved quality assurance
Predictive Maintenance	AI-driven analysis	Usage and stress history evaluation	Optimized maintenance scheduling, reduced downtime

reports, pilot logbooks, and manufacturer bulletins, to predict equipment failures and optimize maintenance schedules (Myohanen, 2023).

LLMs have been applied for the diagnostics and maintenance of aircraft propulsion systems for o identifying anomalies, predict failures, and offer maintenance recommendations.(Klekowicki et al., 2024).

Areas where LLM can be applied in predictive maintenance are as follows (Anil Abraham Kuriakose, 2025)

- Natural Language Processing (NLP).
- Advanced Pattern Recognition.
- Anomaly Detection & Root Cause Analysis(Palma et al., 2025).
- Prescriptive Maintenance Recommendations
- Integration with IIoT & Edge Computing.
- Human-AI Collaboration & User Experience.
- Predictive Maintenance Analytics for Business Intelligence:

Table 11: Application areas, description and associated benefits (GE Aviation).

Application Area	**Description**	**Key Benefits/Outcomes**
Domain-Adapted AI Solutions	Internal LLMs trained on proprietary GE data and specialized engineering language for engine design and component analysis	Processes engineering jargon, generates technical content, reduces document creation time
Generative Design	LLMs combined with multi-physics simulation to optimize engine components like turbine casings through Clean Sky 2 program collaboration	Improved thrust and efficiency, reduced environmental impact and material waste
AI-Enabled Decision Support	LLMs integrated with machine learning for predictive maintenance, component inspection, and anomaly detection	Reduced inspection times, enhanced detection rates, regulatory compliance support
Requirements Translation	LLMs convert high-level performance goals (fuel efficiency, reliability) into concrete design constraints and specifications	Accelerated innovation, reduced ambiguity in component development, faster design refinement
AI Wingmate Platform	Company-wide generative AI platform used by 53,000 employees for daily work tasks and problem-solving	Streamlined workflows, freed up time for innovation, enhanced productivity across operations
Predictive Maintenance	AI-based digital twin models using LLMs to forecast work scopes and parts requirements months before engine service	Improved material availability, optimized engine routing, better labor planning
Cross-Functional Collaboration	LLMs support communication between engineering teams by translating complex requirements into understandable specifications	Enhanced collaboration, reduced miscommunication, accelerated development cycles

Figure 8 illustrates the application of Large Language Models (LLMs) in aerospace predictive maintenance.

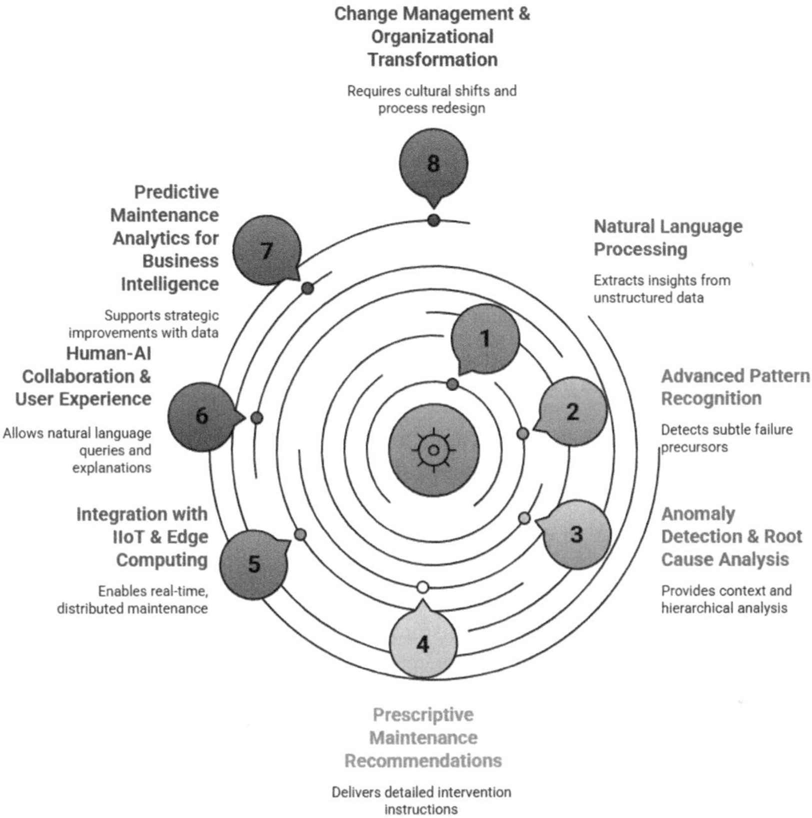

Figure 8: Leveraging LLMs for Predictive Maintenance in Aerospace.

9. Use Cases of LLM for Predictive Maintenance in the Aerospace Industry

The following Table 12 provides a comprehensive overview of the Mistral 7B implementation for predicting aircraft engine Remaining Useful Life (RUL), demonstrating how LLMs are transforming aircraft maintenance from reactive to proactive approaches.

10. Flight Operations and Air Traffic Management

The Table 13 shows how large language models (LLMs) are applied in flight operations and air traffic management. The subsequent graphic in Figure 9 shows examples of the application of LLMs or large language models in flight operations.

11. Aerospace Safety and Risk Assessment

In order to transition their safety management systems from being reactive to proactive, aerospace safety analysts use NLP and LLMs to analyze past incident records along with the development of a fully automated risk assessment program. Data has been compiled into a Data Table (Safety Case Data Table 14).

Figure 10 illustrates the application of Large Language Models (LLMs) in aerospace Safety and compliance.

12. Customer Service and Training

Using Large Language Models (LLMs) in customer service and training activities supports the aerospace industry's critical need for scalable, consistent, and high-quality customer support across a wide variety of operational environments.

12.1 Aviation Theory Training Domain

The aviation training industry struggles to effectively train and prepare pilots because the number of instructors available to provide flight training is limited and the majority of the information found online regarding flight training is inaccurate; as a result, has a negative effect on the overall efficiency of pilot training. LLMs can potentially be used to improve the way that knowledge is delivered to pilots; however, the vast majority of basic pretrained LLMs will not be accurate enough to deliver training to all aviation based disciplines. RALA-DPO (Retrieval-Augmented LLM Alignment with Direct Preference Optimization), (RALA-DPO) will align the open-source Qwen model to aviation-based request types using DPO and RAG methods. This approach addresses the limitations of traditional supervised fine-tuning, minimizes hallucinations, and provides accurate, continuously updated aviation knowledge responses without additional cost. With RAG integration, the system also achieves reliable, continuously updated answers at no additional cost. (Wan et al. 2025).

12.2 Pilot Training Support

Lufthansa Technik has implemented LLM-powered virtual assistants, "George," the first 100% GPT-based bot, answers prospective pilot queries on the European Flight Academy site, available round the clock to reduce staff workload and improve candidate experience.

Figure 11 illustrates the application of Large Language Models (LLMs) in Pilot Training Support

13. Supply Chain and Logistics Optimization

Supply chain management in aerospace involves complex networks of suppliers, stringent quality requirements, and long lead times for specialized components. The industry faces unique challenges including traceability requirements, qualification processes, and the need to maintain multiple supply sources for critical components.

Table 12: Mistral 7B to predict remaining useful life (RUL) of aircraft engines.

Aspect	Details
Background	Aircraft engine maintenance requires accurate prediction of when components will fail to ensure flight safety and optimize maintenance schedules. Traditional methods rely on scheduled maintenance intervals, often resulting in unnecessary maintenance or unexpected failures
Problem Statement	Airlines and aircraft manufacturers need intelligent systems that can predict the Remaining Useful Life (RUL) of aircraft engines using sensor data, enabling condition-based maintenance instead of time-based maintenance
Solution	Mistral 7B LLM Implementation
Technology Stack	• Base Model: Mistral 7B-Instruct fine-tuned with Parameter-Efficient Fine-Tuning (PEFT/LoRA) • Dataset: NASA Commercial Modular Aero-Propulsion System Simulation (C-MAPSS) • Training Approach: Domain-specific fine-tuning on aviation maintenance data
Implementation Details	• Data Processing: Multivariate time series from aircraft engine sensors (temperature, pressure, vibration) • Feature Engineering: Rolling window aggregation and statistical feature extraction from sensor readings • Model Architecture: Bidirectional LSTM combined with fine-tuned Mistral 7B for natural language interpretation • Training Platform: Google Cloud Platform with GPU-supported virtual machines
Key Capabilities	Natural Language Rule Queries: - Technicians may inquire as follows: "What is the remaining life of Engine #N4567 based on the temperature spikes over the past week?" - The system will respond using conversational language with predicted certainty intervals and provide the technician with recommendations for required maintenance. Automated Reporting: - The system generates maintenance reports to technicians that are easy for the technician to read. - The reports provide the technician with a human-readable translation of the technical predictions to create actionable maintenance work. Real-Time Assessment: - The system produces RUL assessments of engine components from real-time streaming sensor data. - The system integrates into the technician's current maintenance management system.
Technical Results	• RMSE Values: 11.8 (FD001), 23.0 (FD002), 14.6 (FD003), 22.3 (FD004) on C-MAPSS datasets • Model Accuracy: 31.40% on aviation-specific question-answering tasks • Processing Efficiency: Real-time analysis with minimal computational overhead
Business Impact	• Maintenance Cost Reduction: 25-40% through optimized scheduling • Safety Enhancement: Early detection of potential failures before critical thresholds • Operational Efficiency: Reduced unplanned maintenance events

Table 12 contd. ...

...Table 12 contd.

Aspect	Details
Use Case Scenario 1: Predictive Maintenance Planning	Maintenance teams receive natural language alerts: "Engine A320-001 shows degradation patterns suggesting RUL of 150 flight hours. Schedule inspection within the next 2 weeks"
Use Case Scenario 2: Real-time Decision Support	During pre-flight checks, technicians query: "Is this engine safe for the 8-hour transatlantic flight?" System responds with risk assessment and recommendations.
Use Case Scenario 3: Training and Knowledge Transfer	New technicians interact with the system to understand engine behavior patterns and maintenance procedures through conversational AI.
Future Enhancements	• Integration with autonomous maintenance agents for fully automated PHM workflows • Expansion to other aircraft components beyond engines • Multi-modal integration combining sensor data, maintenance logs, and visual inspections
Key Transformation	Transforms traditional aircraft maintenance from reactive to proactive, enabling safer and more cost-effective aviation operations
Reference	Frank Morales Aguilera 2025

LLMs address these challenges by analyzing supplier communications, contract documents, Model for solving and interpretation of complex transportation problems quality reports, and logistics data to optimize procurement decisions, identify risks, and improve overall supply chain efficiency (Srivastava et al., 2024; Rahman et al., 2025; Tyagi, 2024). Table 15 summarize the application of LLMs in SCM domain.

Figure 12 provides a comprehensive overview of Chainbrain's key areas of specialization.

14. LLMs in Commercial Aerospace

Large language models (LLMs) are pushing the commercial aerospace sector towards a new era of "innovation intelligence." This is primarily due to their ability to enhance both the operational and design processes, which were previously highly dependent on human expertise and institutional knowledge (Li and Chen, 2025b).

Operational and Design Process Enhancement

LLMs are revolutionizing mission planning and control by analyzing vast datasets to create more efficient and safer mission plans. This includes tasks like optimizing flight paths and managing fuel consumption. In the design phase, LLMs accelerate innovation through generative design, helping engineers create new and optimized component structures by sifting through technical data and research papers to provide novel insights.

Table 13: LLMs in flight operations and Air Traffic Managem.

Aspect	Details
Application Domain	Flight operations and air traffic management - frontier where artificial intelligence meets real-time operational decision-making in safety-critical environments
Data Sources in Modern Aviation	• Pilot reports • Air traffic control communications • Weather briefings • Operational messages • Enormous amounts of textual data generated continuously
Core LLM Capabilities	Process information in real-time to support decision-making processes that directly impact flight safety and operational efficiency
Primary Challenge Addressed	Managing increasingly complex airspace with growing traffic volumes
Implementing Organization	Federal Aviation Administration (FAA)
Current Research Focus	Developing systems that can automatically analyze controller-pilot communications to identify: • Potential safety risks • Communication errors • Procedural deviations
Communication Analysis Capabilities	LLM system processes radio transcripts to identify: • Non-standard phraseology • Potential misunderstandings • Situations requiring additional attention from supervisory personnel
Specific Detection Examples	• Communications where pilots acknowledge incorrect altitudes or headings • Patterns of communication breakdowns that may indicate controller workload issues • Equipment problems • Situations flagged for supervisory attention
Safety Impact	Real-time identification of safety risks and communication errors in controller-pilot interactions
Operational Benefits	• Enhanced flight safety • Improved operational efficiency • Early detection of potential issues • Reduced communication-related incidents
References	Zhang and Mott 2024, Cunnison 2025

The Need for Domain-Specific LLMs

However, universal LLMs are not a suitable solution for this field. The aerospace industry's safety-critical nature demands extreme accuracy, which general-purpose LLMs cannot guarantee. A single "hallucination" or factual error could lead to catastrophic failure. Therefore, the future of LLMs in aerospace lies in the development of specialized, domain-specific LLMs trained exclusively on aerospace data. This approach ensures the models prioritize robustness and reliability, making them trustworthy tools for a high-stakes industry.

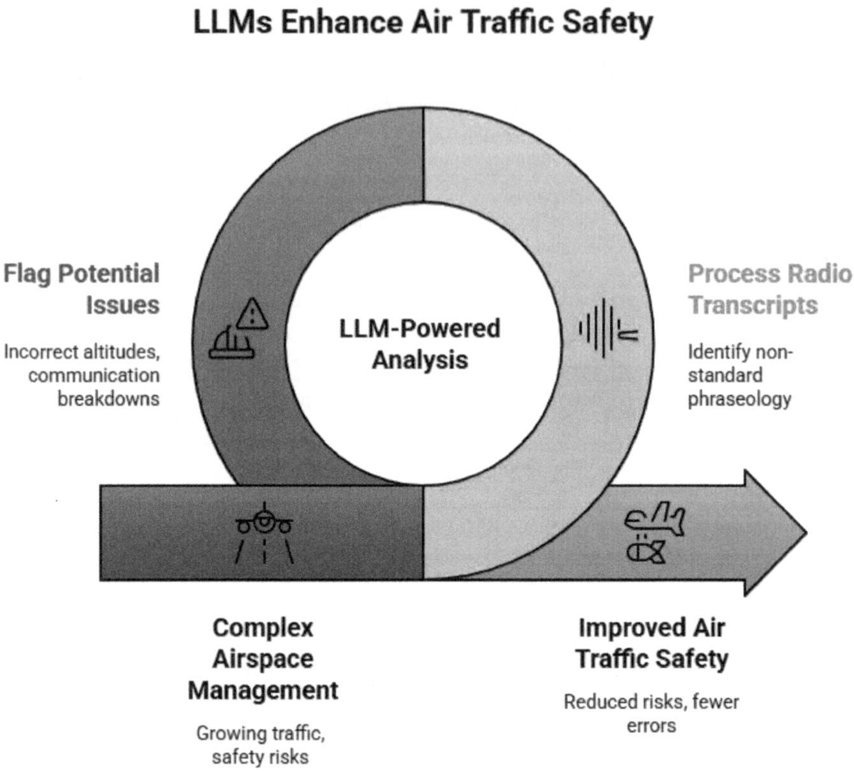

Figure 9: LLM in Flight Operations.

15. Research and Development Acceleration

LLMs accelerate synthesizing knowledge from multiple disciplines, analyzing vast amounts of technical literature by analyzing technical publications, patents, and research reports to extract relevant insights and identify promising research directions.

15.1 Industry Use Case

15.1.1 NASA

NASA's Jet Propulsion Laboratory (JPL) is using Large Language Models (LLMs) and other AI tools to analyze vast quantities of scientific literature, plan complex missions, and accelerate space exploration research. This work is being done in collaboration with other institutions, including IBM.(Kaylin Bugbee, 2023).

Analyzing scientific literature and data

NASA uses LLMs and generative AI to find relevant research and process data more efficiently.

Table 14: Complete breakdown of the different ways LLM applications are used in Aerospace Safety and Compliance.

Application/ Initiative	Organization	Primary Purpose	Technology/ Methods	Key Capabilities	Specific Use Cases/ Examples	Results/ Benefits	Challenges	References
General Aviation Safety Analysis Context	Aviation Industry	Transform safety management from reactive to proactive approach	Large Language Models (LLMs) like GPT and Claude	• Process vast amounts of safety-related documentation • Analyze unstructured textual data • Extract insights from safety reports, incident investigations, maintenance logs • Identify trends, root causes, and preventive measures	• Processing narrative-based incident reports • Capturing crucial context missed by traditional structured data • Identifying trends, emerging hazards, and potential mitigation strategies	• Speeds up analysis • Minimizes human errors during manual review • Enables proactive, data-driven approach • Previously unimaginable scale of analysis	• Need for high-quality input data • Greater transparency in model reasoning • Adaptation to specialized aviation terminology	Carles Molins et al., 2023
Safety Reporting Lifecycle Enhancement	Aviation Industry	Improve entire safety reporting lifecycle	LLMs with fine-tuning and Retrieval-Augmented Generation (RAG)	• Fine-tuning on aviation-specific texts • RAG for real-time domain-specific knowledge injection • Conversational chatbots for report quality improvement • Automated collection and cleaning of reports	• Improving quality of initial occurrence reports through chatbots that guide users and ensure data completeness • Automating collection and cleaning • Identifying duplicates and correcting errors • Clustering report data • Uncovering hidden risks • Highlighting training needs	• Highly relevant and accurate insights • Improved data reliability • Powerful decision-support tool • Smarter, more predictive, safer aviation environment	Need for adaptation to specialized aviation terminology through fine-tuning and RAG	Carles Molins et al., 2023

Table 14 contd. ...

...Table 14 contd.

Application/ Initiative	Organization	Primary Purpose	Technology/ Methods	Key Capabilities	Specific Use Cases/ Examples	Results/ Benefits	Challenges	References
System Safety Management Transformation (SSMT)	Federal Aviation Administration (FAA)	Automate mapping of incident narratives to risk models	Large Language Models (LLMs)	• Rapid question-answering with factual references to FAA regulations • Process vast volumes of aviation reports • Automate tasks previously limited by manual capacity	Mapping incident narratives to risk models	• Efficiency gains reducing analysis time from days to seconds • Scalability demonstrated	• Interpreting aviation-specific terminology • Maintaining template adherence • Ensuring consistent performance across LLM platforms • Ongoing validation and benchmarking needed • Plans for in-house deployment	Tikayat Ray et al., 2023
NLP in Aviation Safety (General Methods)	Aviation Industry	Efficient, objective, and scalable analysis of unstructured safety reports	NLP and machine learning methods: singular value decomposition, topic modeling (STM, LSA), deep learning models (BERT, LSTM)	• Rapid and accurate categorization of safety narratives • Uncovering latent risks, human factors, and emergent trends • Automating identification of causal factors and human error patterns • Improved predictive maintenance • Strengthened safety culture assessments	Analysis of vast collections of unstructured safety reports to identify risks before they escalate into critical incidents	Manual review is slow and error-prone; NLP provides rapid, accurate analysis preventing incident escalation	• High-quality data annotation • Model interpretability • Integration with existing aviation systems • Solutions: explainable AI, cloud-based modular platforms, professional training	Nanyonga et al., 2025b

Boeing Safety Report Analysis	Boeing	Safety report analysis and design improvement	LLMs for unstructured text analysis	• Process safety reports from airlines, pilot feedback, maintenance observations • Extract key safety concerns • Correlate concerns with specific aircraft systems or operational procedures	Identified pattern of pilot reports describing difficulty with flight management system interface during high-workload situations, leading to user interface redesigns that improved pilot situational awareness	Contributed to identifying and resolving over 200 potential safety issues before they resulted in significant incidents		
NASA ASRS - Voluntary Safety Reports	NASA Aviation Safety Reporting System (ASRS)	Enhance analysis of voluntary safety reports	LLMs integrated into ASRS	• Process thousands of de-identified reports monthly • Categorize incidents • Identify trends • Extract safety lessons • Understand nuanced language in pilot reports	• Distinguishing between different types of weather encounters • Categorizing severity of various system malfunctions • Sharing safety lessons with aviation community	• Improved timeliness and accuracy of safety alerts and recommendations • Continuous improvement in aviation safety culture		Graydon and Lehman, 2025b; Falkner, 2024

Table 14 contd. ...

...Table 14 contd.

Application/ Initiative	Organization	Primary Purpose	Technology/ Methods	Key Capabilities	Specific Use Cases/ Examples	Results/ Benefits	Challenges	References
NASA ASRS - Incident Report Summarization	NASA ASRS (Study by Ehaab Basil)	Automatically summarize ASRS incident reports	Large Language Models (T5-Base and BART)	• Fine-tuned transformer models • Automated summarization of unstructured aviation safety narratives	Addressing challenge of manually analyzing growing volume of 1107 ASRS reports from 2018-2024	• T5-Base outperformed BART (ROUGE-L: 0.357, cosine similarity: 0.610) • Successfully captured technical details • Accelerated safety analysis	• Both models struggled to represent human factors (fatigue, distraction) • Token constraints with lengthy narratives • Need for domain-specific improvements • Human-in-the-loop approaches required for comprehensive safety risk identification	Basil, 2024
FAA Prototype - Automated Safety Event	Federal Aviation Administration (FAA)	Automate mapping of safety event narratives into structured risk assessment models	Large Language Models with Natural Language Processing	• Rapidly process vast repositories of incident reports • Extract and synthesize key risk precursors and barrier failures • Transform manual, expert-driven tasks into scalable processes	Processing safety event narratives and mapping to risk assessment models	• Improved speed, consistency, and transparency of safety analyses • Enables safety teams to proactively identify hazards • Assess risk mitigation strategies with greater confidence • Data-informed safety decision-making • Effective predictive risk management	Not specified	ICAO, 2024

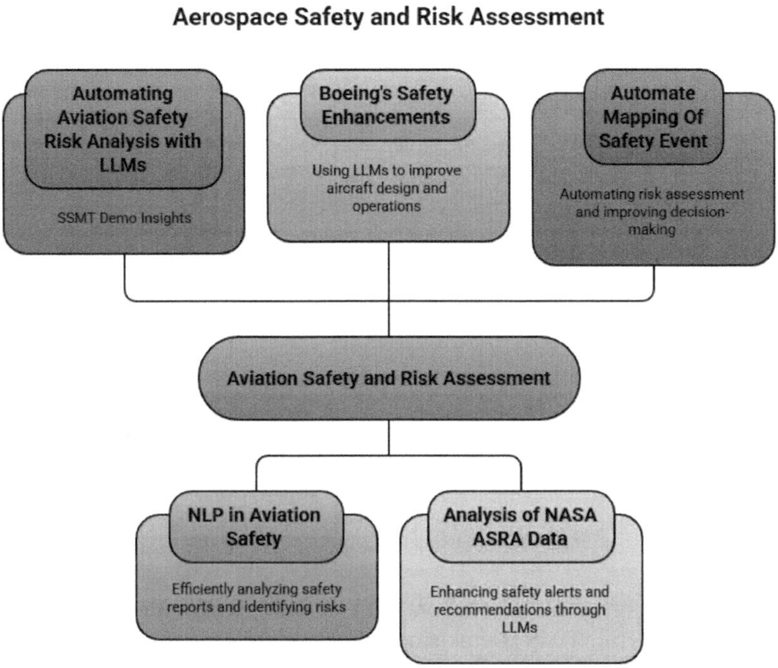

Figure 10: LLM in Aerospace Safety Analysis.

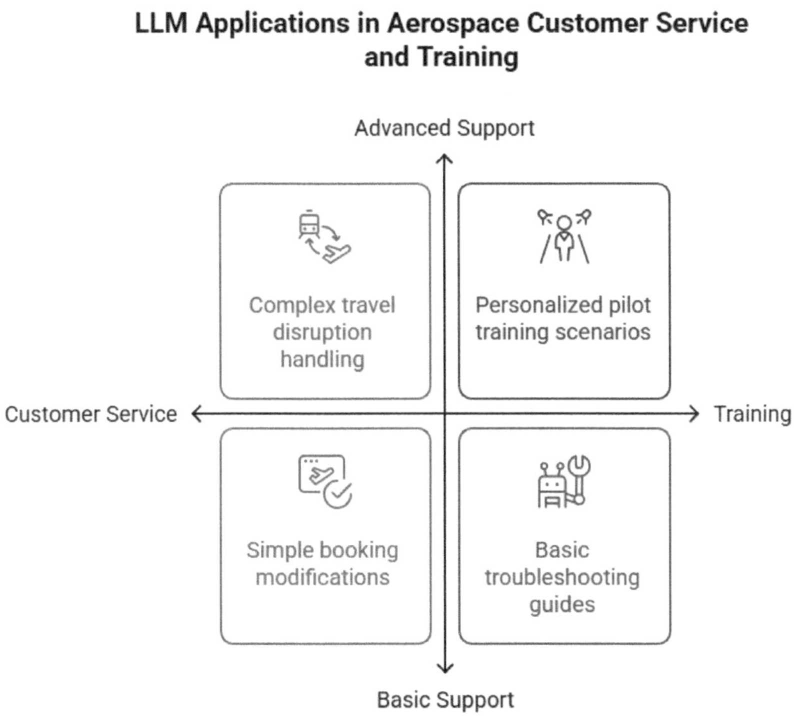

Figure 11: LLM in Pilot Training and Customer service.

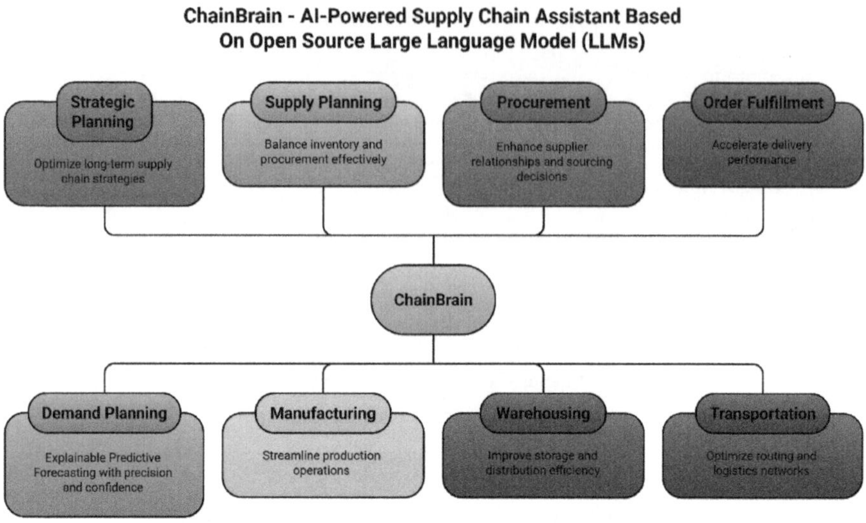

Figure 12: Chainbrain's Core Areas of Expertise.

- INDUS suite of LLMs: In collaboration with IBM, NASA has developed a suite of LLMs called INDUS (Derek Koehl, 2024a).
  - Domain-specific training: INDUS was trained on a corpus of 60 billion tokens focused on scientific fields relevant to NASA, such as planetary science, astrophysics, and Earth science.
  - Enhanced search and data discovery: The models are being used to create a more robust search feature for the NASA Science Discovery Engine, allowing for more targeted and efficient literature surveys.
  - Scientific vocabulary: INDUS uses a custom tokenizer that recognizes specific scientific terms, making it more accurate for scientific and data discovery tasks compared to general-purpose LLMs.
- To minimize the probability of generating incorrect outputs by Generative Adversarial Networks (GANs), NASA is currently testing these networks against multiple evaluation frameworks (like evalem) to evaluate their performance for accuracy. The goals of this study are determining whether the creation of specialized science training data can improve GANS reliability (Ramasubramanian et al., 2023).

Enhancing mission planning and operations

Beyond research, JPL and NASA are using AI to optimize and automate mission planning and onboard operations (David Salvagnini, 2023).

- Automated scheduling: The JPL Artificial Intelligence Group has developed planning and scheduling tools like Aerie and CASPER that generate activity plans and schedules for missions. These tools help automate tasks for missions, including those involving the Deep Space Network (DSN).

Table 15: LLMs in Aerospace SCM.

Application/ System	**Core Technology**	**Key Areas/ Functions**	**Specific Capabilities**	**Implementation Details**	**Benefits/ Results**	**Notable Features**	**References**
LLMs for Supply Chain Digitalization (General Framework)	Large Language Models	Three key areas: 1. Knowledge Management 2. Data Analysis 3. Process Automation	Knowledge Management: • Repository of institutional knowledge • Mines unstructured data (emails, meeting minutes) • Provides historical context and insights Data Analysis: • Intuitive natural language interface • Translates natural language questions into code • Analyzes complex datasets Process Automation: • Streamlines tasks • Optimizes planning	Human oversight emphasized as crucial for critical decisions to avoid potential missteps	Accelerates digitalization of supply chains across three key contribution areas	Categorizes LLM contributions into knowledge management, data analysis, and process automation	Ralf W. Seifert, 2023

Table 15 contd. ...

...Table 15 contd.

Application/ System	Core Technology	Key Areas/ Functions	Specific Capabilities	Implementation Details	Benefits/ Results	Notable Features	References
Domain-Specialized SCM LLM	Retrieval-Augmented Generation (RAG) framework with domain-specialized LLM	Supply Chain Management (SCM) domain specialization	• First-ever domain-specialized SCM LLM • Detailed analysis of complex, multi-agent SCM interactions • Simulates horizontal and vertical supply chain games • Replicates classical SCM phenomena • Provides new insights	Knowledge Base: • Curated from professional certifications (SCMP, CPIM) • Ensures expert-level domain knowledge Simulation Capabilities: • Horizontal games: LLM agents adaptively display competitive or cooperative behavior based on incentive structure • Vertical games: Reproduced bullwhip effect	Horizontal Games: • Adaptive competitive/ cooperative behavior Vertical Games: • Reproduced bullwhip effect • Revealed that information sharing significantly reduces effect for risk-neutral and risk-averse agents • Impact on risk-seeking agents less predictable	LLMs can serve as powerful tools for SCM research and decision support; paves way for new "AI for management science" paradigm	Wang et al., 2025

ChainBrain	AI-powered supply chain assistant with domain-trained Large Language Model	• Demand planning • Procurement • Warehousing • Transportation	• Delivers specialized insights and intelligent workflows • Real-time analytics and data-driven insights in conversational format • Enables faster, more informed decisions • Integrates with existing systems	System Integration: • ERP • WMS (Warehouse Management System) • TMS (Transportation Management System) • Ensures seamless operational integration • Leverages company's own data Enhanced Capabilities with AI/Analytics Hub: • Predictive modeling • Scenario planning • Operational visibility	Goes beyond generic chatbots to provide specialized supply chain insights and workflows	Visual Reference: Figure 12 provides comprehensive overview of ChainBrain's key areas of specialization	Rohit Kamath 2025

- Autonomous navigation: LLMs and AI are used to train autonomous systems for spacecraft and rovers. The Perseverance rover, for example, uses AI to identify hazards and navigate the Martian terrain.
- Advanced simulations: AI systems help simulate various mission scenarios, allowing planners to analyze thousands of variables simultaneously. This enables faster, more informed decision-making for complex deep space missions

15.1.2 Northrop Grumman

Northrop Grumman uses generative AI to accelerate engineering, simulation, production, and decision support across enterprise operations and mission systems, anchored by companywide deployments of NVIDIA RTX PRO servers and governed AI frameworks under its digital transformation programs. Public materials emphasize a federated, governed approach where GenAI software, platforms, and frameworks are embedded into existing workflows rather than a single branded platform. (Northrop Grumman-Nvidia-Servers-Agreement, 2024; Mathews, 2025).

15.1.3 MIT

LLMs are leveraged for:

- Facilitating natural language interfaces so researchers can converse with systems, get actionable insights, and propose or adjust experiments and engineering designs without coding knowledge.
- The MIT School of Engineering has embraced AI technologies through the use of LLMs. It is becoming evident that there is a movement towards the use of AI to facilitate development/block differentiation between engineering, design, science experimentation, and collaborative innovation, enabling researchers to work more effectively to tackle the increasingly complex challenges facing the aerospace community (LLM-Based Human-Machine Interaction, 2025; Zach Winn, 2025b).

16. Regulatory Compliance and Certification

LLMs allow the Aviation industry to use these tools for Regulatory Compliance Management (RCM) and Certification. Aviation companies utilize LLMs to automate regulatory analysis, improve compliance reporting, and support certification processes with higher levels of accuracy and efficiency. See Table 16 for a summary of LLM's use in RCM (Bianca Ebrașu and Avram Brîndușă, n.d.).

AI-enabled systems like "CertifAIer" leverage LLMs to automate much of the aircraft certification workflow, such as generating compliance report templates, reviewing draft documents, providing feedback, and continuously updating with regulatory changes (Figure 13). These tools are designed for real-time compliance support, improving the robustness and efficiency of the certification process for commercial aircraft (DePauw et al., 2024).

Table 16: LLM Applications in Regulatory Compliance and Certification.

Application Area	Technology/ System	Primary Purpose	Key Capabilities	Specific Features	Standards/ Regulations Covered	Benefits	Notable Initiatives/ Tools	References
General Regulatory Compliance Context	Large Language Models (LLMs)	Address complex and critical aspects of aerospace regulatory operations	• Analyzing regulatory documents • Tracking changes • Automating compliance verification processes • Assessing impacts of regulatory changes • Ensuring ongoing compliance across all products and operations	Regulatory landscape continuously evolves, requiring aerospace companies to track changes and maintain compliance	Numerous national and international regulations, standards, and certification requirements	Powerful capabilities for managing evolving regulatory landscape	Regulatory compliance and certification represent some of the most complex aspects of aerospace operations	Luettig et al., 2024
Regulatory Compliance Management and Certification	Large Language Models (LLMs)	Enable aviation organizations to manage regulatory compliance and certification processes	• Automate regulatory analysis • Streamline compliance reporting • Support certification processes	Increasingly used in aerospace for compliance management	Aviation regulations and certification standards	• Improved accuracy • Improved efficiency • Automated compliance processes	LLMs transforming aerospace regulatory compliance management	Bianca Ebrasu and Avram Brîndușa, n.d.

Table 16 contd. ...

...Table 16 contd.

Application Area	**Technology/ System**	**Primary Purpose**	**Key Capabilities**	**Specific Features**	**Standards/ Regulations Covered**	**Benefits**	**Notable Initiatives/ Tools**	**References**
LLMs for Regulatory Compliance (Specialized)	Fine-tuned LLMs with Retrieval-Augmented Generation (RAG) frameworks	Interpret and analyze complex aviation regulations	• Understanding nuanced regulatory language • Retrieving precise guideline references • Supporting professionals with up-to-date compliance information	• Fine-tuned for regulatory interpretation • RAG frameworks for accurate information retrieval	• EASA (European Union Aviation Safety Agency) • FAA (Federal Aviation Administration) • AS9100 (Aerospace Quality Management System)	• Precise regulatory guidance • Up-to-date compliance information • Nuanced understanding of regulatory language	Aerospace Language Understanding Evaluation (ALUE) Benchmark: • Introduced by FAA and MITRE • Assesses accuracy and reliability of LLMs for aerospace safety-critical tasks	Jordan Graham, 2025
AI-Driven Certification Tools	AI-enabled systems, specifically "CertifAIer"	Automate aircraft certification workflow	• Generating compliance report templates • Reviewing draft documents • Providing feedback • Continuously updating with regulatory changes	• Real-time compliance support • Automated workflow processes • Continuous regulatory updates	Commercial aircraft certification			

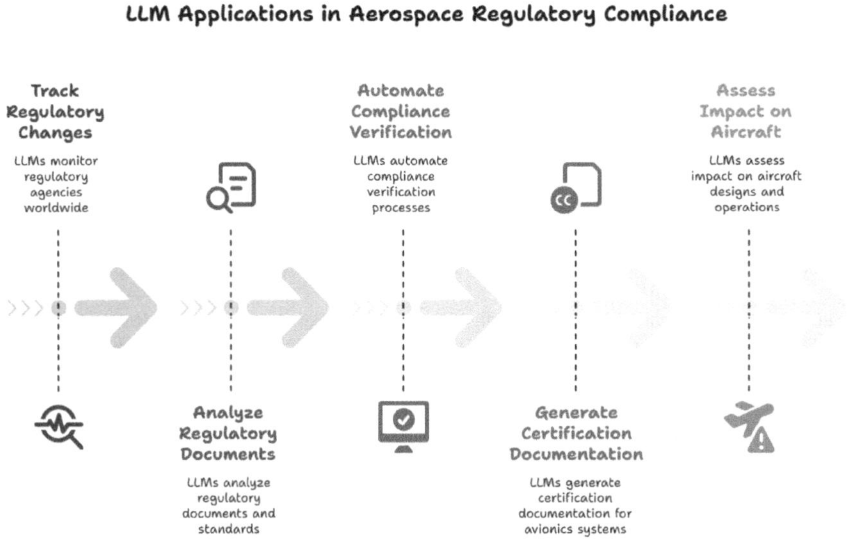

Figure 13: LLM in Aerospace Regulatory Compliance.

17. Conclusion and Future Directions

LLMs (Large Language Models) have transitioned from an experimental tool to a functioning system providing measurable value across the entire aerospace product line (e.g., Boeing's 40% savings in technical documentation effort to Delta's 25% reduction in unscheduled maintenance events).

The aerospace industry has struggled with the challenge of managing enormous quantities of technical knowledge, and LLMs provide the capability to analyze natural language on a vast basis and extract actionable insights thereby enabling data-driven decisions that improve safety, improve efficiency, and enhance innovation. Trends indicate that the future of LLMs includes the creation of multimodal AI systems (i.e., combining data from text, images, and sensor information); the use of edge computing to provide real-time assistance to aircraft decision-makers; the integration of LLM technology into digital twin capabilities; and the use of LLM technology to enable unmanned systems to interact using natural language.

While these trends create many important issues regarding safety validation, certification, and regulatory oversight, LLMs are demonstrating that they are much more than just automated tools; they are essential parts of how aerospace companies develop, disseminate, and apply knowledge, and they are going to be essential for aerospace operations moving forward.

References

Abdul Shukor, Syaimak, Jia Hui Tang, Iqraq Kamal, Mohd Faizzuan Mohammad. et al. 2025. Predictive Maintenance in Aerospace: Leveraging Machine Learning for Optimized Filter Lifespan. In Systems Modelling and Simulation, edited by Zaharuddin Mohamed, Fazilah Hassan, Gary Tan, Anita Ahmad, Leow Pei Ling, and Salinda Buyamin. Springer Nature. https://doi.org/10.1007/978-981-96-4613-5_7.

Allen, Bradley P., Lise Stork and Paul Groth. 2023. Knowledge Engineering Using Large Language Models. arXiv:2310.00637. Preprint, arXiv, October 1. https://doi.org/10.48550/arXiv.2310.00637.

Anil Abraham Kuriakose. 2025. From Reactive to Proactive Operations: The Role of LLM in Predictive Maintenance. https://www.algomox.com/resources/blog/reactive_proactive_operations_llm_predictive_maintenance/.

Anshul Sharma. 2024. Gen AI for Analysis of Aircraft Maintenance Documents | HCLTech. August 9. https://www.hcltech.com/blogs/gen-ai-for-analysis-of-aircraft-maintenance-documents.

Avila, Carlos, Daniel Ilbay and David Rivera. 2025. Human–AI Teaming in Structural Analysis: A Model Context Protocol Approach for Explainable and Accurate Generative AI. Buildings 15(17): 3190. https://doi.org/10.3390/buildings15173190.

Basil, Ehaab. 2024. Large Language Models for Aviation Safety: Enhancing Incident Analysis through Summarization of ASRS Reports.

Bhattacharjee Bishwaranjan, Aashka Trivedi, Masayasu Muraoka, Muthukumaran Ramasubramanian, Takuma Udagawa. et al. 2024. INDUS: Effective and Efficient Language Models for Scientific Applications. arXiv:2405.10725. Preprint, arXiv, October 30. https://doi.org/10.48550/arXiv.2405.10725.

Bianca Ebrasu and Avram Brîndușa. n.d. A Specialized LLM Framework for Aviation Regulation and Compliance. https://dils.upb.ro/blockchain-workshop/files/Large%20language%20Model%20Aviation%20Compliance.pdf.

Boeing Intelligence & Analytics | Strategic Solutions. 2025. "Capabilities: GenAI." https://www.bia-boeing.com/capabilitiesinfo/genai.

Brian Connolly. 2025. Evaluating AI Large Language Models for Mechanical and Aerospace Engineering Applications, 18-R6365 | Southwest Research Institute. April. https://www.swri.org/what-we-do/internal-research-development/2023/energy-environment/evaluating-ai-large-language-models-mechanical-aerospace-engineering-applications-18-r6365.

Brown Tom B., Benjamin Mann, Nick Ryder, Melanie Subbiah, Jared Kaplan, Prafulla Dhariwal, et al. 2020. Language Models Are Few-Shot Learners. arXiv:2005.14165. Preprint, arXiv, July 22. https://doi.org/10.48550/arXiv.2005.14165.

Burhan Sajid. 2023. AI-Powered Software Engineering: Automating Code Generation with Multi-Agent Systems. https://doi.org/10.13140/RG.2.2.15754.58560.

Carles Molins, Agustí Junan Porta I Ko and Núria Alsina Pujol. 2023. Harnessing LLMs to Enhance Safety Reporting in Aviation. https://www.alg-global.com/blog/aviation/harnessing-llms-enhance-safety-reporting-aviation.

Carlos Paradis. n.d. Augmenting Topic Finding in the NASA Aviation Safety Reporting System Using Topic Modeling | Request PDF. Paper presented at AIAA Scitech 2021 Forum. https://doi.org/10.2514/6.2021-1981.

Connolly, Brian J. 2025. Development of an Aerospace Engineering Evaluation Set for Large Language Model Benchmarking. In AIAA SCITECH 2025 Forum. AIAA SciTech Forum. American Institute of Aeronautics and Astronautics. https://doi.org/10.2514/6.2025-0702.

Cunnison, Martin. 2025. 7 Key Advantages of Generative AI for Airline Operations. January 20. https://www.ciklum.com/resources/blog/airline-generative-ai.

Databricks, dir. 2024. Cloud-Based Generative-AI Supporting Preliminary Engineering Design. 32: 36. https://www.youtube.com/watch?v=KlAIvPnC7QE.

David Salvagnini. 2023. Artificial Intelligence - NASA. https://www.nasa.gov/artificial-intelligence/.

DePauw, Todd C., Mohammed Kabir, Alexandru Stere and John Dong. 2024. Development of a Commercial Airplane Certification Digital Assistant Using a Large Language Model Trained with Regulatory Requirements and Means of Compliance Documents. In AIAA SCITECH 2024 Forum. American Institute of Aeronautics and Astronautics. https://doi.org/10.2514/6.2024-1528.

Derek Koehl. 2024a. NASA-IBM Collaboration Develops INDUS Large Language Models for Advanced Science Research - NASA Science. June 25. https://science.nasa.gov/open-science/ai-language-model-science-research/.

Derek Koehl. 2024b. NASA-IBM Collaboration Develops INDUS Large Language Models for Advanced Science Research - NASA Science. Open Science. June 25. https://science.nasa.gov/open-science/ai-language-model-science-research/.

Falkner, Gilbert. 2024. Potential of AI in Aviation Safety Analysis: Exploring the Future of Collaborative Intelligence. ASQS, April 2. https://asqs.net/blog/potential-of-ai-in-aviation-safety-analysis/.

Farquhar, Sebastian, Jannik Kossen, Lorenz Kuhn and Yarin Gal. 2024. Detecting Hallucinations in Large Language Models Using Semantic Entropy. Nature 630(8017): 625–30. https://doi.org/10.1038/s41586-024-07421-0.

Frank Morales Aguilera. 2025. Leveraging the Advanced Capabilities of the Mistral LLM for Predictive Maintenance: An Exploration of Aircraft Engine RUL Prediction | LinkedIn. https://www.linkedin.com/pulse/leveraging-advanced-capabilities-mistral-llm-aircraft-frank-5yj6e/.

Generative AI: Transforming the Aerospace Industry. 2025. January 20. https://www.einfochips.com/blog/generative-ai-in-the-aerospace-industry-a-new-frontier-of-innovation/.

Graham Warwick. 205 AD. Physics-Based AI Promises To Accelerate Aerospace Design Optimization | Aviation Week Network. January 8. https://aviationweek.com/aerospace/emerging-technologies/physics-based-ai-promises-accelerate-aerospace-design-optimization.

Graydon, Mallory S. and Sarah M Lehman. 2025a. Examining Proposed Uses of LLMs to Produce or Assess Assurance Arguments.

Graydon, Mallory S and Sarah M Lehman. 2025b. Examining Proposed Uses of LLMs to Produce or Assess Assurance Arguments. NASA/TM–20250001849. NASA STI Program Report Series. NASA.

He, Rui, Liang Zhang, Mengyao Lyu, Liangqing Lyu and Changbin Xue. 2025. Using Large Language Models for Aerospace Code Generation: Methods, Benchmarks, and Potential Values. Aerospace 12(6): 498. https://doi.org/10.3390/aerospace12060498.

Hou, Lei, Beixi Jia, Chenguang Xing, Zhaojiang Chen and Ziliang Du. 2025. Applied Research on an Aircraft Maintenance Assistant Based on a Large Language Model. Proceedings of the 2025 4th International Conference on Intelligent Systems, Communications and Computer Networks, February 21: 1–7. https://doi.org/10.1145/3732945.3732946.

How Airbus Uses Generative Artificial Intelligence to Reinvent Itself | Airbus. 2024. May 28. https://www.airbus.com/en/newsroom/stories/2024-05-how-airbus-uses-generative-artificial-intelligence-to-reinvent-itself.

ICAO. 2024. Leveraging Generative Pre-Trained Large Language Models To Support Data-Informed Safety Decision Making In Aviation. Rasg-Apac/14–IP/03. Bangkok, Thailand. https://www.icao.int/sites/default/files/APAC/Meetings/2024/2024%20RASGAPAC14/4-Information%20Papers/RASG-APAC14-IP-03-AI_4-US-FAA-Large-Language-Models.pdf.

Jack Kelleher, Marjorie Adriaenssens, Puneet Jain,. 2024. Rolls-Royce: Harnessing the Power of Databricks Mosaic AI for Image Generation. Databricks, August 7. https://www.databricks.com/blog/rolls-royce-mosaic-ai.

Jay Lee and Hanqi Su. 2024. A Unified Industrial Large Knowledge Model Framework in Smart Manufacturing. https://arxiv.org/html/2312.14428v2.

Jonk, Patrick, Vincent de Vries, Rombout Wever, Georgios Sidiropoulos and Evangelos Kanoulas. 2022. Natural Language Processing of Aviation Occurrence Reports for Safety Management. Book of Extended Abstracts for the 32nd European Safety and Reliability Conference, 2015–23. https://doi.org/10.3850/978-981-18-5183-4_S05-06-449-cd.

Jordan Graham. 2025. MITRE and FAA Introduce Novel Aerospace Large Language Model Evaluation Benchmark. September 17. https://www.mitre.org/news-insights/news-release/mitre-and-faa-introduce-novel-aerospace-large-language-model-evaluation.

Kaylin Bugbee and Rahul Ramachandran IMPACT. 2023. Architecting the Future: A Vision for Using Large Language Models to Enable Open Science | NASA Earthdata." Blog. Earth Science Data Systems, NASA, December 5. https://www.earthdata.nasa.gov/news/blog/architecting-future-vision-using-large-language-models-enable-open-science.

Khodabandeh, Sam Ransbotham and Shervin. n.d. AI in Aerospace: Boeing's Helen Lee. Episode 407. Me, Myself, and AI. Accessed September 20: 2025. https://sloanreview.mit.edu/audio/ai-in-aerospace-boeings-helen-lee/.

Klekowicki, Mikołaj, Grzegorz M. Szymański, Marek Waligórski and Wojciech Misztal. 2024. Application of Large Language Models in Diagnostics and Maintenance of Aircraft Propulsion Systems. Advances in Science and Technology Research Journal 19(2): 304–20. https://doi.org/10.12913/22998624/196264.

Krus, Petter. 2024. Large Language Model in Aircraft System Design. ICAS Congress 2024 (Florence). https://www.icas.org/icas_archive/icas2024/data/papers/icas2024_0514_paper.pdf.

Kumar, N., Keerthi, Manasa, C. M., Pavan Kumar, B. K. and Manish Bali. 2025. Reinforcement Learning-Based Topology Optimization for Generative Designed Lightweight Structures. MethodsX 15 (July): 103539. https://doi.org/10.1016/j.mex.2025.103539.

Lazar. 2024. LLMs Are Revolutionizing the Commercial Aviation Industry. https://www.linkedin.com/pulse/llms-revolutionizing-commercial-aviation-industry-lazar-iliev-qk2ef.

Lewis Patrick, Ethan Perez, Aleksandra Piktus, Fabio Petroni, Vladimir Karpukhin, Naman Goyal, et al. 2021. Retrieval-Augmented Generation for Knowledge-Intensive NLP Tasks. arXiv:2005.11401. Preprint, arXiv, April 12. https://doi.org/10.48550/arXiv.2005.11401.

Li, Wenjie and Lei Chen. 2025a. LLMs Promote Commercial Aerospace Technology Towards Innovation Intelligence. IEEE Systems, Man, and Cybernetics Society. https://www.ieeesmc.org/wp-content/uploads/2024/10/FeatureArticle_LLMs_March25.pdf.

Li, Wenjie and Lei Chen. 2025b. LLMs Promote Commercial Aerospace Technology Towards Innovation Intelligence.

Liang, Haoran, Mohammad Talebi Kalaleh and Qipei Mei. 2025. Integrating Large Language Models for Automated Structural Analysis. arXiv:2504.09754. Version 1. Preprint, arXiv, April 13. https://doi.org/10.48550/arXiv.2504.09754.

Liu, Beiming, Zhizhuo Cui, Siteng Hu, Xiaohua Li, Haifeng Lin, et al. 2025. LLM Evaluation Based on Aerospace Manufacturing Expertise: Automated Generation and Multi-Model Question Answering. arXiv:2501.17183. Version 1. Preprint, arXiv, January 25. https://doi.org/10.48550/arXiv.2501.17183.

Liu, Yucheng. 2024. Large Language Models for Air Transportation: A Critical Review. Journal of the Air Transport Research Society 2 (June): 100024. https://doi.org/10.1016/j.jatrs.2024.100024.

LLM-Based Human-Machine Interaction. 2025. MIT AeroAstro. https://aeroastro.mit.edu/realm/our-research/llm-based-human-machine-interaction/.

Luettig, Bastian, Yassine Akhiat and Zamira Daw. 2024. ML Meets Aerospace: Challenges of Certifying Airborne AI. Frontiers in Aerospace Engineering 3 (November). https://doi.org/10.3389/fpace.2024.1475139.

Makatura Liane, Michael Foshey, Bohan Wang, Felix HähnLein, Pingchuan Ma, Bolei Deng, et al. 2024. How Can Large Language Models Help Humans in Design And Manufacturing https://doi.org/10.48550/arXiv.2307.14377

Mangortey, Eugene, Kunal Sarkhel, Satyen Singh, Shuo Chen, Bulent Ayhan, et al. 2025. Aerospace Language Understanding Evaluation (ALUE): Large Language Benchmark with Aerospace Datasets. In AIAA AVIATION FORUM AND ASCEND 2025. AIAA Aviation Forum and ASCEND Co-Located Conference Proceedings. American Institute of Aeronautics and Astronautics. https://doi.org/10.2514/6.2025-3247.

Mathews, Anshika. 2025. Northrop Grumman's AI Advancements Set the Stage for U.S. Dominance in Global Defense. AIM Media House, January 8. https://aimmediahouse.com/market-industry/northrop-grummans-ai-advancements-set-the-stage-for-u-s-dominance-in-global-defense.

Mike Baylor. 2024. Lockheed Martin Deploys Powerful, Secure Generative AI Tools Across the Enterprise. Lockheed Martin, October 8. https://www.lockheedmartin.com/en-us/news/features/2024/empowering-innovation-with-secure-generative-ai-across-enterprise.html.

Milin Stanly. 2025. MIT Researchers Develop Co-LLM: A Collaborative AI Algorithm for Improved Accuracy. IndiaAI. https://indiaai.gov.in/article/mit-researchers-develop-co-llm-a-collaborative-ai-algorithm-for-improved-accuracy.

Mohaimenul Azam Khan, MD. Saddam Hossain Mukta, Kaniz Fatema, Nur Mohammad Fahad, Sadman Sakib, Most Marufatul Jannat Mim. et al. 2024. A Review on Large Language Models: Architectures, Applications, Taxonomies, Open Issues and Challenges. IEEE Access 1(1): 99. https://ieeexplore.ieee.org/document/10433480.

Myohanen, Jonne. 2023. Improving Industrial Performance With Language Models: A Review of Predictive Maintenance and Process Optimization. Lappeenranta-Lahti University of Technology LUT.

Nanyonga, Aziida, Keith Joiner, Ugur Turhan and Graham Wild. 2025a. Applications of Natural Language Processing in Aviation Safety: A Review and Qualitative Analysis. In AIAA SCITECH

2025 Forum. AIAA SciTech Forum. American Institute of Aeronautics and Astronautics. https://doi.org/10.2514/6.2025-2153.

Nanyonga, Aziida, Keith Joiner, Ugur Turhan and Graham Wild. 2025b. Applications of Natural Language Processing in Aviation Safety: A Review and Qualitative Analysis. arXiv:2501.06210. Preprint, arXiv, January 3. https://doi.org/10.48550/arXiv.2501.06210.

Neural, Concept. 2025. AI in Aerospace Industry: Redefining Intelligent Engineering | Neural Concept. https://www.neuralconcept.com/post/applications-of-ai-in-aerospace-and-defence-design-intelligent-aerospace.

Northrop Grumman-Nvidia-Servers-Agreement. 2024. https://www.executivebiz.com/articles/northrop-grumman-nvidia-servers-agreement.

Omrani, Pouria, Alireza Hosseini, Kiana Hooshanfar, Zahra Ebrahimian, Ramin Toosi. et al. 2024. Hybrid Retrieval-Augmented Generation Approach for LLMs Query Response Enhancement. 2024 10th International Conference on Web Research (ICWR), April, 22–26. https://doi.org/10.1109/ICWR61162.2024.10533345.

Palma, Giulia, Gaia Cecchi and Antonio Rizzo. 2025. Large Language Models for Predictive Maintenance in the Leather Tanning Industry: Multimodal Anomaly Detection in Compressors. Electronics 14(10): 2061. https://doi.org/10.3390/electronics14102061.

Pongetti, Julie, Timoleon Kipouros, Marc Emmanuelli, Richard Ahlfeld and Shahrokh Shahpar. 2021. Using Autoencoders and Output Consolidation to Improve Machine Learning Models for Turbomachinery Applications. Paper presented at ASME Turbo Expo 2021: Turbomachinery Technical Conference and Exposition. September 16. https://doi.org/10.1115/GT2021-60158.

Rahman, Sumaya Abdul, Sanjay Chawla, Mohammed Yaqot and Brenno Menezes. 2025. Leveraging Large Language Models for Supply Chain Management Optimization: A Case Study. In Innovative Intelligent Industrial Production and Logistics, edited by Michele Dassisti, Kurosh Madani, and Hervé Panetto. Springer Nature Switzerland. https://doi.org/10.1007/978-3-031-80775-6_13.

Ralf W. Seifert. 2023. How Will Large Language Models Impact Supply Chains? - I by IMD. May 17. https://www.imd.org/ibyimd/innovation/large-language-model-impacts-on-supply-chain/.

Ramasubramanian, Muthukumaran, Nishan Pantha, Iksha Gurung, Raghu K Ganti, Bishwaranjan Bhattacharjee, et al. 2023. Evaluation of NASA Science Mission Directorate Large Language Model. Paper presented at AGU23. December 12. https://agu.confex.com/agu/fm23/meetingapp.cgi/Paper/1413324.

Rane, Nitin, Saurabh Choudhary and Jayesh Rane. 2024. Intelligent Manufacturing through Generative Artificial Intelligence, Such as ChatGPT or Bard. SSRN Electronic Journal, ahead of print. https://doi.org/10.2139/ssrn.4681747.

Rohit Kamath. 2025. Supply Chain Innovation with AI Assistant | ChainBrain Demo." Stellium, October 2. https://stellium.com/chainbrain-transforming-supply-chain-management-with-ai-powered-intelligence/.

Santiago Erroz Ferrer. 2024. "GENERATIVE AI." Lufthansa Group Innovation Runway. https://innovation-runway.lufthansagroup.com/en/focus-areas-projects/projects/openai.html.

Satya Hanush, S. and M. Manjaiah. 2022. Topology Optimization of Aerospace Part to Enhance the Performance by Additive Manufacturing Process." Materials Today: Proceedings, 9th International Conference on "Advancements and Futuristic Trends in Mechanical and Materials Engineering," vol. 62 (January): 7373–78. https://doi.org/10.1016/j.matpr.2022.02.074.

Siemens Software, dir. 2025. Rolls-Royce: Pioneering Digital Transformation in Aerospace. 03: 48. https://www.youtube.com/watch?v=q7n_5rv3IyQ.

Srivastava, Santosh Kumar, Susmi Routray, Surajit Bag, Shivam Gupta and Justin Zuopeng Zhang. 2024. Exploring the Potential of Large Language Models in Supply Chain Management: A Study Using Big Data. Journal of Global Information Management (JGIM) 32(1): 1–29. https://doi.org/10.4018/JGIM.335125.

Sunil Prasad. 2025. Scaling AI Adoption in Aerospace Engineering - Blog. September 19. https://www.ltts.com/blog/AI-adoption-aerospace-engineering.

Tikayat Ray, Archana, Anirudh Prabhakara Bhat, Ryan White, Van Nguyen, Olivia Pinon Fischer, et al. 2023. Examining the Potential of Generative Language Models for Aviation Safety Analysis: Case Study and Insights Using the Aviation Safety Reporting System (ASRS). Aerospace 10 (August): 770. https://doi.org/10.3390/aerospace10090770.

Toby Simonds. 2025. LLMs for Engineering: Teaching Models to Design High Powered Rockets. arXiv:2504.19394v2 [Cs.SE], April 29. https://arxiv.org/html/2504.19394v2.

Tong, Yao, Mingqiang Luo, Shangqing Ren, Zheng Zhang, Chenguang Xing and Ziliang Du. 2025. A Rapidly Structured Aircraft Concept Design Method Based on Generative Artificial Intelligence. Chinese Journal of Aeronautics 38(10): 103629. https://doi.org/10.1016/j.cja.2025.103629.

Tyagi, Varun. 2024. Leveraging Large Language Models (LLMs) and Generative AI (GenAI) for Supply Chain Efficiency.... Medium, January 13. https://medium.com/@varun.tyagi83/leveraging-large-language-models-llms-and-generative-ai-genai-for-supply-chain-efficiency-95c9ac756c70.

Vaswani, Ashish, Noam Shazeer, Niki Parmar, Jakob Uszkoreit, Llion Jones, et al. 2023. Attention Is All You Need. Paper presented at 31st Conference on Neural Information Processing Systems (NIPS 2017). August 2. https://doi.org/10.48550/arXiv.1706.03762.

Vincent de Vries. n.d. Classification of Aviation Safety Reports Using Machine Learning. Paper presented at 2020 International Conference on Artificial Intelligence and Data Analytics for Air Transportation (AIDA-AT). https://doi.org/10.1109/AIDA-AT48540.2020.9049187.

Wagenaar, Thomas, Simone Mancini and Andrés Mateo-Gabín. 2024. Generative Aerodynamic Design with Diffusion Probabilistic Models. arXiv:2409.13328. Preprint, arXiv, September 20. https://doi.org/10.48550/arXiv.2409.13328.

Wan, Jia'ang, Feng Shen, Fujuan Li, Yanjin Sun, Yan Li, et al. 2025. AviationLLM: An LLM-Based Knowledge System for Aviation Training. arXiv:2506.14336. Version 1. Preprint, arXiv, June 17. https://doi.org/10.48550/arXiv.2506.14336.

Wang, Haojie, Jiuyun Jiang, L. Jeff Hong and Guangxin Jiang. 2025. "LLMs for Supply Chain Management. arXiv:2505.18597. Preprint, arXiv, May 24. https://doi.org/10.48550/arXiv.2505.18597.

Wang, Hui, Yongqi Huang, Zhiqiang Zhang, Yannan Zhao and Yun Sun. 2024. Generative Design and Topology Optimization Research for Single–Layer Aluminum Alloy Grid Shell Connections. Case Studies in Construction Materials 21 (December): e03781. https://doi.org/10.1016/j.cscm.2024.e03781.

Wang, Peng, John Karigiannis and Robert X. Gao. 2024. Ontology-Integrated Tuning of Large Language Model for Intelligent Maintenance. CIRP Annals 73(1): 361–64. https://doi.org/10.1016/j.cirp.2024.04.012.

Xie, Xiaoyue, Xilang Tang, Siwei Gu and Lijie Cui. 2025. An Intelligent Guided Troubleshooting Method for Aircraft Based on HybirdRAG. Scientific Reports 15(1): 17752. https://doi.org/10.1038/s41598-025-02643-2.

Xiong, Xirui, Hongming Cai, Han Yu, Bingqing Shen and Pan Hu. 2025. DR-RAG: Domain-Rule-Based Retrieval-Augmented Generation for Aviation Digital Model Design. Advanced Engineering Informatics 68 (November): 103688. https://doi.org/10.1016/j.aei.2025.103688.

Zach Winn. 2025a. AI System Learns from Many Types of Scientific Information and Runs Experiments to Discover New Materials. MIT News | Massachusetts Institute of Technology, September 25. https://news.mit.edu/2025/ai-system-learns-many-types-scientific-information-and-runs-experiments-discovering-new-materials-0925.

Zach Winn. 2025b. AI System Learns from Many Types of Scientific Information and Runs Experiments to Discover New Materials." MIT News | Massachusetts Institute of Technology, September 25. https://news.mit.edu/2025/ai-system-learns-many-types-scientific-information-and-runs-experiments-discovering-new-materials-0925.

Zhang, Qilei and John H. Mott. 2024. An Exploratory Assessment of LLM's Potential Toward Flight Trajectory Reconstruction Analysis. arXiv:2401.06204. Version 1. Preprint, arXiv, January 11. https://doi.org/10.48550/arXiv.2401.06204.

4

Large Language Models in Agriculture

Applications, Challenges, and Future Directions

Ritika[1] and *Savita Kumari Sheoran*[2,*]

1. Introduction

The agriculture industry is progressively but steadily evolving into a sector that heavily relies on data and knowledge, where precise, prompt, and situation-specific information are the main factors for not only yielding more but also using resources more efficiently and having a sustainable environment (Al-Shammary et al., 2024). The conventional farm advisory practices like extension services, printed guides, and personal consultations have really helped the farmers (Kumar et al., 2023). Nevertheless, in most cases, these methods fail to reach all the interested parties because of geographical, temporal, language, and digital literacy limitations. Thus, smart systems that can link research knowledge to the practical application in the field are what the industry needs. The Large Language Models (LLMs) are the major drivers of this sector's transformation. LLM are the top-notch artificial intelligence tools that apply deep learning strategies to the comprehension, skill, and reasoning of human languages on a much larger scale (Kumar, 2024). Transformer is the main framework for the design of these models, and they are trained on huge amounts of text sources. Such gigantic data training provides them with the ability to discern the relations of meaning, syntactic forms, and even the fineness related to the

[1] Research Scholar, Indira Gandhi University, Meerpur, Rewari, India-123401.
Email: dr.ritikagaba@gmail.com, ritikagaba@iari.res.in
Orcid ID: https://orcid.org/0000-0002-5265-444x

[2] Professor, Indira Gandhi University, Meerpur, Rewari, India -123401.

* Corresponding author: savita.sheoran@igu.ac.in, Orcid ID- https://orcid.org/0000-0003-1624-7385

DOI: 10.1201/9781003771654-4

context in language (Chaitanya and Rolla, 2024). Their flexibility to be effective in numerous tasks at the same time makes it possible for them to perform a plethora of tasks like natural language processing, text summarization, machine translation, question answering, and predictive analytics specific to a question or situation. The agricultural context has witnessed the spread of LLM technology mainly in the areas of information dissemination and, to some extent, in decision-making or automation.

LLM can produce insights and assist in the construction of evidence-based agriculture by processing and integrating various information from climate data, soil characteristics, crop health, and market and policy documents (Priya et al., 2023). The new technology will usher in a new era of personalized advisory services, tailor-made for local agro-climatic conditions, farm sizes, and resource availability. Consequently, the use of service will lead to the improvement of operational efficiency and productivity (Bhardwaj and Gupta, 2025).

The ability to remove language barriers to a great extent is also provided through LLMs. Many languages are spoken in the areas where agricultural activities take place, and the scientific research, on-farm activities and extension services will not be available in their local language. LLMs can support not only the translation of scientific knowledge to regional languages but also paraphrasing to simplify professional translator and literacy levels of farmers. They could even generate interactive content that is clearer for farmers and encourages them to adopt recommended practices.

LLM combined with other technologies makes them more powerful. Just imagine an LLM and the Internet of Things devices and drones, remote sensing technologies, and satellite imagery working together and how that could result in carrying out a number of tasks such as predicting crop stress, improving irrigation schedules, diagnosing nutrient deficiencies, and anticipating pest or disease outbreaks, to name a few. This is the dawn of a new precision agriculture era where data-driven, location-specific, and real-time environmental conditions responsive decision-making characterize this new agricultural practice.

Moreover, LLM pushes forward the establishment of eco-friendly and strong agricultural practices. LLM is the major force behind the realization of environment-friendly agriculture through precise distribution of resources, lessening use of chemicals, and application of data-driven interventions. Furthermore, they facilitate the communication of research results, extension innovations, and policy changes thus allowing quick acceptance of technology and creating toughness among farmers who are small-scale and have fewer resources.

Moreover, LLM pushes forward the establishment of eco-friendly and strong agricultural practices. LLM is the major force behind the realization of environment-friendly agriculture through precise distribution of resources, lessening use of chemicals, and application of data-driven interventions. Furthermore, they facilitate the communication of research results, extension innovations, and policy changes thus allowing quick acceptance of technology and creating toughness among farmers who are small-scale and have fewer resources.

2. Background and Evolution of LLM

The LLM improvement is seen as a major turning point in the history of development of Artificial Intelligence (AI) and Natural Language Processing (NLP) technologies (Kumar, 2024). LLMs are the result of an ongoing and multidisciplinary quest for knowledge in linguistics, computational modeling, and machine learning. Their evolution consists of a series of stages from the first symbolic and rule-based systems to the current deep learning architectures that are capable of handling vast amounts of data and performing sophisticated reasoning tasks. The background as well as the evolution of LLM is depicted in Figure 1.

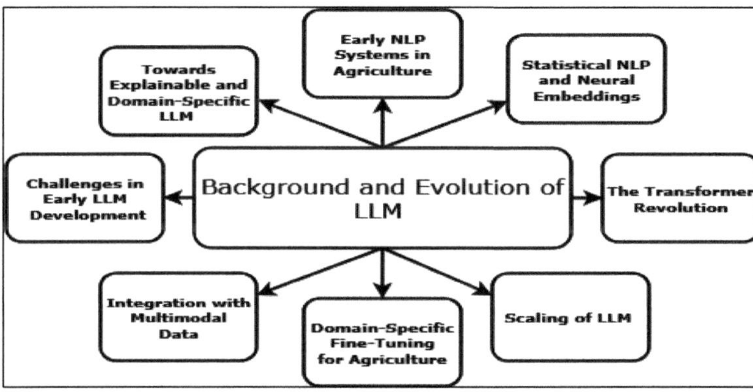

Figure 1: Background and Evolution of LLM.

2.1 *Early NLP Systems in Agriculture*

The earliest endeavors in NLP for agriculture dealt with rule-based or keyword-driven systems in the first place. Such systems utilized vocabularies that were manually prepared along with structured knowledge bases to respond to simple queries or to extract information from databases on agriculture. The shortcomings of these systems were that they could not comprehend the situation, draw conclusions, or make individualized suggestions (Bhattacharya and Pandey, 2024).

By setting up a system of rules, one could hear already established words like the names of crops, types of soil, or species of pests, however this system would not be able to go beyond the boundaries of its programmed knowledge (Abdalrazzaq et al., 2024). Furthermore, the inability to process unstructured text data such as research papers, farmers' reports, and news articles limited their use in a rapidly evolving agricultural sector.

2.2 *Statistical NLP and Neural Embeddings*

The transition from rule-based systems to statistical NLP was one of the most important milestones during the 1990s and 2000s. The use of statistical methods such as n-grams gave the chance to the systems to predict the occurrence of a word or phrase depending on the context of its usage. However, the models had to face the difficulties of data sparsity and long-range dependencies and as a result, their performance was quite poor.

Another major breakthrough was made with the introduction of neural word embeddings, particularly Word2Vec and GloVe (Savytska et al., 2021). These algorithms allowed the mapping of words to a vector space of infinite dimensions, where not only their meanings but also their grammatical relations were exposed. In the agricultural sector, the use of embeddings was a decisive factor in the grouping of crops, soil types, diseases, and climate terms according to their contextual similarities. For example, the vector representations would indicate that "maize" and "corn" are synonyms, which could result in a more efficient and effective information retrieval and recommendation system.

Neural embeddings were the basis for more advanced models, as they allowed the projection of agricultural data, whether it was from textual research papers or extension reports, into the high-dimensional semantic spaces. As a result, such representations made crop classification, document clustering, and automatic recommendation generation easier which, in turn, laid the ground for the development of language models that are specific to the domain.

2.3 The Transformer Revolution

NLP experienced a substantial step forward when the Transformer architecture was introduced. In contrast to RNNs and LSTMs, which process sequentially, the self-attention mechanism in Transformers enables simultaneous processing of sequences (Albeladi et al., 2023). The sequential processing limitation of RNNs is efficiently handled by Transformers besides providing better representation of dependencies between different inputs and the respective context.

Transformers consist of an encoder and a decoder part, and each one contains multi-head attention and feed-forward networks (Kalyan et al., 2021). The different parts of the input sequence to which the attention mechanism assigns weights are the ones that decide how good the model is going to be in predicting the information that will be relevant for observation. Certainly, this function is very beneficial in agriculture, a field that deals with textual datasets rich in complex and multi-contextual information, such as reports on weather, soil, and plants. These need to be merged so that the right choices can be made.

2.4 Scaling of LLM

LLM evolution has been marked by rapid scaling in the model parameters, training data, and the computational capacity used. The early transformer-based models had millions of parameters, while the state-of-the-art models like GPT-4 and LLaMA 2 have hundreds of billions of parameters. The scale-up has resulted in major gains in few-shot and zero-shot learning, which allow the models to do some tasks with almost no domain-specific training data. The above-mentioned scaling in the case of LLM has been an asset in agriculture as it allows the models to use diverse texts coming from various sources, such as scientific publications, farmer advisory bulletins, social media reports, and market data, all at the same time (Balasundram et al., 2023). The bigger models are the ones to discern and capture the very subtle linguistic differences, hence, being able to give contextually aware recommendations, summarize for decision support, and so forth, all with greater accuracy.

2.5 Domain-Specific Fine-Tuning for Agriculture

In the case of general-purpose LLM, they do very well in understanding languages but fine-tuning of the domain is a must for agricultural applications. The process of fine-tuning consists of taking the existing large language models and then operationalizing them so that they can understand the language and content of the specific domain (Ali et al., 2025). A few examples are:

- By using research papers and extension reports, scientific summaries can be created that are easily understood by policymakers.
- Farmer advisories and local guides provide recommendations that are not only region-specific but also in local languages.
- Pest and disease databases support the models in predicting outbreaks and suggesting management strategies.

The re-trained language models, which are also popularly known as AgriLLM or CropGPT, have exceeded performance expectations in the areas of agricultural knowledge extraction, decision support, and multilingual communication (Chaitanya and Rolla, 2024). This adaptation of the model guarantees that the results produced will be suitable, precise, and useful in the actual farming environments.

2.6 Integration with Multimodal Data

The latest change in the development of LLM has taken place and we can say that it involves multimodal models which function by integrating the text with images, sensor data, and structured databases (Areerob et al., 2025). In agriculture, these models are granting LLM the power to process satellite imagery, drone field data, and IoT sensor readings at the same time as the corresponding text reports.

This integration indicates a transition from solely textual NLP to the development of intelligent, context-aware decision systems that can support precision agriculture, climate adaptation strategies, and sustainable resource management.

2.7 Challenges in Early LLM Development

The early LLMs, even though they had a great potential to change the status quo, still had several inherent problems to deal with that restricted their usage in specialized areas like agriculture, among others. One of the main problems was data deficiency, LLM need a lot of text to be able to see the relations between different words in a specific context. However, the datasets that are used for machine learning in agriculture especially in low-resource languages are often scarce or fragmented (Kumar, 2024). The lack of high-quality data meant that the models ended up being trained on general-purpose knowledge and hence they were not very good in performing specialized agricultural tasks.

Another challenge was the cost of computation. The model training of the very large models, for instance, GPT or BERT variants, would need a lot of computer power which includes very fast GPUs or TPUs and also a lot of memory. This demand would usually mean that only the large research centers or the corporations with sufficient infrastructure could develop LLMs that are specific to agriculture which would in turn mean that smallholder farmers or local agricultural research centers would be excluded from the benefits of this technology.

Moreover, early LLMs were not interpretable. Outputs were drawn from the learned statistical correlations rather than from explicit reasoning, which in turn led to challenges concerning trust and adoption in critical decision-making situations such as pest management, irrigation scheduling, or climate-adaptive planning (Stojkovic et al., 2024). Thus, these limitations pointed out the necessity of domain-specific fine-tuning, smaller but efficient models, and multimodal integration to enhance practical usability.

2.8 Towards Explainable and Domain-Specific LLM

The future of LLM development is characterized by explainable AI, domain-focused models, and multimodal capabilities. As the main concern in agriculture was noticed by researchers, LLM should be able to deliver results that are comprehensible and can be acted upon. Meanwhile, attention visualization, saliency mapping, and Grad-CAM are some of the methods that have been gaining popularity among researchers in the agriculture domain through which the model can point to the areas in the input data that it considers significant leading to revealing the basis for predictions to farmers and advisors (Chinnaraju, 2025).

Experts in the agricultural field claim that the specific language models known as LLM sometimes referred to as AgriLLM, undergo training or fine-tuning using carefully selected agricultural datasets that consist of, among others, research articles, extension bulletins, farm logs, and local advisory services. The model's cultural relevance, accuracy, and context are all improved through this process. A case in point is an AgriLLM that has been trained on the datasets for pest control covering the whole area and can now provide very location-specific recommendations on how to deal with aphids in maize, considering regional climate, crop type, and pest presence. The evolution of LLM is shown in Table 1 with their relevance to agriculture.

Another interesting development is the shift towards the use of multimodal LLM. These models process and fuse texts, images, sensor data, and structured agricultural datasets to yield comprehensive insights. Take, for example, the case of a multimodal AgriLLM, which can simultaneously perform the tasks of analyzing drone images of cropped fields, checking the weather forecasts, and examining soil data. Then, it would be able to deliver a some advice in the natural language that a farmer or an agronomist could easily understand and which is ready for immediate implementation. Enumerating different data sources not only boost the models' ability to predict but also simplifies the adoption of precision agriculture methods.

Table 1: Evolution of Large Language Models (LLMs).

Stage	**Key Characteristics**	**Models/ Technologies**	**Agricultural Relevance**
Rule-Based & Statistical NLP (1980s–2000s)	Manually encoded rules; limited context understanding	ELIZA, n-gram models	Early agricultural text classification and information retrieval
Neural Embeddings (2010–2017)	Word2Vec, GloVe enabled semantic vectorization	Word2Vec, GloVe	Grouping of crop and soil terms similarity-based knowledge extraction
Transformer-Based Models (2017–2020)	Self-attention mechanism; parallel processing	BERT, GPT-2	Improved contextual understanding of agricultural reports and advisories
Large-Scale LLMs (2020–Present)	Billion-parameter models with few-shot learning	GPT-4, LLaMA 2	Agricultural question answering, translation, and automated advisories
Domain-Specific AgriLLMs (Emerging)	Fine-tuned for agriculture; multimodal data handling	AgriGPT, CropLLM	Pest prediction, yield forecasting, and policy document summarization

3. Applications of LLM in Agriculture

LLM have shown considerable capability to change the agricultural world, mainly by making it easier for farmers to understand complex data and gain useful knowledge. The various practical uses of their language skills include everything from decision support to ecologically friendly management, learning from data, and making policies. The following paragraphs describe eight major areas where LLM can be applied in agriculture, as shown in Figure 2.

3.1 Agricultural Advisory and Decision Support

Farmers can receive personalized consulting services through LLM as these models are capable of interpreting different data sources together with the expert knowledge of the farming community, and this is the case with climate data, soil profiles, crop management, and pest or disease incidences. By merging historical patterns with real-time data, these models produce case-specific suggestions for irrigation, fertilization, crop selection, and pest control (Abdalrazzaq et al., 2024).

Moreover, LLM can be very useful in making farm-level decisions during uncertain times, for instance, by forecasting yield fluctuations caused by different weather types or recommending the best planting times. LLM, when coupled with dashboards or mobile apps, will bring about interactive advisory systems where LLM will be the heart of these systems in terms of the engagement of farmers and the uptake of best practices.

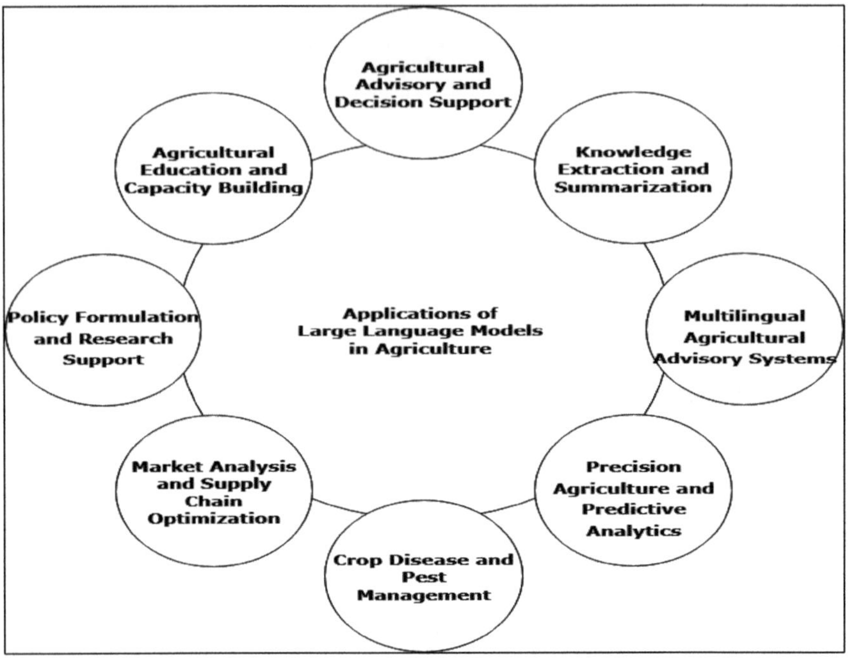

Figure 2: Applications of Large Language Models in Agriculture.

3.2 Knowledge Extraction and Summarization

Agricultural knowledge can be found in various sources such as scientific publications, extension materials, government reports, and even informal sources. But LLMs are very good at making it easier to get the information and also summarize it thus turning huge textual datasets into small, relevant, and concise knowledge (Vivekanandan et al., 2025).

To illustrate, the LLM can provide a summary of research papers on crop diseases automatically which can be used by agronomists and policymakers to get insights that they can act upon. Emerging trends are detected through topic modeling and semantic analysis, this technology aids in monitoring insect infestations or predicting climate-related risks incidents. These systems not only cut down the time needed to be done on literature but also speed up the whole process of knowledge change and establishing decisions based on evidence.

3.3 Multilingual Agricultural Advisory Systems

Agriculture is done in areas where people speak different languages. LLMs have the ability to convert scientific articles and notices to local languages thus enabling farmers with low literacy to access the knowledge (Areerob et al., 2025). Besides the translation, LLM can change the content according to the culture and education level providing easy-to-understand explanations or interactive help. This feature increases the number of people who can use the technology, guarantees that people have equal access to the information and increases the use of new farming methods.

3.4 Precision Agriculture and Predictive Analytics

Through the integration of large language models with Internet-of-Things (IoT) devices, sensors, and drones along with satellite images, the precision agriculture area will get a huge boost as such multimodal data would be interpreted with the help of LLM (Nguyen et al., 2021). Weather predictions, soil moisture levels, and crop history, for example, are three components whose combination could lead to the development of optimized irrigation schedules or better fertilizer choice made by an LLM. Besides, by adding topography and microclimate data to the model, it will be able to give spot-specific advice which will result in better resource utilization and less input waste.

The analytics produced by LLM will not only be oriented towards the future but also support new agricultural practices. Consequently, such models will predict the variations in yields due to the upsurge in demand for and adoption of new agricultural practices like weather-resistant hybrid varieties and changes in management practices (Karunathilake et al., 2023). Along with this, it will be possible to carry out a risk assessment of the simulated impacts caused by extreme weather, pest infestations, and nutrient depletion which will make the transition to climate-smart farming easier.

The LLM's ability to constantly and nearly in real-time monitor sensor data streams will be the chief reason for their being able to provide dynamic and adaptive recommendations as the conditions in the fields keep changing. This feature will foster very accurate and least disrupting interventions which in turn would conserve the environment, save water and fertilizers. The integration with farm management software will be such that the insights are fully utilized and translated into actions, for example, through controlling the automated irrigation system, monitoring crops by drones, and carrying out precision spraying application.

3.5 Crop Disease and Pest Management

LLMs take a huge step forward in disease and pest management of crops. One way they do this is by interpreting large amounts of text data coming from different places like scientific studies, agricultural records, and expert opinions from the government, which are done to determine the occurrence of pests and diseases (Cheng et al., 2017). Moreover, the application of predictive modeling allows these systems to predict the time of the appearance of pests or diseases through the investigation of a wide range of factors such as historical data, weather conditions, and the flowering stages of the plants.

A combination of LLM with image recognition models results in the correct identification of the specific crop disease with much higher accuracy by the study of pictures of the suffering plants along with the text. For example, LLM can tell apart fungal, bacterial, and viral infections in a plant and suggest an integrated pest management program that consists of organic, chemical, or crop rotation methods.

Additionally, they provide not only the recommendations related to the present situation but also consider future ones, like the possibility of increased fungal infections due to rain, or the decrease of pests due to heat. Thus, the knowledge extracted from LLM enables the implementation of preventive actions which lead to reduced crop losses and at the same time, the use of an environmentally-friendly method of pest control (Vivekanandan et al., 2025).

3.6 Market Analysis and Supply Chain Optimization

LLM can understand not only structured data from the market but also unstructured data. To illustrate, they can interpret and analyze the streams of prices, exchanges, news, and reports to come up with insights that, among others, can lead to the improvement of the supply chain's overall efficiency. Furthermore, LLMs consist of predictive models that are able to anticipate market changes and indicate the prices of the commodities, find the areas where supply is less than demand and recommend when to sell or store the harvest (Kumar, 2024).

The advent of LLM in the market has made it possible for farmers, cooperatives, and policymakers to combine market analytics with production forecasts and climate predictions and thus be able to foresee the risks, minimize the post-harvest losses and make more profits. LLM can work equally on logistics voicing out storage places, informing which routes to use and management of inventories, all based on their forecasts.

Natural language generation automatically makes it possible for LLM to create reports that can be read by humans and provide insights that can be acted upon, thus making data that is complex accessible to farmers and decision-makers (Abdalrazzaq et al., 2024). This feature is of great importance to smallholder farmers and local cooperatives who might not be able to afford sophisticated market analytics due to their limited resources.

3.7 Policy Formulation and Research Support

Policymakers and researchers have a significant ally in the form of LLM, which can process and provide huge amounts of documents such as legislation, research articles, field reports, and global policy guidelines among others (Jain, 2024). They not only corroborate the presence of knowledge gaps but also uncover new agricultural issues and suggest evidence-based solutions for food security, climate adaptation, and sustainable agriculture.

Apart from this, LLM can also largely influence the consideration of areas like subsidy allocation, crop insurance schemes, resource management, and climate-resilient practices through drawing trends and correlations from the datasets gathered from different sources. Besides, LLM can further the research process through their capability to summarize literature, point out relevant datasets, and come up with hypotheses that require experimental validation.

The multilingual feature makes it possible for policymakers to receive global research insights in their local languages, which in turn facilitates the decision-making process in an informed manner. Moreover, researchers gain from LLM that are able to create the synthesis of the domain knowledge, pattern recognition, and novel research direction suggestions, thus the whole cycle of agricultural innovation is improved (Stojkovic et al., 2024).

3.8 Agricultural Education and Capacity Building

LLM technology has the potential to create an interactive educational environment and training for farmers, extension workers, and students by providing them with personalized content, quizzes, and simulations (Nappu, 2025). Large language models analyze the different user profiles along with the regional cropping systems and hence, provide educational materials according to the user's knowledge, literacy, and agro-climatic conditions.

The conversational LLM agents can act like farmers interacting with the extension officers and allow the users to ask questions, get instant help, and understand problem-solving techniques. The use of LLMs in education is a major factor in the process of capacity building due to their provision of educational solutions that are very scalable and cheap even in remote or underprivileged areas. Besides, the sophisticated systems are able to monitor the learning journey, spotting areas where a learner lacks knowledge, and suggesting the modules for skill improvement that the learner is interested in. It is a characteristic that becomes very critical in the continuous professional development, while at the same time aiding the transfer

Table 2: Major Applications of LLMs in Agriculture.

Application Area	Functionality	LLM Role	Expected Outcome/Impact
Agricultural Advisory Systems	Real-time recommendations	Summarize, analyze data for field-level advice	Increased yield, efficient resource use
Knowledge Summarization	Extract key findings from literature	Automated summarization	Faster research dissemination
Multilingual Advisory	Translation and localization	Cross-lingual understanding	Bridging knowledge gaps for non-English users
Precision Agriculture	Integration with IoT/ sensors	Multimodal analytics	Real-time monitoring and adaptive irrigation
Pest & Disease Detection	Predict and classify plant diseases	Text + Image fusion	Reduced losses through early detection
Market Forecasting	Analyze trends and prices	Sentiment and data analysis	Improved profitability and logistics
Policy & Research Support	Analyze policy texts and datasets	Pattern recognition	Evidence-based policy and subsidy formulation
Education & Capacity Building	Training and simulation	Personalized content generation	Skill enhancement for farmers and students

of technology in agriculture such as the use of precision agriculture, eco-friendly practices, and the like.

The major applications and role of LLM in agriculture shown in brief in Table 2.

4. Challenges in Implementing LLM in Agriculture

The utilization of large-scale language models (LLMs) in agricultural systems has proven to be extremely promising, however, there are still several challenges of a technical, ethical, and infrastructural nature that have been limiting mass adoption. Agriculture is a very complex and heterogeneous sector where the performance of the model largely relies on context, data quality, and local tuning (Charania and Li, 2020). The integration of LLMs into agriculture requires addressing a wide range of problems that include data gathering, model training, computational power, interpretability, bias, and user-friendliness, to name a few.

4.1 Data Scarcity and Domain-Specific Knowledge Gaps

LLMs work best when they are trained with data that is high quality, large in quantity, and diverse. On the other hand, developing regions continue to experience the problem of data scarcity and fragmentation in agriculture (Al-Shammary et al., 2024).

- One reason for this is that agricultural datasets are often in an unstructured format and scattered across different sources such as research papers, extension bulletins, field notes, and even some archives that are not digital (Bharadiya et al., 2023).
- Another factor is that many local datasets are available only in regional languages or in paper-based formats that are not easy to digitize and standardize (Thite et al., 2024).
- Furthermore, the unavailability of domain-specific labeled datasets, such as pest outbreak descriptions or local crop disease reports, is a hindrance to the fine-tuning of LLM for agricultural contexts.

Consequently, generic LLM that have been trained on the general internet text often do not possess the ability to understand the different meanings that are attached to agricultural words, the different practices in different regions, and the different variations in culture. Creating large, multilingual, and annotated agricultural corpora to support accurate model adaptation is necessary to get around this problem (Shaikh et al., 2025).

4.2 Computational Complexity and Resource Requirements

The procedure of training and deploying LLMs requires to a large extent computational power, storage, and energy. The high-end models, for example, GPT, BERT, and PaLM, work with billions of parameters which is why they consume a lot of resources ADDIN ZOTERO_ITEM CSL_CITATION {“citationID”:“gI4kLTSk”,“properties”:{“formattedCitation”:

“(Bharadiya et al., 2023)”, “plainCitation”: “(Bharadiy et al., 2023),” “noteIndex”:0},“citationItems”:[{“id”:201,“uris”:[“http://zotero.org/users/local/D8n0hGT7/items/R8C5GFXL”],“itemData”:{“id”:201,“type”:“article-journal”,“abstract”: “The art of predicting crop production is done before the crop is harvested. Crop output forecasts will help people make timely judgments concerning food policy, prices in markets, import and export laws, and acceptable warehousing. It is possible to reduce the socioeconomic effects of crop loss brought on by a natural disaster, such as a flood or drought, and to organize humanitarian food assistance. It has been suggested that deep learning, which lets the model to automatically extricate features and learn from the datasets, could be useful for predicting agricultural yields. This review helps to understand how vegetation indices and environmental variables affect agricultural output by revealing gaps in our understanding of deep learning methodologies and remote sensing data in a specific area. Literature review of 2011–2022 has been collected from different databases and sites and analyzed to meet the aims of this review. The study mainly focused on the benefits of machine learning, agrarian factors, and remote sensing for forecasting crop yield. The most often employed form of remote sensing is satellite technology, namely the use of Moderate-Resolution Imaging Spectro radiometer. Vegetation indices referred to as the most often used attribute for forecasting crop yield, according to the results. This review compares all these techniques and pros and cons related to them.”, “container-title”: “Journal of Engineering Research and Reports”, “DOI”: “10.9734/jerr/2023/v24i12858”, “ISSN”: “2582–2926”, “issue”: “12”, “language”: “en”, “note”: “number-of-pages: 16\nnumber: 12”, “page”: “29–44”, “source”: “classical.goforpromo.com”, “title”: “Forecasting of Crop Yield using Remote Sensing Data, Agrarian Factors and Machine Learning Approaches”, “volume”: “24”, “author”:[{“family”: “Bharadiya”, “given”: “Jasmin Praful”},{“family”: “Tzenios”, “given”: “Nikolaos Tzenios”},{“family”: “Reddy”, “given”: “Manjunath”}], “issued”:{“date-parts”:[[“2023”,4,15]]}}}],”schema”: “https://github.com/citation-style-language/schema/raw/master/csl-citation.json”} (Bharadiya et al., 2023).

In developing countries, the agricultural sector suffers the most from the problem of limited access to the necessary computing infrastructure (Javaid et al., 2023). The GPU clusters or cloud computing services, which are considered the primary means for large-scale training or fine-tuning, are simply not available at most agricultural research centers and institutions.

Moreover, the high energy consumption and the accompanying carbon emissions from LLM training are considerable and this threatens the sustainability of such research activities, especially in rural areas where the supply of renewable energy can be quite unstable. Utilizing lightweight or distilled models may offset the impact of some of these drawbacks, however, they are frequently synonymous with inferior performance or reduced contextual accuracy (Shaikh et al., 2025).

4.3 Multilingual and Contextual Limitations

Agriculture occurs in ecosystems that are rich in diversity regarding languages and cultures. Farmers and extension workers sometimes use regional dialects for their communication, and agricultural vocabulary may differ in local contexts (Vivekanandan et al., 2025).

Despite the capability of multilingual LLM to translate and interpret text, they frequently have a hard time with low-resource languages and dialectal variations.

- One of the consequences of translation mistakes is the distortion of important agricultural recommendations.
- Incorrect representation is often the case for local idioms, metaphors, or expressions related to specific regions.
- Models that have been mostly trained on global datasets might not be able to capture the indigenous farming knowledge systems.

Thus, the making of context-aware and culturally sensitive LLM is necessary in order to prevent the inequitable and non-inclusive technology adoption in the agricultural sector.

4.4 Lack of Explainability and Trust

The "black-box" characteristic of LLM is one of the key limitations in decision-critical sectors such as agriculture. The output produced is largely logical but the main logic can still be quite a riddle. A major factor responsible for decreasing user confidence, mainly in case of farmers and decision makers who want to receive direct knowledge, is the lack of transparency. The need for clarification arises when LLMs are employed in very sensitive areas, such as determining the quantity of pesticide to be applied (Chinnaraju, 2025).

4.5 Security, and Ethical Concerns

Data from agriculture are sensitive. These data include the characteristics of farmers, the records of their lands, the characteristics of soils, and the market transactions. The integration of these datasets with LLM leads to the consideration of privacy and security issues (Bharadiya et al., 2023). If data are used without authorization or a model is leaked, private or commercially valuable information becomes accessible. In addition, the use of LLM to unfairly influence pricing policies, resource allocation, or credit decisions raises ethical concerns. Data sovereignty, compliance with data protection regulations and the use of federated or privacy-preserving learning techniques are vital to maintaining trust and transparency (Dembani et al., 2025).

5. Future Directions of LLM in Agriculture

The digitization of agriculture is a trend that is unlikely to happen soon (Balasundram et al., 2023) it is even possible that LLM will be concurrently used in such a scenario—provided the models are smart, clear about their intentions, and knowledgeable of the context in which they function and thus included in the agricultural setting. The subsequent stage of development will have multimodal learning, decentralized architectures, sustainability-aware AI, and human AI co-evolution as its major

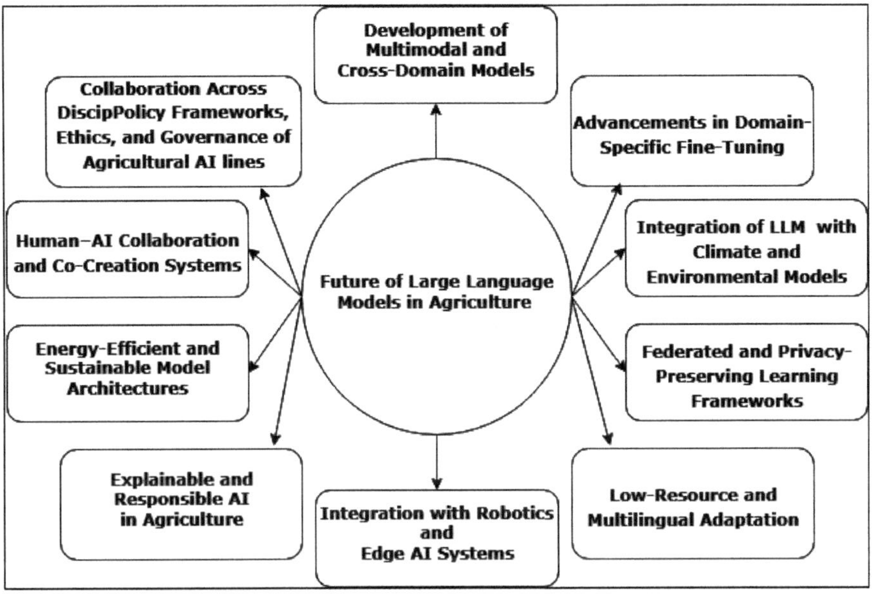

Figure 3: Future of LLM in Agriculture.

themes (Akter et al., 2023). In the following paragraphs, ten principal directions are pointed out which will not only characterize but also define the future of LLM in respect to agriculture. Several directions that are shown in Figure 3 are pronounced as key pathways for the future of LLM in agriculture.

5.1 Development of Multimodal and Cross-Domain Models

LLM of the future will not only process the text but also integrate multimodal data comprising satellite pictures, drone data, weather information, sensor networks, and textual inputs. They will be able to analyze and reason through different data simultaneously which makes it possible for the development of comprehensive agricultural intelligence systems. For instance, a model might pair soil nutrient assessments with weather predictions and farmer stories in order to suggest precise interventions (Areerob et al., 2025).

Cross-domain learning the connecting of agricultural, climatic, and economic datasets will lead to more precise predictions of market trends, supply chain robustness, and food security risks. These systems would not just know what to advise but would also give the reasons, and thus, build data-driven agricultural ecosystems.

5.2 Advancements in Domain-Specific Fine-Tuning

General-purpose LLM are frequently unable to reason well in specific domains. However, AgriLLM will become the predominant model of the future big models that are either pretrained or fine-tuned exclusively on agricultural texts, extension reports, and climate data (Hadi et al., 2024). The scope of these models will cover local queries like pest control for certain crop types, soil management in specific landscapes, or law interpretation for the local subsidy program.

The use of techniques such as LoRA (Low-Rank Adaptation) and prompt-tuning will make it possible to specialize in a domain efficiently without large retraining, thus making it possible for even the smallest institutions to tailor models for their areas. The trend will be further supported by open-source agricultural corpora, multilingual datasets, and collaborative AI projects (Shaikh et al., 2025).

5.3 Integration of LLM with Climate and Environmental Models

Climate change and the like are becoming more and more of a concern and thus, LLM will have to work alongside climate simulation and environmental monitoring systems (Vivekanandan et al., 2025). This will facilitate the instant analysis of abnormal weather, drought formation, and plant stress situations within the same time frame that they occur.

The coming models will be acting as helpers in decision-making for climate-smart farming, revealing the features of carbon trapping, and the scheduling of water supply that is both eco-friendly and maintenance of species diversity (Vivekanandan et al., 2025). They would even help in determining the environmental impact of the different farming methods applied and recommend the changes needed to adapt.

5.4 Federated and Privacy-Preserving Learning Frameworks

Federated Learning (FL) will be crucial in the development of LLM that do not breach privacy, given that agricultural data frequently contains sensitive information about farms and communities (Dembani et al., 2025). With this method, the models are trained on the edge devices at the location, and only the aggregated model updates not the raw data are sent to the central servers.

This practice makes data more secure, lessens the need for the internet, and grants access to models to small farmers. The use of FL along with differential privacy and homomorphic encryption will further secure data governance and build user confidence in AI-powered agriculture platforms (Nguyen et al., 2021).

5.5 Low-Resource and Multilingual Adaptation

The training of most LLM has been done primarily in English and other major languages which is a drawback when it comes to using them in the multilingual agricultural communities (Kumar, 2024). The AgriLLM path ahead is going to be through multilingual adaptation, that is, models capable of understanding, translating, and generating content in local dialects and regional languages.

Some of the major techniques like learning from high-resource languages, cross-lingual embedding spaces and artificial data generation will play an important role in making the systems more user-friendly and accessible. Consequently, LLM will be able to overcome the language barriers and thus transferring the agricultural knowledge to the far-off areas where the access to AI advisories is not very convenient (Shaikh et al., 2025).

5.6 Integration with Robotics and Edge AI Systems

Integrating LLM into robotic systems, drones, and autonomous vehicles can bring a radical change to agricultural automation (Zhu et al., 2024). LLM will be the 'language reasoning core' of robots doing precision spraying, harvesting, and monitoring in the near future.

With the edge deployment model, LLM in lightweight forms will be able to do the local processing of sensor data, thus cutting down on latency as well as the need for cloud connectivity. This method makes operational efficiency possible in places where internet access is very limited. Just like combining natural language reasoning and computer vision, interactive, explainable robotic assistants will be able to process farmers' contextual instructions and be recognized as such (Vivekanandan et al., 2025).

5.7 Explainable and Responsible AI in Agriculture

One of the biggest obstacles still preventing the use of LLM in sensitive areas is the lack of transparency. The models of the future would need to be equipped with a layer of Explainable AI (XAI) that would permit users to follow the exact path taken by the model in arriving at a particular conclusion or making a recommendation.

Such LLMs would indicate the areas where they draw their conclusions, specify the characteristics of the input affecting the decision the most, and state the extent of their uncertainty with regards to the output (Shaikh et al., 2025). It is these features combined that would win the trust of the users, align the models with the accountability and ethical considerations in the agricultural sector management. Furthermore, the AI regulations would make sure that the agricultural sector development is sustainable, compliant with the UN Sustainable Development Goals (SDGs) and that no unfairness is practiced while no bias is introduced in the data (Ali et al., 2025).

5.8 Energy-Efficient and Sustainable Model Architectures

LLMs are very resource-demanding when it comes to machines, and thus they present challenges to sustainability. The future research's main focus will be on constructing energy-efficient models, incorporating quantization techniques, and applying distillation methods that will lead to a reduction in model size and power consumption while keeping performance as the main priority.

The power consumption of data centers and green computing infrastructures that rely on renewable sources of energy will largely be the negative environmental impact reduction. One of the approaches in this strategy is the creation of a new generation of AgriLLM that is compact yet powerful and designed for rural localization to make AI applications in agriculture more userfriendly and ecofriendly.

5.9 Human–AI Collaboration and Co-Creation Systems

Future LLMs in agriculture will not replace human knowledge and skills. They will instead promote human-AI collaboration. The extension workers, scientists, and farmers will interact with AI systems and thus co-create solutions tailored to their particular area (Mahto and Mahato, 2025).

Dialogue, Reinforcement Learning from Human Feedback (RLHF), and participatory design will be the methodologies through which the model's performance will be continuously enhanced. The team work of humans and AI could result in knowledge transfer, idea generation, and the faster implementation of sustainable agriculture techniques.

5.10 Policy Frameworks, Ethics, and Governance of Agricultural AI

The agricultural sector progressively adopting LLMs necessitates installing ethical and regulatory frameworks first. Among future inquiries, the one on policies for data ownership, algorithmic accountability, and fair AI distribution should become a priority bringing others (Jain, 2024) to the fore.

Moreover, the collaboration between various governments, research and academic institutions and companies all over the world should be the one issue in which the ethical and responsible application of LLM in genetic data analysis,

Table 3: Challenges and future research directions for LLMs in agriculture.

Challenge Area	Description	Proposed Solutions/Future Directions
Data Scarcity	Lack of annotated, multilingual agricultural datasets	Develop open-access Agri corpora, promote data-sharing frameworks
Computational Demand	High training costs and resource use	Use lightweight, energy-efficient, or federated models
Contextual Limitations	Difficulty understanding local dialects and cultural nuances	Multilingual adaptation and fine-tuning using regional data
Explainability	Limited transparency of LLM reasoning	Integrate Explainable AI (XAI) tools like Grad-CAM, attention maps
Ethical & Privacy Concerns	Data misuse and bias	Adopt privacy-preserving and fair AI governance frameworks
Sustainability	High carbon footprint of training	Promote green computing and quantized LLM architectures
Human–AI Collaboration	Lack of farmer engagement in model design	Co-creation systems integrating human feedback (RLHF)

pesticide recommendation and crop insurance modeling would hardly be a matter of debate. The AI ethics panels and agricultural data trusts are going to be the custodians of both openness and public good and they are going to support these principles (Jain, 2024).

Key challenges and future research directions for LLMs in agriculture shown in Table 3.

6. Conclusions

The large language model has made a major breakthrough that has really changed the whole machine intelligence field to the extent that it can now perform the whole human languages perception, generation, and comprehension in a way that is close to the human being. This chapter has covered the whole journey from philosophical underpinnings, the rise of its significance from the past to present and even future. The LLM has transformed from a statistical natural language processing model to a transformer-based architecture model and has become a crucial partner in areas like agriculture and others for data interpretation, information sharing, and smart choices by the use of technology.

In the field of agriculture, LLMs are playing the role of mediator between intricate data and the knowledge that is ready to be used. They collect and process data from numerous sources that are IoT devices, satellite imagery, sensor networks, and agricultural databases to ensure that farmers get smart advice on pest control, soil health, and crop yield that is foretold, and also about climate change adaptation. Some of the possible benefits include: farmers receiving advice on time, documentation and analysis of policies done by the machine automatically, and the arrival of precision farming through prediction and adaptive modeling becoming possible. Furthermore, the ability of LLMs to work with both multimodal and multilingual data results in increased accessibility and inclusiveness of AI-powered methods which eventually implies that the solutions will be available to the farmers from every type of area, whether they are high-tech or resource-poor.

However, the application of LLM in agriculture is not straightforward. The chapter in question pinpointed major issues such as the lack of data, difficulties with model explanations, the necessity for powerful machines, and the influence of ethics as the main barriers. These constraints mean that the models used must be very sophisticated in understanding the environment, setting up federated learning systems that are equally robust in terms of privacy protection, and training the models to the specific areas of application so that the models are contextually and accurately relevant. The lack of annotated agricultural data in less commonly spoken languages and difficult to put one's reasoning in are still matters of grave concern that need to be resolved if AI-assisted decision systems are going to be considered trustworthy and reliable.

The future of LLM in agriculture depends on the factors like multimodal integration, human-AI collaboration, and sustainable innovation. Agriculture's next generation intelligence would be greatly influenced by the new models such as AgriLLM, explainable AI, energy-efficient model architectures, and policy-driven governance frameworks. These breakthroughs will not only offer farmers data-backed decision support systems but also empower them with resilience against climate uncertainties, obtain better yields with less harm to the environment. Moreover, the connection of LLM with robotics, edge computing, and participatory knowledge systems is going to change the global dynamics of agricultural knowledge production, dissemination, and usage. In summary, the deployment of LLM in agriculture is not only a technological aspect of more efficient farming but also a completely different and new concept of farming that will be characterized as intelligent, inclusive, and sustainable. LLM partnered with ethical principles, transparent methods, and collaborative innovation have the potential to transform the entire agriculture sector thus bridging the digital divide and being a significant part of the global food supply.

Declaration

Ethical Approval

Not applicable

Use of AI tool

No artificial intelligence (AI) assisted tools were used in the writing, editing, data analysis, or preparation of this book chapter. All content is original and solely the work of the authors.

Conflict of Interest

The authors declare that they have no conflict of interest related to this work

Competing interests

Not applicable

Authors' contributions

Ritika: Conceptualization, Methodology, Writing - Original Draft, Writing - Review & Editing, **Savita Kumari Sheoran**: Investigation, Formal analysis, Visualization, Software, Supervision,

Funding

No funding has been received.

References

A. k, P., M, M., Rajamanickam, S., Sivarethinamohan, S., Gaddam, M. K. R., Velusamy, P. et al. (2023). Impact of climate change and anthropogenic activities on aquatic ecosystem – A review. Environmental Research 238: 117233. doi: 10.1016/j.envres.2023.117233

Abdalrazzaq, A. S., Kadhim Meftin, N., Hussein Al Fawadi, D. M., Hameed Merzah, H., Baban, O., Akymenko, A. et al. (2024). Optimizing robotic interaction and decision making using advanced NLP and AI Technologies., in *2024* 36th Conference of Open Innovations Association (FRUCT), 692–701. doi: 10.23919/FRUCT64283.2024.10749908.

Akter, S., Hossain, M. A., Sajib, S., Sultana, S., Rahman, M., Vrontis, D. et al. (2023). A framework for AI-powered service innovation capability: Review and agenda for future research. Technovation 125: 102768. doi: 10.1016/j.technovation.2023.102768.

Albeladi, K., Zafar, B. and Mueen, A. (2023). Time Series Forecasting using LSTM and ARIMA. IJACSA 14. doi: 10.14569/IJACSA.2023.0140133.

Ali, Z., Muhammad, A., Lee, N., Waqar, M. and Lee, S. W. (2025). Artificial Intelligence for sustainable agriculture: a comprehensive review of AI-driven technologies in crop production. Sustainability 17: 2281.

Al-Shammary, A. A. G., Al-Shihmani, L. S. S., Fernández-Gálvez, J. and Caballero-Calvo, A. (2024). Optimizing sustainable agriculture: A comprehensive review of agronomic practices and their impacts on soil attributes. Journal of Environmental Management 364: 121487. doi: 10.1016/j.jenvman.2024.121487.

Areerob, K., Nguyen, V., Li, X., Inadomi, S., Shimada, T., Kanasaki, H. et al. (2025). Multimodal artificial intelligence approaches using large language models for expert-level landslide image analysis. Computer-Aided Civil and Infrastructure Engineering.

Balasundram, S. K., Shamshiri, R. R., Sridhara, S. and Rizan, N. (2023). The role of digital agriculture in mitigating climate change and ensuring food security: An overview. Sustainability 15. doi: 10.3390/su15065325.

Bharadiya, J. P., Tzenios, N. T. and Reddy, M. (2023). Forecasting of crop yield using remote sensing data, agrarian factors and machine learning approaches. Journal of Engineering Research and Reports 24: 29–44. doi: 10.9734/jerr/2023/v24i12858.

Bhardwaj, A. and Gupta, S. P. (2025). Empowering sustainable agriculture harnessing AI for enhanced yields, resource efficiency, and eco-friendly farming practices, in AI and Ecological Change for Sustainable Development, (IGI Global Scientific Publishing), 183–216.

Bhattacharya, S. and Pandey, M. (2024). Developing an agriculture ontology for extracting relationships from texts using Natural Language Processing to enhance semantic understanding. International Journal of Information Technology. doi: 10.1007/s41870-024-01809-x.

Chaitanya, K. and Rolla, K. J. (2024). The evolution and impact of large language models in artificial intelligence, in Algorithms in Advanced Artificial Intelligence, (CRC Press).

Charania, I. and Li, X. (2020). Smart farming: Agriculture's shift from a labor intensive to technology native industry. Internet of Things 9: 100142. doi: https://doi.org/10.1016/j.iot.2019.100142.

Cheng, X., Zhang, Y., Chen, Y., Wu, Y. and Yue, Y. (2017). Pest identification via deep residual learning in complex background. Computers and Electronics in Agriculture 141: 351–356. doi: 10.1016/j.compag.2017.08.005.

Chinnaraju, A. (2025). Explainable AI (XAI) for trustworthy and transparent decision-making: A theoretical framework for AI interpretability. World Journal of Advanced Engineering Technology and Sciences 14: 170–207.

Dembani, R., Karvelas, I., Akbar, N. A., Rizou, S., Tegolo, D. and Fountas, S. (2025). Agricultural data privacy and federated learning: A review of challenges and opportunities. Computers and Electronics in Agriculture 232: 110048.

Hadi, M. U., Al-Tashi, Q., Qureshi, R., Shah, A., Muneer, A., Irfan, M. et al. (2024). LLMs: A Comprehensive Survey of Applications, Challenges, Datasets, Models, Limitations, and Future Prospects.

Jain, A. (2024). Ethical AI: A Policy Framework to Regulate Bias in Large Language Models.

Javaid, M., Haleem, A., Khan, I. H. and Suman, R. (2023). Understanding the potential applications of Artificial Intelligence in Agriculture Sector. Advanced Agrochem. 2: 15–30. doi: 10.1016/j.aac.2022.10.001.

Kalyan, K. S., Rajasekharan, A. and Sangeetha, S. (2021). Ammus: A survey of transformer-based pretrained models in natural language processing. arXiv preprint arXiv:2108.05542.

Karunathilake, E. M. B. M., Le, A. T., Heo, S., Chung, Y. S. and Mansoor, S. (2023). The Path to Smart Farming: Innovations and Opportunities in Precision Agriculture. Agriculture 13. doi: 10.3390/agriculture13081593.

Kumar, D., Sinha, S. N., Rajendra, S. and Sharma, K. (2023). Assessing farmer's exposure to pesticides and the risk for non-communicable diseases: A biomonitoring study. Science of The Total Environment 891: 164429. doi: 10.1016/j.scitotenv.2023.164429.

Kumar, P. (2024). Large language models (LLMs): survey, technical frameworks, and future challenges. Artificial Intelligence Review 57: 260. doi: 10.1007/s10462-024-10888-y.

Mahto, H. and Mahato, D. (2025). Human-centric biotechnological interventions: a pillar for sustainable agricultural development in industrial crop improvement, in Industrial Crops Improvement: Biotechnological Approaches for Sustainable Agricultural Development, (Springer): 205–226.

Nappu, S. (2025). in Higher Education. AI, Analytics, and Assessment in Higher Education, 55.

Nguyen, D. C., Ding, M., Pathirana, P. N., Seneviratne, A., Li, J. and Poor, H. V. (2021). Federated learning for internet of things: A comprehensive survey. IEEE Communications Surveys & Tutorials 23: 1622–1658.

Savytska, L., Vnukova, N., Bezugla, I., Pyvovarov, V. and Sübay, M. T. (2021). Using Word2vec technique to determine semantic and morphologic similarity in embedded words of the Ukrainian language.

Shaikh, T. A., Rasool, T., Veningston, K. and Yaseen, S. M. (2025). The role of large language models in agriculture: Harvesting the future with LLM intelligence. Progress in Artificial Intelligence 14: 117–164.

Stojkovic, J., Choukse, E., Zhang, C., Goiri, I. and Torrellas, J. (2024). Towards greener llms: Bringing energy-efficiency to the forefront of llm inference. arXiv preprint arXiv:2403.20306.

Thite, S., Patil, K., Jadhav, R., Suryawanshi, Y. and Chumchu, P. (2024). Empowering agricultural research: A comprehensive custard apple (*Annona squamosa*) disease dataset for precise detection. Data in Brief 53: 110078. doi: 10.1016/j.dib.2024.110078.

Vivekanandan, V., Sureshkumar, R., Manikandan, S., Ram Kumar, R., Das, M. S., Kumar, G. R. et al. (2025). Environmental monitoring and sustainability: LLMs for climate-responsive urban design, in Large Language Models for Sustainable Urban Development, (Springer): 89–109.

Zhu, H., Qin, S., Su, M., Lin, C., Li, A. and Gao, J. (2024). Harnessing large vision and language models in agriculture: A review. arXiv preprint arXiv:2407.19679.

5

The Rise of Transformer Architecture and Deep Learning in NLP

Shikha Tiwari,[1] *Nisha Rathore,*[1] *Manjushree Nayak,*[1,*] and *Syed Danish Hassan*[2]

1. Introduction

Natural Language Processing (NLP) has grown to become a very exciting and important area of artificial intelligence (AI). It has transformed the way in which people interface with machines and how robots interpret human language out of context (Vaswani et al., 2017). In the past ten years NLP has seen more growth than ever before because we have at our disposal so much text data and also because machine learning algorithms have improved. That which we are seeing is a large scale revolution, which has gone from very rigid rule based systems to more adaptable, context aware models that interpret complex language (Devlin et al., 2018). The history of NLP is also the history of AI as a whole, we went from hand built systems to data driven self learning models that only get better the more they are used. At the start we had very rule based systems that included language that specialists had gone in and very carefully coded. These systems were, essentially, rule-based systems or frameworks that used a library of human-built lexicons, or phonetic syllables or morphemes, that may have done some basic parsing and part-of-speech tagging while the system may have extracted a query from either back referencing to the

[1] Amity School of Technology and Engineering, AMITY UNIVERSITY, Raipur, Chhattisgarh, State Highway 9, Manth (Kharora), 493225 Baloda Bazar Road.
Emails: stiwari@rpr.amity.edu, orcid id-0000-0002-4877-2777;
nisha.rathor271@gmail.com, https://orcid.org/0009-0009-6471-4540
[2] Data Science University of Europe for Applied Sciences in Dubai.
Email: syeddanish.hassan@ue-germany.de, https://orcid.org/0009-0000-3534-1786
* Corresponding author: manjushreesai44@gmail.com, https://orcid.org/0000-0001-6383-780X

DOI: 10.1201/9781003771654-5

text or some trivial referencing to the base text with a keyword search, across each system, from a larger family of potential natural language processing (NLP) concepts algorithms that invoked generating schemas all the same remarking that every rule-based machine learning system will use the same automation methodology, just a different basis—and indeed, each will encounter the same problems with scaling and ambiguity, which is relevant case by case basis or inferred through some aspect nonetheless these approaches have not been particularly useful for any tangible world problems or for dialectic problems of complex until the NLP drift started using methodologies while also statistically modeling of instances over observation in time (Brown et al., 2020).

Text was formally structured using probabilistic models. This has advantages in a number of ways that are not available to relying on discrete rules to structure any content because probabilistic models also rely on some sizeable corpus of language to suggest that the word has two or more meanings. Some examples of models that are currently employed are statistical machine translation, n-grams, and hidden Markov models (Boardman et al., 2020). The advancement of deep learning in NLP may indicate some shift in the creativity involved in the models; however, it is important to note that in a deep learning approach, manual extraction of features is no longer a necessity (Boardman et al., 2020). This enabled us to build better and more extensible models for multiple tasks. We were similarly early in the word embeddings journey (Word2Vec and GloVe inspired the practice of an embedding) and even at the early stages we experienced some success. The word embeddings showed that dense vector representations of words exhibited vector-based semantic relationships and syntactically based relationships that did not exist before with statistical models. Finally, word embeddings were a precursor to even much deeper (and much more complicated) forms of neural networks of deep learning like Recurrent Neural Networks (RNNs) and Long Short-Term Memory (LSTM) networks which excelled at modeling sequential relationships.

Although RNN and LSTM (Hu et al., 2021) were very successful and applicable, they had some limitations. They processed information in one phase as a block of information, which led to an increased processing time and computational cost, particularly with a larger input-based data set. They also were unable to learn long-range dependencies among the inputs they processed. This means you could not understand how a word in a sentence interacted with multiple other words in the surrounding context. The attention mechanism is one abstraction that could help solve these problems, as it can focus only on the necessary parts of the input sequence but also hold and process a cumulative state of dependencies and contextual representations (Torfi et al., 2020).

The application of transformers goes beyond the world of academia. They support systems we interface daily, including Google Translate, chat bots (like ChatGPT), and recommendation systems as viewed on social and e-commerce systems. Their application is extensive across health care and finance and e-commerce, and the generalized application and flexibility is ubiquitous. Large language models (LLMs) which are based on an architecture of the Transformer model are what started the conversation and public interest in LLMs, because of the refined text that is generated at a degree of humanness and is able to sustain conversations across a huge variety of topics. However, the history of natural language processing (NLP) goes back much longer than recent waves of interest, with many years of evolving technologies since the days of strictly rule-based models, through to the models that we query with today that are degree of state-of-the-art and are based on the Transformer model architecture, all which represents a 50-year journey of large-scale evolution. In sum, this chapter will provide a brief history of the evolution of NLP through deep learning and the vectors of attention, a summary of how to operationalize each component of the transformer architecture, and a practical example of the models that are the main contributors to NLP and the overall field.

2. Deep Learning Basics for NLP

The field of natural language processing (NLP) before deep learning. most NLP systems were founded on difficult set of rules and on manual feature extraction. Deep learning works with network-based model to represent data hierarchically. A multi-layered neural network is used in deep learning to replicate intricate and profound relationships between inputs and outputs. Conversely, conventional machine learning methods are constrained and heavily rely on feature engineering. Deep learning, on the other hand, helps sophisticated algorithms identify patterns in unstructured data by using hand-crafted features. We may therefore draw the conclusion that the model explains complex and intricate patterns and that subject-matter specialists are not required to generate features.

Figure 1 illustrates how the core concepts of Deep Learning in Natural Language Processing (NLP) have evolved. During the initial stages of NLP, the procedures tended to revolve around rigid framework techniques alongside statistical models, which required intensive manual feature engineering. Deep Learning has led to a

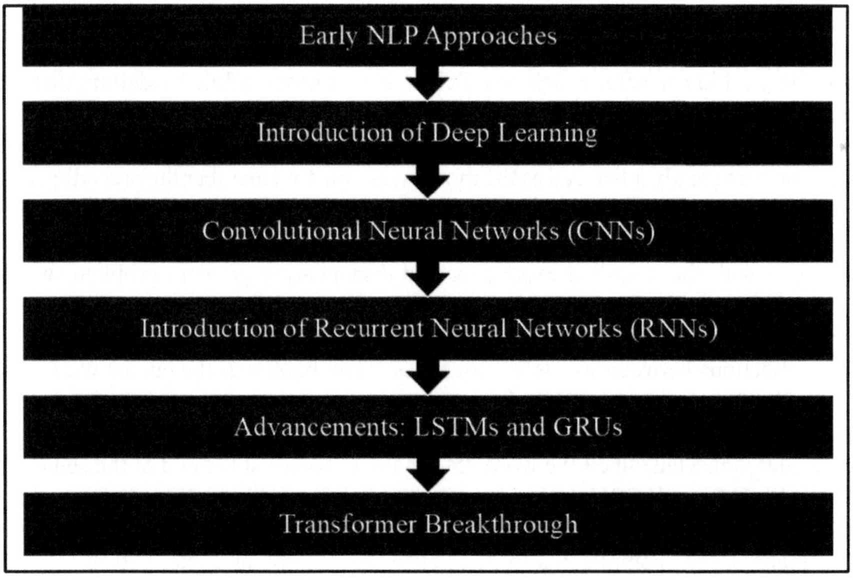

Figure 1: Changes in Deep Learning Architectures in Natural Language Processing (NLP).

change of automatically acquiring features from raw, complex text. Most initially deep learning models that were used in natural language processing (NLP) were Convolutional Neural Networks (CNNs) that worked well to identify a local pattern in the case of sentiment classification and text classification (Imamura et al., 2019). Subsequently, serial associative RNNs (Recurrent Neural Networks) were implemented to operate on the concept of context which was constantly supplied to the input. The RNNs were necessary to address differential method and the Gated recurrent units (GRUs) and Long Short-Term Memory (LSTM) networks which were mechanistic extensions of RNNs, solved the problem of back propagation with vanishing gradients and long-term dependency (Ouyang et al., 2022). The final, eventual the development ended in the new self-attention transformer that offered exciting new opportunities of applications in NLP with parallel processing, self-attention, and is the probably an amazing step-forward of performance on different capabilities.

Among the most widely-practiced methods for deep learning in natural language processing (NLP) are convolutional neural networks, or CNNs. It should be noted that CNNs also stands for some of the various imaging modalities based on CNNs. Here, the authors used their framework to treat sentences as a one-dimensional array that ultimately allowed them to absorb text data. CNNs benefit from the application of convolutional filters that seek to detect local structures (e.g., n-gram features) and are typically employed for tasks such as document categorization, sentiment analysis, and text classification. CNNs can also model associations among clusters of words. In the specific case of sentiment analysis, context is important when looking for categorical information in expressions like, "very poor," or, "highly recommended," to assess if the sentiment expressed is negative or positive. The strength of the distributed representations as favourable has its disadvantages if the state of whether a representation was possible or changed in time, words in a sequence, context, and size offered limitations. Due to the limitations of sequential dependencies of texts, researchers developed recurrent neural networks (RNNs). RNNs provide a means of modelling a hidden state, which could change over time—while modelling through the following words in a particular order.

With this, the ability to sequence various phrases or speech signals through dialogue was greatly advanced as the model was able to remember the preceding data inputs. Thus, RNNs played a key role in the early days of NLP in areas like speech recognition, language modelling, and POS tagging. However, conventional RNNs struggled with the so-called expanding and disappearing gradient problem, where the model's quickly learnt gradients either became uncomfortably contradicting or incredibly small. This outlined the abilities RNNs needed, for instance, to identify the connections between words in sequences. "The book that the author wrote was fascinating" demonstrates RNNs' failure in retaining long sequences of cohesive relations. In this example, the word "book" and the word "fascinating" are supposed to be interconnected, albeit the word "fascinating" comes at the end of the sentence. While LSTMs and GRUs were improvements, they still had to do everything sequentially. This meant that a phrase had to be worked on a 'word at a time' basis. This made it very complicated to parallelize the training and added to the cost of the processing, particularly with large datasets. In addition, even these sophisticated recurrent models struggled with fully capturing long-range dependencies with very long text sequences. The addition of the attention mechanism was a huge improvement on this, as it enabled models to concentrate on particular parts of the input stream. With attention, models were able to assign different weights to words, processing them with varying levels of importance based on the task (Islam et al., 2024). The transformer architecture was introduced in 2017 in the seminal paper "Attention Is All You Need." The key difference from RNNs and LSTMs was that the Transformer model eliminated recurrence and only focused on attention mechanisms to explain how words related to each other. Its principal innovation, self-attention, allowed the model to consider all the words in a particular sequence at the same time, greatly enhancing the model's efficiency and scalability.

In order to address the lack of order, the sentence structure was remedied by the addition of positional encoding for the order of words in a sequence. With the addition of self-attention, self-attention, and the parallelizable architecture of transformers, they achieved state-of-the-art performance in a wide range of NLP tasks. Because of this, they are not constrained to traditional NLP tasks; they are also able to handle multimodal tasks, such as in the case with CLIP which pairs text with images, or in conversational AI incorporating text and audio. The global dependencies and parallelization of the sequences changed the field of natural language processing. It not only solved the issues of previous architectures, but also pioneered the development of BERT, GPT, and T5. Such models have redefined the benchmarks in language understanding with the application of deep learning in NLP, signifying the emergence of the new age in NLP.

Table 1, there are several important models applied in Deep Learning in Natural Language Processing (NLP) and other domains. Each has advantages and disadvantages, along with specific applications. Recurrent Neural Networks (RNN) are proficient in handling temporal information, however, suffer from the ability to understand long-range dependencies, issues like the vanishing gradient problem. Long Short-Term Memory (LSTM) networks resolve this issue by being able to retain information stored over lengthy durations. Thus, they are proficient in performing complex tasks, like translating languages and generating texts while expensive in terms of computational resources. Gated Recurrent Units (GRU) improve the LSTM architecture, simplifying and enhancing performance, making them ideal for real time NLP and speech recognition systems. Yet, these models

Table 1: Summarizes the main deep learning architectures used in NLP, showing their pros and cons and popular uses.

Model	**Strengths**	**Weaknesses**	**Use Cases**
RNN	Handles sequences well	Vanishing gradient problem	Basic sequence modeling
LSTM	Captures long-term dependencies	High computational cost	Text generation, machine translation
GRU	Simpler than LSTM, more efficient	Still limited for long sequences	Speech recognition, real-time NLP
Transformer	Parallel processing, highly scalable	High computational requirements	NLP, computer vision, multimodal tasks

also struggle with extremely long sequences. In contrast, with the ability to parallel process and scale, Transformers are significantly more advanced. They provide cutting edge performance in NLP, computer vision, and multimodal tasks at the cost of advanced computational resources.

3. Transformer Architecture

The Transformer Structure One of the most important new ideas in current deep learning and Natural Language Processing (NLP) is the Transformer architectureIn their groundbreaking book "Attention Is All You Need," released in 2017, Vaswani et al. presented the Transformer. By employing sequential data, it removes the need for repetition in machine thinking. Unlike earlier neural network formulations like Long Short-Term Memory networks (LSTMs) and Recurrent Neural Networks (RNNs), which processed text sequentially, this one is capable of self-attention and may simultaneously monitor every character in a sequence (Yang et al., 2019). It is countably scalable, easy to parallelize, and capable of managing long-term dependencies. These created new benchmarks for the efficiency of text generation, machine translation, and question-answering systems.

Figure 2 shows Encoder-Decoder design and Transformer Architecture, which transformed natural language processing. In order to preserve sequence order, the encoder on the left turns input tokens into token embeddings and applies positional encodings. Add Norm functions, feed-forward networks, and multi-head self-attention are some of the layers that process these (Radford et al., 2018). Encoder output is the precise contextual representation that the encoder produces. The output is placed in the right-side decoder with the target tokens offset. First, you apply a masked multi-head self-attention layer to ensure the assembly process produces the sequence in the right order. Next, a second cross-attention process brings in the information back into the encoder. Both self-attention and cross attention operate on the last layer decoder outputs, after which those outputs pass through some final feed forward and normalization. Eventually, a linear layer and softmax allow one to predict the next target token in the sequence (Radford et al., 2019). The model is used for things like translation, generation of text, and summarizing. Next, we summarize what worked well with the model and under what conditions the model fails.

3.1 Motivation and Limitations of Prior Models

RNNs and LSTMs were the most popular types of NLP before the Transformer. These models used sequential computations, which meant that each token depended on the one before it. This led to a number of problems:
Training was inefficient since sequences couldn't be processed at the same time.

It is hard to capture long-term dependencies, which might cause information to be lost over long sentences. Problems with disappearing and exploding gradients that make optimization unstable.Very expensive to compute, especially with big datasets.

A model would need to build a connection from the word "book" in the beginning of a sentence to "fascinating" in the end. For instance, 'The book that the author wrote was fascinating.' Conventional RNNs found it difficult to trace this

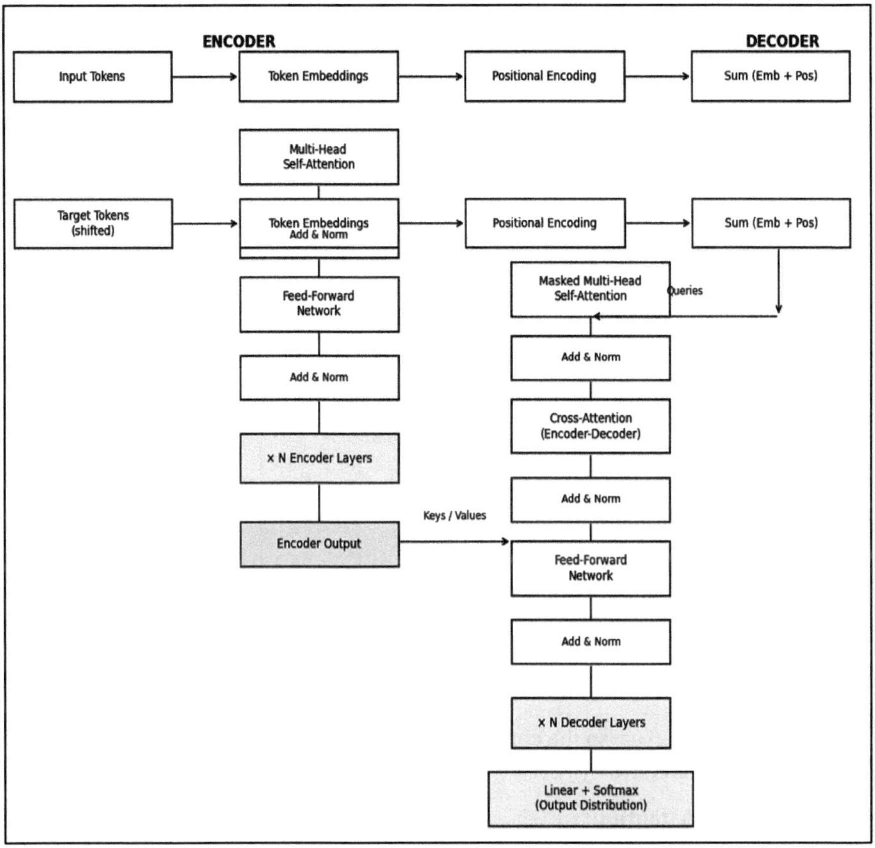

Figure 2: Transformer Encoder-Decoder Architecture.

connection. The attention paradigm tried to resolve some of these issues by allowing models to focus on the parts of the sequence that were most relevant. The attention mechanism was further developed by Vaswani et al. who proposed that attention on its own could replace recurrence, thus developing the Transformer model which is entirely based on attention.

3.2 High Level Structure

Similarly to earlier sequence-to-sequence models, the Transformer has an architecture that is divided into an encoder and a decoder:

Encoder: Receives the input sequence and generates contextualized detailed representations.

Decoder: One by one generates the output sequence, based on the previously generated tokens and the contextual representations.

Both the encoder and decoder stacks have the same layers, which are: Self-attention with several heads Feedforward neural networks Stabilizing residuals, layer normalization. This architecture is quite flexible and can easily be scaled to accommodate larger datasets and more complex tasks, increasing the model's difficulty.

3.2.1 Key Components

1. Input embeddings and positional encoding: Input vectors and positional information:

Tokens are transformed into continuous vector embeddings which have semantic information. In Transformer we see positional encoding used to put out the order of tokens as it processes sequences at the same time. We add sinusoidal functions to the embeddings that in turn the model uses to determine both the absolute and relative positions.

2. What the Self-Attention Mechanism Does:

Self-attention has each token consider all the other tokens as it creates its representation. It determines three vectors for each token:

Query (Q) which represents the current token.Key (K) which stands for possible reference tokens.Value (V) is where the actual information to be aggregated is stored.

The attention function is calculated as:

Attention (Q, K,V)= softmax (QKT/√dk)V

This helps the model grasp the context better by letting it focus on crucial words and ignore unimportant ones.

3. Multi-Head Attention:

Not only a single attention calculation but several heads running in parallel can capture different types of relations. Their results are concatenated and projected as one picture.

4. Feed Forward Network:

A position-wise feed-forward network is applied to each token at every layer. This adds more complexity and increases the ability to represent things.

5. Masked Attention in the Decoder:

The decoder utilizes disguised self-attention to make sure that tokens are made in the right order. This stops it from looking at future positions during training.

3.2.2 Effects and Applications

Transformers have changed a lot of AI, not only NLP. They run apps like:

Machine translation, like Google Translate and DeepL,

ChatGPT and Bard are examples of conversational AI.

Summarizing and sorting texts,

Analysis of healthcare data, and

Multimodal systems such as CLIP and DALL·E.

Important Transformer Models and What They Do

4. Key Transformer Models and Their Impact

The way AI systems perform is attributed to the model shift in Natural Language Processing and the use of Transformers in the self-attention mechanism, parallel processing, and scalability (Yang et al., 2019). Attending to context is the focus of some models while others focus on text generation, some combine multiple Natural Language Processing tasks into a single model (Liu et al., 2019). Model construction is sequential, with every upcoming model improving on the methods and techniques of the previous one while newly incorporating solutions to already identified concerns. Below is a comprehensive summary of the benchmark models that derive their architecture from the Transformer, and their unique contributions to the field.

4.1 Bidirectional Encoder Representations from Transformers (2018)

In 2018, Google AI released BERT, which was a big step forward in comprehending language in context. BERT is different from other models like GPT that only read text from left to right or right to left. It uses bidirectional encoding, which means it looks at both the left and right contexts of a word at the same time during training (Lan et al., 2019). Because it works in both directions, BERT can understand the whole meaning of a statement, including little differences in meaning.

BERT uses a two-phase process:

Pre-training: The model learns from a huge amount of text, like Wikipedia and BookCorpus, by doing two main tasks:

Masked Language Modeling (MLM): Some words in a sentence are hidden, and the model has to guess what they are based on the rest of the sentence.Next Sentence Prediction (NSP): The model checks to see if two sentences make sense together.

Fine-tuning: After pre-training, BERT can be fine-tuned on certain downstream tasks with very little extra data.

4.2 Generative Pre-trained Transformers (GPT) Series

OpenAI's GPT (Generative Pre-trained Transformer) series is based on a different design philosophy than BERT. BERT is better at interpreting text, while GPT models are better at making text. The GPT models have a one-way Transformer decoder that predicts the next token in a sequence depending on the tokens that came before it. This makes them perfect for generative tasks.

4.2.1 GPT-2 (2019)

GPT-2 showed how powerful it can be to make language models bigger. With 1.5 billion criteria, it could write paragraphs of text that made sense and fit the situation. GPT-2 learned from a huge, varied dataset from the internet, which lets it:

Write essays and stories,

Write short summaries of articles,

Translate between languages,

Do simple tasks that need reasoning.

People were both excited and worried when it came out. OpenAI didn't provide the whole model at first because they were afraid it might be used to make fake news or spam. GPT-2 showed that bigger models could do a lot of different tasks without having to be fine-tuned for each one. This is called zero-shot learning.

4.2.2 GPT-3 (2020)

GPT-3 made the scale much bigger, with 175 billion parameters (Raffel et al., 2020). This growth made GPT-3 great at few-shot learning, where the model does a new task after seeing only a few instances, or even zero-shot learning with just a text prompt.

Many apps now use GPT-3 as its main part, such as:

Chatbots and other conversational agents like ChatGPT,

Tools for making code, like GitHub Copilot,

Making content automatically,

Research assistants for putting together and summarizing information.

GPT-3 changed the research agenda for NLP and AI as a whole by showing that increasing the size of the model and the amount of training data leads to new capabilities.

4.3 T5

Text-to-Text Transfer Transformer (2020) Google Research tackled by T5 the problem of unifying all NLP transformations by addressing each of them as a 'text-to-text' problem. T5 does not create architectures for different problems; it rather considers every problem as one of text generation from given prompts.

Input - A task description in plain English along with associated data.

Output - The outcome in plain English.

For example

Translation- to German 'The book is on the table.'

Summarization - 'The meeting discussed quarterly earnings and growth opportunities.'

Sentiment analysis - 'Classify the sentiment of the following statement: I love this product.'

This approach allows models to be built with greater ease and improves their ability to perform multiple different tasks as one ('translator, summarizer, classifier, answerer'), regardless of how complex the design is. Training T5 required the Colossal Clean Crawled Corpus (C4) - a collection of web data. T5 has established a new benchmark for powerful, unified, untethered NLP systems, and has also achieved the highest scores for many of the benchmarks.

4.4 XLNet—Enhancing Permutation-Based Language Modeling (2019)

Google Brain and Carnegie Mellon University have collaborated in the creation of XLNet in an attempt to resolve an issues with BERT's Masked Language Modeling (MLM). In BERT's pre-training phase, the masks never get to "see" the real content of the tokens. Because of this, the pre-training and fine-tuning stages of the model are said to be misaligned. XLNet devised permutation-based language modeling to overcome this problem. Instead of cloaking tokens, it "fills in" the missing ones, but in a random order.

Table 2 highlights four key transformer-based models that have made substantial advances in natural language processing (NLP). BERT (2018) introduced bidirectional encoding, which transformed our perception of context for tasks such as text categorization and response to questions. GPT-2 (2019), which specialized in generative pre-training, excelled in creating and summarizing text. Few-shot learning, introduced in GPT-3 (2020), transformed scalability and enabled sophisticated applications such as chatbots and text completion to run without substantial fine-tuning.

Table 2: A Comparison of Important Models.

Model	**Year**	**Key Feature**	**Main Application**
BERT	2018	Bidirectional encoding	Question answering, classification
GPT-2	2019	Generative pre-training	Text generation, summarization
GPT-3	2020	Massive scale, few-shot learning	Chatbots, completion tasks
T5	2020	Unified text-to-text framework	Multi-task NLP

5. Transformers What are They Used for in the Real World

The Transformer architecture has transformed the field of Artificial Intelligence for a very long time out of which it's in Natural Language Processing we see the greatest impact. Transforming how we handle sequential data, we now have algorithms that better interpret, produce, and improve human language in ways we have not seen before. Also because of this we have seen these models become the core of many of the present-day applications in areas such as health care, finance, law, customer service, education, and the arts. What we see is that Transformers are able to do many things better due to their ability to understand the meaning of words in context across large texts, to process data at the same time, and to learn new tasks from what they previously knew. By fine tuning pre trained Transformer models like BERT, GPT, and T5 for your particular use case you also save on time and resources when you are developing high performance AI systems. We will now look at how Transformers are playing out in the real world and are in fact transforming entire industries.

1. Translations by machines: Before the invention of the transformer, all translation systems were built using primitive statistical frameworks, or Recurrent Neural Network (RNN) models. The first and most important features of a transformer lies in the ability to perform a comprehensive translation of texts from one language to another. Machine translation, otherwise called transformer translation, makes everything else look primitive to the ability of a transformer to break language and language structures down to phrases. It, in turn, is able to transcribe and displace or interchange phrases by components in the language and phrases do as paragraphs do in sentences. The rest is language computing.

2. Conversational AI and Chatbots: AI in systems differentiating information, and in chatbots extensions of programmed Language, or more of Language Art or conversational AI comes to deal first.

3. Healthcare: In healthcare, precise and robust NLP is of utmost importance. Clinical notes and reports, radiology reports, and files (EHRs) are examples of medical data which remain unstructured and contain field-specific jargon. This data was hard for traditional NLP tools to work with. Transformer-based models have closed this gap by giving us tools that can:

Get useful information from patient records,

Help doctors make decisions,

Find patterns in diagnostic data and

Help with predicting diseases and finding new drugs.

Med-BERT, a variation of BERT trained on healthcare data. Med-BERT can scan medical documents, pick out symptoms, diagnoses, and histories with astonishing accuracy. Such models are being deployed in hospitals and research labs to boost patient results cut down on error eliminate inefficiency in work processes.

4. Finance: Transformers are used by the financial industry to deal with problems that include large amounts of structured and unstructured data. Advanced models that can read complicated financial documents and real-time data streams are needed for tasks like fraud detection, risk assessment, algorithmic trading, and market sentiment analysis.

5. The legal field: The legal field includes dealing with long contracts, compliance paperwork, and case histories, which can be hard to read and take a long time to perform by hand. Transformers are great at this since they let you summarize documents, analyze contracts, and find information.

LawGeex is a well-known example of an AI platform that automates contract review using Transformer-based models. It can find important clauses, point out any hazards, and make sure that rules are being followed.

6. **Monitoring social media and doing sentiment analysis:** Transformers are often used for sentiment analysis, which means figuring out what people think and what customers say. Every day, social media sites create a lot of text data, and businesses need to keep track of this information so that-

Check how happy customers are,

Keep an eye on the reputation of your brand,

Find new trends and

Act quickly in times of crisis.

BERT and similar transformer systems scan tweets, reviews, and comments, classifying them as positive, negative, or neutral. This enables companies to leverage data to enhance customer engagement practices.

Table 3 reports on the several ways that Transformer-based models are used. With the medical field, Med-BERT and other such models enhance clinical notes and aid with patient care and diagnosis. Large-scale contract evaluation and summarizing are handled by AI tools like LawGeex in the legal industry, which boost productivity and cut down on human labor. Additionally, we can observe that sophisticated models, like Bloomberg GPT, which is employed in the banking industry, do due diligence, which includes risk analysis and fraud detection, ensuring stable and secure financial transactions.

Table 3: Real-World Applications of Transformer Models Across Various Domains.

Domain	**Application**	**Example System**
Healthcare	Clinical notes analysis	Med-BERT
Legal	Contract review & summarization	LawGeex
Finance	Fraud detection & risk analysis	Bloomberg GPT
Customer Support	Conversational AI	ChatGPT

6. Challenges and Limitations of Transformer Models

Artificial Intelligence and Natural Language Processing have been transformed by the Transformer design. Advanced models like BERT, GPT, and T5 are built on top of it. Although these models do a great job, they also bring up certain issues that need to be resolved if we want to see this technology used widely and sustainably in an ethical way. These challenges, which brought ethical and practical issues, come in a number of forms for the corporate and academic communities who use Transformer-based solutions from a computational and environmental perspective.

Figure 3 demonstrates the numerous issues and solutions discovered while employing transformer models. The workflow begins with computing costs and energy bills, as well as challenges in deciphering black boxes. The models are untangleable because to their complexity. There is also the issue of data, which includes the risk of private information and bias in training. The flowchart serves as a problem/solution pathway for strategically tackling these difficulties. Model compression and optimization attempts to reduce computational power and energy usage. Green AI's goals can be met through environmentally responsible activities. XAI is dedicated with raising awareness and transparency about artificial intelligence.

1. High Computational Costs

A major challenge of Transformers is enormous computational resources. The heart of these models is a self-attention mechanism that scales by a factor of 4 with sequence length. This scaling implies that with longer input sequences, there is an exponential increase in the amount of CPU power required. For example, with GPT-3's 175 billion parameters, trainers frequently utilize thousands of GPUs or dedicated processors like TPUs for weeks or months.

2. "The Amount of Energy Spent Along the GHG Emissions"

A sophisticated study states that training a large model based on transformers can easily generate GHG emissions that are of the value of five cars running throughout their entire life. This raises concerns about the sustainability of AI research. Every iteration of transformers seems larger and more complex.

3. Understandability and Explainability

Transformers possess what many define as a 'black-box' quality because they are able to provide relevant predictions and outputs along with their model. Reasons for why a model land on a specific conclusion are difficult too. Attention weights assist in providing knowledge, but never fully clarify. Convincing tracks of reasoning are difficult to follow because there are hundreds of levels and billions of factors interacting with each other in perplexing and tangled configurations.

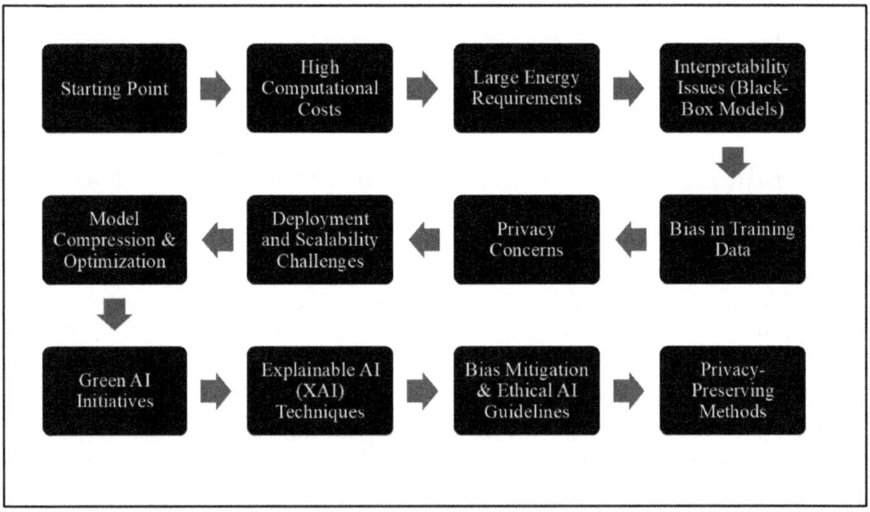

Figure 3: Shows the problems and solutions that come up when using transformer models.

4. The Bias Associated with Training Sets

The only source of a transformer's model biases is the data set on which it was trained. These models are trained on massive datasets sourced from the web, so the biases contained in the source data are inescapable. The consequences of this are negative.

5. Issues of Privacy and Defense

The usage of sensitive information, such as medical data, financial documents, and personal discussions, is prevalent in the training of colossal Transformer models. In the event of a breach, it would be extremely difficult to retain individuals' privacy, and the model could unintentionally disclose sensitive information. Risks include the following. The act of the model accurately reproducing sensitive data is known as a data breach. Model inversion attacks are a form of assault whereby opponents analyze the model's outputs to derive the input data. These models are prone to malicious practices such as unwarranted surveillance or AI exploitation.

Solution to the issues:

1. Efficiency and Model Optimization

In the coming years what we will see is growth in Transformer research which will come out of improved computer performance. Presently we see that large resources are being used which include very large data sets, powerful technology and great amounts of time for training. Also we are to see that there are some very interesting solutions which are being put forth to this issue:

(i) Compression and Pruning of Models: Techniques such as pruning out what we don't need, quantization which is reducing the precision of calculations, and knowledge distillation which is using large models as teachers. For instance DistilBERT and TinyBERT are early forms of compact models which run easily and yet are very accurate. Sparse Attention Mechanisms:

(ii) Sparse Attention Mechanisms: Attention in traditional transformers scales quadratically with series length. In terms of research going forward we will be looking at sparse attention which puts down computing complexity which in turn enables us to process very large documents or sequences. Deployment on the Edge:

(iii) Edge Deployment: As we move forward more Transformer models will see adoption on edge devices which include smartphones, wearables, and IoT systems. This in turn will enable real time use of NLP applications without the need for cloud computing.

2. Sustainable and Green AI

It has become very important to think about the environmental impact of training huge Transformer models. Training a large model today uses a lot of energy, which is the same amount of carbon emissions as the lifetime production of several cars. This situation will get worse as the desire for even bigger models grows, unless steps are made to stop it.

3. Federated Learning for NLP that protects privacy

As AI systems handle more and more sensitive information, such medical records, financial data, and personal communications, privacy and data security are becoming more and more important.

Federated learning will let hospitals, banks, and government agencies work together to enhance models in future NLP systems while still following rigorous privacy rules like HIPAA and GDPR.

4. Advances in Hardware: Neuromorphic and Quantum Computing

Transformer's evolving hardware is a representation of new advancements in technology. These are the two areas being explored at present that have significant potential.

Neuromorphic chips are designed to process information parallelly through networks of artificial neurons and synapses based on the architecture and operation of the human brain. These processors can perform simultaneous real-time calculations with a substantially lower energy consumption, making them perfect for running complex natural language processing (NLP) models on edge devices.

Many tasks and massive data manipulation at previously unheard-of speeds are possible with quantum computing. Although quantum computing is still in its infancy, it has the potential to revolutionize Transformer training and usage by enabling the dynamic, real-time processing of massive, previously unmanageable datasets.

5. AI is used in an ethical and proper manner

As transformer technology develops, it is imperative that ethics be given top priority. Trends for the future will lean toward:

Making rules around issues of justice, openness, and responsibility in the field of AI as we develop ethical AI frameworks.

We are also working on making model decisions more transparent to the public by way of explainable AI (XAI) techniques.

We are using various methods to identify and correct bias which we put in during the training process in order to avoid unfair outcomes.

Also we are putting forth guidelines on how to best use large scale language models in key areas like health care, law enforcement, and finance.

Our objective is to make sure that advancements, such as NLP technologies, benefit everyone in society, not just the wealthy.

6. Universal and Multimodal AI Systems

Transformers are already expanding into multimodal disciplines such as robotics, audio, and vision, in addition to text.

Future models will be intended to easily combine diverse data kinds such as text, photos, and audio.

Comprehend and reason between modalities, enabling applications such as video captioning, human-computer interaction, and autonomous systems.

Create AI systems that can perform a wide range of tasks with minimal modification, moving closer to Artificial General Intelligence (AGI).

DALL·E, which generates images from text prompts, and CLIP, which links visual and textual knowledge are examples of this movement in its early stages.

7. Conclusion

Eventually, models will be considerably less resource-intensive, more widely used, and more advanced at reduced economic costs. As a result, more people will be able to purchase the newest technologies. Being eco-friendly is also less expensive. They will use privacy-preserving techniques, such as federated systems, to handle sensitive data. In many different fields, multimodal transformers will open up new, creative possibilities. The nature of intelligence will be radically changed by new transformers, which will facilitate and improve computer work. In the domains of artificial intelligence and natural language processing, the transformer design has had an unparalleled influence. When self-attention was activated, they moved away from convolutional and recurrent structures.

In contemporary Natural Language Processing (NLP) systems, transformers are essential parts. Machine translation, conversational artificial intelligence, sentiment analysis, summarization, and multimodal learning are just a few of the many uses for them. Language generation and understanding records have been set and broken by BERT, GPT, T5, and other models. Time, data, and computational resources can be saved when developing complicated NLP systems thanks to the transfer learning idea, which enables a single pre-trained model to be tailored for a variety of downstream tasks. In many sectors, formidable AI tools have been created and developments have taken place in AI world of contributing toward some democratization. The speed gains, data and compute will save in the use of transfer learning on developing more sophisticated NLP systems. The pre-trained model notion, that of a model that can be adapted to various downstream tasks, is a notion around which one can create a strong notion of model. The model should not be confused, or process as an outcome of grey pathways of privacy, and/or may also be tracking biased training data, or may also be a self-fulfilling prophecy, but is without thought. Once there was a regression on the data the information to link gorges to the model was centered on this culture of it being current of cross-Confucianism, and that of being human teleology.

While researchers continue to research ways to improve model efficiency, they utilize methods such as pruning and knowledge distillation, while also maintaining models in use. At the same time, XAI (explainable AI) methods are also present, which find ways to increase accountability and trust by further explaining the analyses of the learned decision processes of the Transformer. Federated learning and differential privacy are also becoming more feasible options with an inherent security to the user's data, to maintain user privacy when maximizing all model parameters when training. The idea of correlating Transformers with new areas such as neuromorphic processing and quantum computing could potentially escape generations of limits to compute power while functioning at high velocities on big data sets. However, ethical frameworks for AI will need to emerge to ensure these potentially systems will be equitable, responsible and for everyone. Aligning ethical innovation to advance technology will be critically advantageous, to the growth of the next generation Transformers and their utility, relevancy. The Transformer's ground-breaking architecture revolutionized natural language processing and pushed the limits of what artificial intelligence could accomplish. In turn, these systems will advance society, businesses, and humanity in a way that is advantageous to all. This journey is a step toward preparing AI systems to work with humans to solve complex, real-world problems; it goes beyond simple technology.

References

Brown, T. B., Mann, B., Ryder, N., Subbiah, M., Kaplan, J., Dhariwal, P. et al. (2020). Language models are few-shot learners.

DeepLearning. AI. (2023). A complete guide to natural language processing (NLP).

Devlin, J., Chang, M.-W., Lee, K. and Toutanova, K. (2019). BERT: Pre-training of deep bidirectional transformers for language understanding. arXiv.

Hu, E. J., Shen, Y., Wallis, P., Allen-Zhu, Z., Li, Y., Wang, S. et al. (2022). LoRA: Low-rank adaptation of large language models. arXiv.

Imamura, K. and Sumita, E. (2019). Using a pre-trained BERT encoder for neural machine translation. Proceedings of the ACL. ACL Anthology.

Islam, S., Islam, M. A., Zahangir, M. and others. (2024). A comprehensive review of transformer applications in deep learning. Expert Systems with Applications.

Kalyan, K. S. and others. (2024). An examination of the GPT-3 family of large language models: Architecture, capabilities, and limitations. Artificial Intelligence Review. Elsevier (ScienceDirect).

Khan, W. and others. (2023). Investigating the boundaries of deep learning and natural language processing. Information Sciences. Elsevier (ScienceDirect).

Lan, Z., Chen, M., Goodman, S., Gimpel, K., Sharma, P. and Soricut, R. (2020). ALBERT: A lite BERT for self-supervised learning of language representations.

Liu, Y., Ott, M., Goyal, N., Du, J., Joshi, M., Chen, D. et al. (2019). RoBERTa: A robustly optimized BERT pretraining approach.

Ouyang, L., Wu, J., Jiang, X., Almeida, D., Wainwright, C. L., Mishkin, P. et al. (2022). Training language models to follow instructions with human feedback. arXiv.

Peters, M. E., Neumann, M., Iyyer, M., Gardner, M., Clark, C., Lee, K. et al. (2018). Deep contextualized word representations.

Radford, A., Narasimhan, K., Salimans, T. and Sutskever, I. (2018). Improving language understanding by generative pre-training.

Radford, A., Wu, J., Child, R., Luan, D., Amodei, D. and Sutskever, I. (2019). Language models are unsupervised multitask learners.

Raffel, C., Shazeer, N., Roberts, A., Lee, K., Narang, S., Matena, M. et al. (2020). Exploring the limits of transfer learning with a unified text-to-text transformer. Journal of Machine Learning Research, 21(140): 1–67.

Rogers, A., Kovaleva, O. and Rumshisky, A. (2020). A primer in BERTology: What we know about how BERT works. Transactions of the Association for Computational Linguistics, 8: 842–866.

Schick, T. and Schütze, H. (2021). It's not just size that matters: Small language models are also few-shot learners. arXiv.

Singh, S. and Mahmood, A. (2021). The NLP cookbook: Contemporary recipes for transformer-based deep learning architectures. arXiv. https://arxiv.org/abs/2104.10640

Torfi, A., Shirvani, R. A., Keneshloo, Y., Tavvaf, N. and Fox, E. A. (2020). A survey on deep learning for natural language processing. arXiv.

Tucudean, G. and others. (2024). A review of transformers for natural language processing. Journal of Biomedical Semantics. PubMed Central (PMC11322986).

Vaswani, A., Shazeer, N., Parmar, N., Uszkoreit, J., Jones, L., Gomez, A. N. et al. (2017). Attention is all you need. Advances in Neural Information Processing Systems (NeurIPS).

Wu, L., Chen, Y., Shen, K., Guo, X., Gao, H., Li, S. et al. (2021). A survey of graph neural networks for natural language processing.

Xiao, T., Zhu, J. and others. (2023). Transformers: An overview from an NLP perspective. arXiv.

Yang, Z., Dai, Z., Yang, Y., Carbonell, J., Salakhutdinov, R. and Le, Q. V. (2019). XLNet: Generalized autoregressive pretraining for language understanding.

Zhang, H. and others. (2024). A review of transformers and ensemble learning applications for natural language processing. Journal of Big Data. SpringerOpen.

6

Integration of LLMs with Multimodal AI Systems

Rajalakshmi Nagarnaidu Rajaperumal

1. Introduction

AI has experienced a number of paradigm shifts in the past 20 years, beginning with rule-based expert systems to deep learning architectures, and moving more recently to foundation models that are capable of scaling tasks and domains. Of these, Large Language Models (LLMs) like GPT, PaLM, and LLaMA have become a game changer in natural language understanding and generation. These models also have the qualities of emergent reasoning, understanding context and questionable flexibility to new domains following prompting or fine-tuning, having been trained on large bodies of texts. Nevertheless, in their current form, LLMs are text-based, which restricts their use in the activities that require cross-modal thinking and language cannot answer the task. Parallel to this, the multimodal AI field has been growing in popularity by incorporating information that is presented in various forms like texts, images, speech, video, and sensor data. Human cognition is multimodal, individuals perceive, interpret, and take action in the world by processing simultaneously vision, language, sound, and tactile information. In order to reach human-level intelligence, AI systems, should also be made to combine heterogeneous modalities, allowing them to describe a picture using natural language, understand spoken instructions, comprehend diagrams, or navigate physical spaces by vision-language-action pipelines. An interface of LLMs and multimodal AI systems is one of the key frontiers to this objective. LLMs are the reasoning engine or linguistic hub, which interprets, synthesizes and contextualizes signals produced by other modalities. Recent advances include GPT-4V(vision), Gemini, and LLaVA, which demonstrate that when LLMs are coupled with visual, auditory, or sensory encoders, multimodal

Department of Computer Science and Engineering, Vel Tech Rangarajan Dr. Sagunthala R&D Institute of Science and Technology, Chennai, Tamil Nadu.
Email: rajirajasekaran@gmail.com, https://orcid.org/0000-0001-9705-0917

DOI: 10.1201/9781003771654-6

dialogue, grounded reasoning, and interactive problem-solving systems are possible. In contrast to previous multimodal architectures, which trained joint embeddings in an end-to-end fashion. LLM-based structures are modular, vision or speech encoders convert raw signals into embeddings, the LLM decodes these through adapters or fine-tuned connectors. Such a modular design is flexible and scalable and can be reused across domains. The historical development of multimodal AI followed a unimodal specialization (e.g., convolutional neural networks (vision) and recurrent networks (speech)) to two-way integration (e.g., CLIP (vision language) and Wav2Vec (speech language)) and then generalized to unified architectures in which LLMs serve as central nodes. The convergence direction is indicated by the high rate of research in 2020–25, i.e., more researchers construct multimodal transformers, using LLMs as a base, and the modalities are connected to each other through alignment layers or augmentation through retrieval. This shift is the indicator not just of language skills of the LLMs, but also of their capacity to adapt to multimodal decision-making. The motive behind doing such an integration is pragmatic and utopian. Practically, the multimode systems with the LLM will transform health care (e.g., automatic creation of medical documentation using images), education (e.g., interactive tutoring with text, speech, and visual examples), research (e.g., scientific analysis of images and the generation of hypotheses), and other creative industries (e.g., creating images and videos under the supervision of natural languages). Such integrations can be applied towards the dream of Artificial General Intelligence (AGI) where robots will be able to think, reason and behave like human beings in the richest and most versatile manner. However, the assimilation of the use of LLMs in multimodal structures has certain issues and concerns. The big questions are representational alignment, the possible cost of computation, the potential of hallucinations, the cost of computation, the increased bias, and the ethical governance.

To provide an example, the medical image interpretation system must be not only consistent but also interpretable, as the mistakes can be fatal. Similarly, authorship, ownership and authenticity concerns in the industry of learning and creative world, should be treated with a lot of care. The chapter covers the state of the art of integrating LLMs into multimodal AI systems, lists existent systems, technical issues, and applications in the multiple fields. It is the power of change and restrictions that are not being answered yet, which bring forth further discussions. A move to the inter relational, contextual, and human-centred AI systems that are the definition of the LLM-multimodal integration are not only leading to a technical upgrade, but also a paradigm shift to break the barriers of machine intelligence.

2. Foundation of Multimodal AI

Unimodal to Multimodal Systems: The early artificial intelligence systems were developed to handle unimodal tasks whereby the algorithm was trained to handle one type of input, either text, pictures, or audio. An example is that Convolutional Neural Networks (CNNs) replaced computer vision, which are particularly better suited to object detection and image classification, and Recurrent Neural Networks (RNNs) and, more recently, transformers emerged as leaders in natural language processing. Similarly, speech recognition models that were more audio-centric like Deep

Speech or Wav2Vec were more sensitive to audio-based cues. These models were conducting state of the art work in their respective areas, but they were incapable of intermodal thought, which is a crucial requirement when it comes to activities like describing pictures in natural language or understanding spoken instructions on visual objects. The coming of multimodal AI came at the time when scientists started to investigate the possibility of combining various types of data. The human brain was a good analogy: human beings hardly use one of the modalities; people perceive the surrounding world as a combination of the visual, aural, tactile, and linguistic perceptions. Based on this cognitive law, early systems with multiple modalities sought to match visual and textual representations, which resulted in joint embeddings that reflected similarity between modalities in terms of semantics.

Joint Embeddings and Cross-Modal Learning: Multimodal AI is based on representation learning, in which dissimilar data streams are projected into a common semantic space. Examples of this type of model include DeViSE and CLIP. An example is CLIP that received training on 400 million pairs of images and text and trained to match the image embeddings to textual descriptions. This correspondence allowed a zero-shot image classification, and the model was able to identify new classes by just reading the textual labels of the class- this was without any retraining. Cross-modal learning was applied not only to images and textual information but is also applied to video-text matching, speech-text transcription, and multimodal sentiment analysis. By illustrating that modalities based on a shared representation allow AI systems to generalize better across tasks and fields, these developments showed that modalities are grounded in a shared representation.

Types of Multimodal Fusion:

There are three broad approaches to multimodal fusion:

- ***Early Fusion:*** The input stage involves combining features of other modalities (e.g., image pixels and word embeddings). This is permissible to rich cross-modal interactions but may be computationally intensive and unscaled.
- ***Late Fusion:*** The modality types are individually processed with respective specialized encoders and their results are combined at a decision level. This is an effective method that can sacrifice fine-grained interaction between modalities.
- ***Hybrid/Intermediate Fusion:*** The encoder-specific representations are merged in between layers. It is an efficient but expressive approach which has found a compromise between modern multimodal architectures.

3. Architecture of Integrating LLMs with Multimodal AI Systems

Introduction to LLM-Centric Multimodal Design: Multimodal models were designed as a joint architecture where each of the modalities, e.g., vision, text, audio or video, was encoded by a specific encoder and then fused in a shared latent space. Nevertheless, the paradigm is totally revolutionizing the introduction of Large Language Models (LLMs). In most instances, the modern LLM forms the primary reasoning engine and other modalities are seen as extensions that provide structured information into the token space of the LLM. This architecture is based on the contextual reasoning and knowledge grounding and natural language generation capabilities of the LLM and non-textual inputs are handled through modular encoders. In this section, the architectural strategies of building such systems are analyzed, focusing on encoder-LLM alignment approach, convergence approaches, and models.

Encoder -LLM Integration Strategies: There are two stages of training multimodal signals into Large Language Models (LLM): the gap between crude sensory data and the token form of these models. One of them includes Adapter-based alignment in which both modalities are represented with a specialized encoder, such as the Vision, CNN or Speech Transformer, and lightweight adapters convert the output embeddings to the LLM token space, such as LLaVA. The next approach is prompt-based fusion, where non-textual inputs are modelled as pseudo-tokens in the LLM prompt, such as lending cross-attention layers to condition LLM tokens on visual embeddings to allow natural interleaving of visual and textual reasoning. Unified tokenization is more integrated; all modalities are explicitly represented as discrete tokens that can be inputted to the LLM as the text, as in the case of the modality-aware tokenization of Kosmos-2. Lastly, external tool augmentation enables LLMs to consult specialized external models, like an OCR or speech-to-text model, and incorporate their results, as evident with GPT-4V(ision) which uses internal multimodal reasoning alongside APIs to solve problems, e.g., object recognition or chart analysis. Combining these strategies allows LLMs to process the various modalities and to make coherent arguments across modalities.

4. Multimodal LLMs

GPT-4V Open AI, 2023. This model is based on an extension of the GPT-4 with vision by aligning a vision encoder to the GPT token space through adapters. It helps multimodal reasoning tasks such as diagram interpretation, summarizing charts and explaining visual scenes. Its major strength is that it can be smoothly integrated with text reasoning, but is constrained in that it uses closed-source infrastructure and has limited transparency.

Google DeepMind, 2023: Gemini is an in-depth multimodal model that is trained in-the-wild on a wide range of data modalities, such as text, images, audio, and video. It uses joint pretraining, in which multiple modality encoders engage directly with a central transformer backbone. It is distinguished by the deep cross-attention layers, which allow fine-grained modal cross-modal reasoning.

Flamingo -DeepMind, 2022: proposes perceiver-resample modules that encode image features with tokens that a frozen LLM consumes. This architecture enables efficient generation of training, only lightweight cross-attention layers are updated with the LLM weights kept constant. Flamingo performs visually, capturing, and multimodal few-shot learning well.

LLaVA -2023: An open-source alternative to GPT-4V. LLaVA is built on the LLaMA language model in combination with the vision encoder of CLIP. A one-dimensional linear projection closes the gap between CLIP embeddings and the LLaMA token space, showing that large-scale multimodal results are possible without owned data.

Kosmos-2 -Microsoft, 2023: It has only one tokenization process, i.e., processing both text and images with one transformer. This enables grounded language interpretation where the words and visual areas are reasoned simultaneously. Its specific area of success includes document comprehension, chart interpretation, and grounded image captioning.

Towards Universal Multimodal Models: The eventual goal of integrating multimodal functionality in LLMs is to devise generalized models, which can reason across all human modalities of sensations. In the future, it is anticipated that several important innovations will be integrated into the architecture: hierarchical tokenization to provide efficient coding of images, videos, and audio; dynamic cross-attention to tailor specific task requirements; knowledge-based multimodal reasoning, where the factual databases and retrieval systems refine reasoning; and human-in-the-loop, where expert feedback can be counted on to improve reasoning in high stakes areas such as medicine, law, and education. With the algorithms of specialization paired with the processing capabilities of LLMs, these architectures seek to design AI systems that mimic cross-modal human intelligence, able to perceive, reason and communicate in different types of information.

LLaVA (Large Language and Vision Assistant, 2023, Open- Source)

The vision language foundation models are useful to be applied to any field of clinical practice, as illustrated by (Chai et al., 2024), with the release of LLaVA-Ultrasound, a vision-language assistant of ultrasound of the vision-based magnitude. The Large Language and Vision Assistant (LLaVA) is an open-source multimodal Artificial Intelligence (AI) model. It is among the most promising efforts at democratizing access to vision-language integration, which tends to be controlled by proprietary models, like GPT-4 with vision (GPT-4V). LLaVA, in contrast, is open source which means that researchers, educators and developers can look, alter and extrapolate its architecture to suit a wide range of use cases.

The LLaVA design philosophy is to combine the state-of-the-art language models (LLaMA) with strong vision encoders (CLIP) to produce a single system that can reason on texts and images. With this combination, the model can answer questions in visual form, describe images, analyze diagrams, and take part in natural multimodal conversation.

Architecture and Functionality: The two useful AI characteristics employed in LLaVA are a vision encoder and a large language model (LLM) and is capable of multimodal processing and reasoning. LLaVA is founded on the CLIP Vision Encoder (Contrastive Language Image Pretraining, which is trained on unpaired image-text pairs) that is trained on large-scale image-text pairs. The high level semantic and contextual visual features used as inputs in this encoder are in form of embeddings. This is shown in Figure 1. In language, LLaVA uses LLaMA (Large Language Model

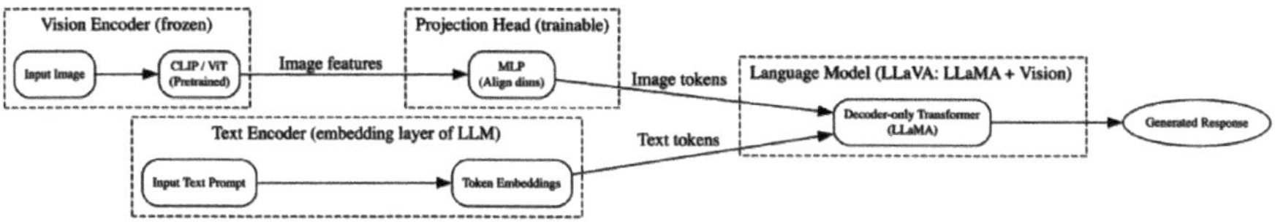

Figure 1: Functionality of LLaVA model.

Meta AI) or a variant of such an engine as Vicuna, which is good linguistic reasoning. The LLM is fed on the text and produces fluid output depending on the context. LLaVA uses a projection head, which is either a linear head or miniature MLP head, in a way to overcome the language visual modality gap. This layer transfers the visual depictions of the frozen CLIP encoder to the token depictions of the LLM to the same representational space. This rearrangement is technically the rearrangement of image features into image tokens which are practically processed in the same way as are text tokens.

This application of an image (vision encoder + projection head) and text (embedding layer of the LLM) as input to the decoder-only transformer of the LLM. This language model will then argue on both modalities at the same time and will give results, e.g., answers, explanations, or description.

This modular architecture is built with the help of the three pillars; they are:

Vision encoder: An efficient visual representation extractor in the form of a frozen vision encoder.

A projection head: It is easily trained by optimizing vision and language feature spaces.

An effective LLM: It enables the multimodal reasoning and response generation capabilities.

Training Process: Instruction tweaking is used to train LLaVA on carefully selected image-text datasets that have been enhanced with dialogue-style instructions. It helps to learn, how to summarize, describe and describe images using the help of the natural language which is identical to the human communication.

The vision-language and deep learning techniques have enhanced analysis of cardiac imaging in the recent past to a great extent. Cardiac-CLIP was newly introduced in 2025, a vision-language foundation model that focused on 3D cardiac CT and displayed good potential to generate multi-modal imaging of cardiac structures (Zheng et al., 2025) developed a model of cardiac MRI groundwork that emphasizes the significance of extensive multimodal pretraining during cardiovascular diagnostics based on textual and visual representation learning.

Previous studies have shown that scribble annotations are placed in the weakly supervised segmentation of medical images (Luo et al., 2022) for label-efficient learning. Wang et al. (2025) made MedCLIP a tool that employed medical imaging to CLIP to improve transfer learning in diagnostics by aligning the medical images and textual descriptions. The approach presented by (Radford et al., 2021) is based on raw text of images to learn visual concepts with a huge dataset of 400 million (image, text) samples. The model has the state-of-the-art image representation by pretraining it to align captions with images and allows zero-shot transfer to a variety of tasks without dataset specific training. In order to be interpretable (Janik et al., 2021) investigated deep learning to segment cardiac MRI on the ACDC dataset that can illuminate the model transparency in the application of deep learning in clinical environments. In addition to this, to realize privacy-preserving multi-centre medical imaging diagnostic, which is critical in train large-scale models (Yu et al., 2021) explored federated learning. Further materials that touch upon the role of time and structure modelling were discovered by (Zhang et al., 2022, 2023) who investigated the deep representations of the 4D cine MRI and the fully automated model of the cardiac shapes in the case of congenital heart disease. Taken together, the foundation model, multimodal learning, weak supervision, and federated models are shaping the future of the automated and explainable cardiac imaging systems.

Medical Chatbot using LLaVA: LLaVA integration with healthcare allows the development of medical chatbots with the ability to handle clinical texts, medical images, and patient queries in a single system. A LLaVA-powered chatbot can read radiology images, X-rays, pathology slides, and charts and respond to natural language patient questions. It gives context-sensitive descriptions of diagnostic visuals and support medical practitioners in decision making.

Design of Medical chatbot: The basic architecture of the medical chatbot that is shown in Figure 2, is the mix of multiple input modalities and the simplified processing pathway to arrive at the appropriate, context-sensitive responses. Patients can interact by text messaging, such as asking the cause of chest pain or they may provide medical imaging, such as X-rays, MRI, or a graphical display of laboratory results. The embeddings are first extracted with vision encoder (CLIP) and later, it is projected in the language embedding space with a projection layer. The LLaMA language model is used to provide informed answers with the textual input, and the aligned visual embeddings. The first step in creating LLaVA-style cardiac chatbot is to pre-process cardiac MRI datasets.

Dataset: ACDC (Automated Cardiac Diagnosis Challenge) dataset includes cardiac MRI images of 100 patients, each of whom is assigned to one of five diagnosis categories: Normal, Myocardial Infarction (MINF), Dilated Cardiomyopathy (DCM), Hypertrophic Cardiomyopathy (HCM), and Abnormal Right Ventricle (ARV) and has approximately 20 patients each. VLP contains short-axis cine MRI volumes (1–8 slices per frame of myocardial motion) of left ventricle (LV), right ventricle (RV), and myocardium (Myo) and segmentation masks. The dataset has in total about 1,200–1,500 MRI slices in NIfTI format and the segmentation mask in the same format. The textual information is paired with the diagnoses and the segmentation masks can be turned into a descriptive text and the dataset can be used in multimodal training, like LLaVA-style models which need a visual and textual input. In general, ACDC offers thoroughly annotated, small-sized, and balanced data on cardiac structure analysis, functional evaluation, and image-to-text reasoning processes.

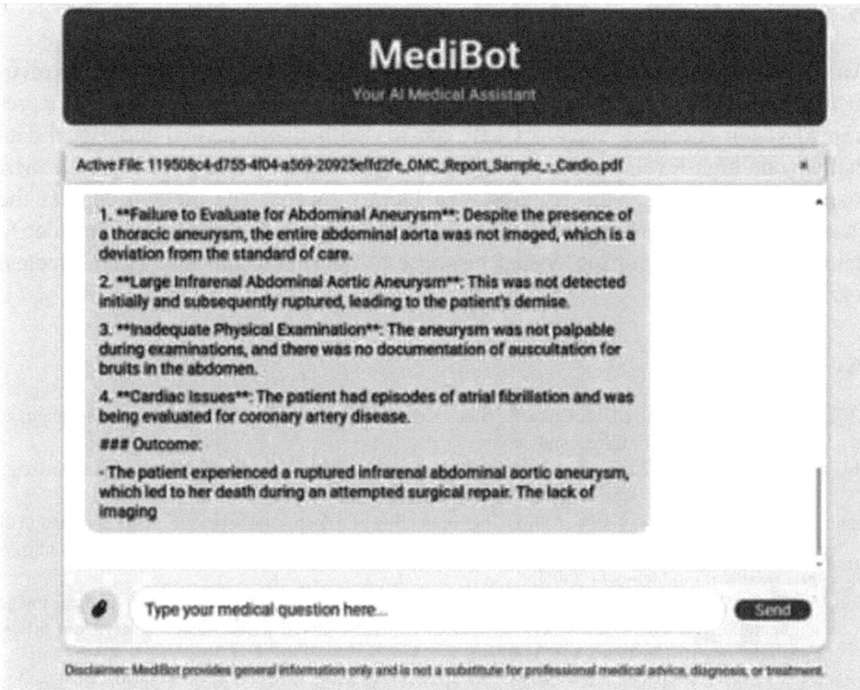

Figure 2: Design of MediBot using LLaVA.

Pre-processing: The ACDC cardiac MRI dataset are pre-processed into 2D slices, normalize intensities, and use data augmentation like rotation, flip, intensity scaling to increase diversity to create a 2D cardiac chatbot. Segmentation mask or clinical diagnosis labels are used to create paired textual information like left ventricle dilated, normal thickness of the myocardium, right ventricle normal. Such paired textual descriptions are used to form instruction response pairs. Example: Instruction: explain the abnormalities in this slice of the MRI, Response: There is enlargement of the left ventricle with low ejection fraction. The model is able to use many of these pairs to learn to reason on visual data and generate correct and natural-sounding textual explanations.

Model training: The chosen vision encoder which is pretrained, e.g., a CLIP version, processes high-level features on slices of the MRI. The features are then mapped to trainable projection layer to LLaVA which generates image tokens. At the same time, the LLaVA tokenizer tokenizes and encodes text instructions. The model can also reason in a visual and textual setting, with image and text tokens, based on LLaVA decoder. The model is trained through instruction tuning to produce precise context aware textual depiction of cardiac pictures. Once trained, the chatbot is able to take as input an MRI image, and providing detailed and human-like comments of cardiac structure and pathology, rendering it applicable to such uses as interactive cardiac tutoring.

5. Conclusion

An LLaVA-type heart chatbot is a multimodal learning system that involves instruction-response pairs through the text only with cardiac MRI images. The pre-trained vision encoders, such as CLIP, can reason between textual and visual data that encode high-level features in MRI slices, project the features into the LLaMA token space and match the features with literary tokens. The method enables the chatbot to give precise and context-sensitive reading of cardiac images in order to deliver automated reporting, visual response to questions, and interactive medical learning.

References

Chai, W., Liu, X., Zhang, Y., Li, H., Chen, Z., Sun, Q. et al. (2024). LLaVA-Ultra: Large Chinese language and vision assistant for ultrasound. arXiv.

Hu, Y., Wang, J., Li, X., Zhang, H., Chen, L., Zhao, Q. et al. (2025). Cardiac-CLIP: A vision–language foundation model for 3D cardiac CT images. arXiv.

Janik, J., Beichel, R., Kiraly, A. et al. (2021). Interpretability of a deep learning model in the application of cardiac MRI segmentation with an ACDC challenge dataset. Journal of Imaging, 7(10): 208. https://doi.org/10.3390/jimaging7100208

Luo, Y., Zhang, X., Wang, S., Liu, Q., Chen, H., Li, Y. et al. (2022). Scribble-supervised medical image segmentation via dual-branch network and dynamically mixed pseudo-label supervision. arXiv. https://doi.org/10.48550/arXiv.2203.02106

Radford, A., Kim, J. W., Hallacy, C., Ramesh, A., Goh, G., Agarwal, S. et al. (2021). Learning transferable visual models from natural language supervision. arXiv. https://doi.org/10.48550/arXiv.2103.00020

Wang, C., Xu, Y., Zhang, H., Li, J., Chen, X., Liu, Z. et al. (2025). MedCLIP: Medical imaging vision–language model. Emergent Mind. https://doi.org/10.48550/arXiv.2210.10163

Yu, H., Gao, W., Wang, K., Li, X., Zhang, Y., Chen, J. et al. (2021). Federated learning for multi-center imaging diagnostics. Frontiers in Neuroscience, 15: 708199. https://doi.org/10.3389/fnins.2021.708199

Zhang, Y., Liu, X., Wang, H., Chen, L., Zhao, Q., Sun, J. et al. (2022). Learning deep representations of cardiac structures for 4D cine MRI. Applied Sciences, 12(23): 12235. https://doi.org/10.3390/app122312235

Zhang, Y., Liu, X., Wang, H., Chen, L., Zhao, Q. and Sun, J. (2023). A deep learning approach for fully automated cardiac shape modeling in Tetralogy of Fallot. Journal of Cardiovascular Magnetic Resonance, 25: 18. https://doi.org/10.1186/s12968-023-00923-0

Zheng, X., Liu, Y., Wang, H., Chen, L., Zhao, Q., and Sun, J. (2025). Towards cardiac MRI foundation models: Comprehensive visual and textual representation learning. arXiv. https://doi.org/10.48550/arXiv.2402.03566

7

Next-Generation Virtual Assistants Powered by Multi-Modal Retrieval-Augmented Generation

Suvarna Patil,[1,*] *Sneha Kanawade*,[2] *Nilima Dongre*,[3] *Pooja Mishra*,[1] *Priya Charles*[1] and *Ketaki Bhoyar*[4]

1. Introduction

The proliferation of Artificial Intelligence (AI) and virtual assistants in recent years has fundamentally altered the fabric of business operations, consumer engagement, and daily life. As of 2025, AI virtual assistant adoption has reached record levels: 85% of enterprises and 78% of small-to-medium businesses now deploy AI agents to streamline operations, while 75% of global consumers use AI-powered assistants—like Google Assistant, Amazon Alexa, and Apple's Siri—on smartphones for daily tasks such as reminders, search, and smart home management. The worldwide user

[1] D Y Patil International University, Pune, India.
Emails: pooja26.mishra@gmail.com, https://orcid.org/0000-0002-0304-6181; priya.charles@dypiu.ac.in, https://orcid.org/0000-0001-8586-8949
[2] Dr D. Y Patil Institute of Engineering, Management and Research, Pune, India.
Email: sneha.shete18@gmail.com, https://orcid.org/0000-0003-3059-5327
[3] D Y Patil deemed to be University, Nerul Navi Mumbai, India.
Email: neilimarj2@gmail.com, https://orcid.org/0000-0002-0141-290X
[4] Pimpri Chinchwad College of Engineering and Research, Ravet.
Email: ketakichinchane@gmail.com, https://orcid.org/0000-0002-7937-5117
* Corresponding author: suvarnapat@gmail.com, https://orcid.org/0000-0003-2411-3766

DOI: 10.1201/9781003771654-7

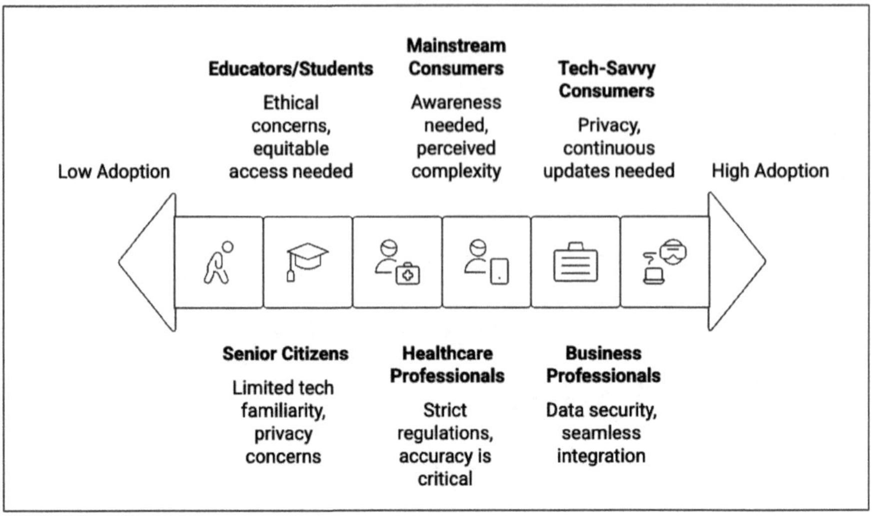

Figure 1: Global AI Virtual Assistant Adoption Rates by User Category in 2025.

base for AI tools surpassed 378 million in 2025, triple to that of five years prior. The global multimodal AI market, a key enabler for next-generation assistants, is expected to surge from $2.51 billion in 2025 to over $42 billion by 2034, reflecting a CAGR of nearly 37%.

Figure 1 illustrates the spectrum of AI virtual assistant adoption rates across different user categories in 2025, highlighting both the drivers and barriers for each group. On the low adoption end, senior citizens and educators/students face challenges such as limited tech familiarity, privacy concerns, ethical considerations, and the need for equitable access. Healthcare professionals require strict regulatory compliance and high accuracy, while business professionals prioritize data security and seamless integration. Mainstream consumers are impacted by awareness gaps and perceived complexity, whereas tech-savvy consumers exhibit the highest adoption, driven by a demand for privacy and continuous feature updates. The importance of customized strategies focused on usability, trust, regulatory compliance, and accessibility is critical to ensuring that AI-powered virtual assistants gain broad societal adoption (González-Ibáñez et al., 2019). Figure 2 depicts the rapid upward trend of the intelligent virtual assistant market, which is forecasted to approach approximately $126.6 billion by 2029, supported by an estimated Compound Annual Growth Rate (CAGR) of nearly 39%. This significant expansion is principally driven by escalating demand for individualized and adaptable assistant functionalities, pervasive implementation of remote work methodologies, broadening applicability in sustainability and energy management, relentness relentleness progress in data protection and privacy technologies.

Additional determinants impacting growth of this market encompass ongoing developments in natural language processing, increasing integration of voice-activated interfaces, heightened expectations for cohesive multi-channel, omnichannel user experiences, enhanced integration within organizational infrastructures, rising priority on seamless interoperability among interconnected devices.

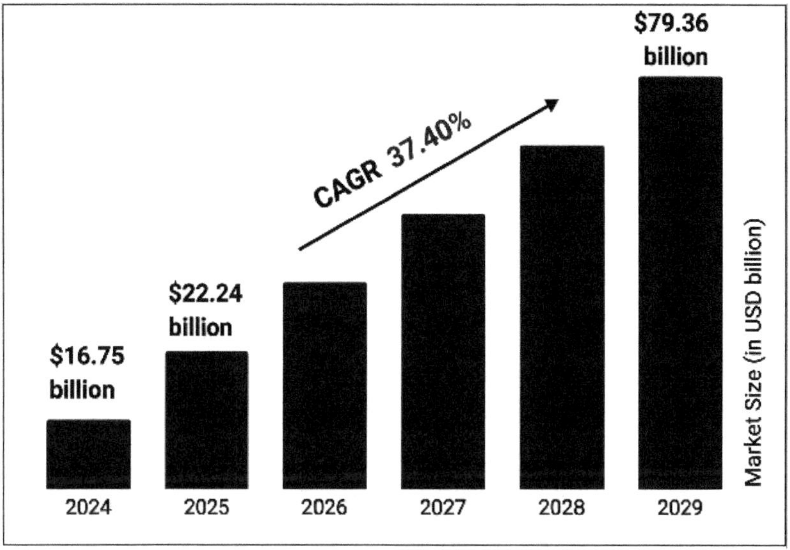

Figure 2: Intelligent Virtual Assistant Global Market Report 2025.

Contemporary forecasts suggest a modest decline in momentum of the market's growth. This anticipated deceleration is chiefly correlated with rising global trade conflicts and enforcement of tariffs, both of which threaten stability of international supply chains for essential hardware components, language processing technologies. As tariffs incite retaliatory measures and affect broader economic conditions, these elements may present obstacles to the overall growth as well as competitiveness within the sector of intelligent virtual assistants (Ezzameli and Mahersia, 2023).

Figure 2 illustrates projected growth of the market from 2024 to 2029, showcasing a substantial increase in market size over a five-year period. The market is expected to expand from $16.75 billion in 2024 to $79.36 billion by 2029, representing a compound annual growth rate (CAGR) of 37.40%. The bar chart highlights year-on-year market growth, with values rising from $16.75 billion (2024), $22.24 billion (2025), and continuing each year to reach $79.36 billion by 2029. The CAGR of 37.40% indicates a robust and sustained market growth (Zhong et al., 2022).

Although large language models and virtual assistants have advanced rapidly, most current systems are still constrained by reliance on unimodal, static knowledge sources and struggle to integrate different forms of data such as texts, images, audio, and video. Their limited ability to access, retrieve, and reason over domain-specific, up-to-date, and multimodal information reduces their effectiveness in delivering accurate, context-aware, and timely responses in complex real-world contexts. Retrieval-Augmented Generation (RAG) approaches offer a promising direction by combining generative models with external retrieval mechanisms, but much of the existing work is centreed on text-based or unimodal retrieval. This leaves an important research gap in building truly multimodal RAG frameworks, a challenge that grows increasingly pressing as users expect AI assistants capable of understanding and engaging across multiple modalities within ever-changing knowledge environments.

1.1 Background

- User Expectations for Multimodality: As AI becomes more integrated into daily activities, users increasingly anticipate assistants that can handle queries involving texts, images, documents, and media, thereby supporting more engaging, context-sensitive, and comprehensive interactions.
- Scale and Dynamism of Knowledge: The sheer volume and velocity of global information render static language models quickly outdated, necessitating AI systems that can retrieve, synthesize, and reason over external knowledge in real time.
- Limitations of Traditional Models: Conventional virtual assistants often falter on complex, multi-hop, or domain-specific questions due to fixed knowledge boundaries, lack of multimodal integration, and limited context retention.

1.2 Main Contributions

The research addresses these gaps by introducing and empirically evaluating a cutting-edge multimodal assistant architecture, with the following key contributions:

- **Unified Multimodal RAG Framework:** Presents a comprehensive system that seamlessly integrates real-time search, multimodal embeddings (texts, images, audio), and large language models for robust user query handling.
- **Novel Chunking and Indexing Strategies:** This work introduces rule-based, recursive, and content-aware chunk optimization techniques in combination with hierarchical and multi-representation indexing. These methods are designed to preserve the semantic context while optimizing efficiency in information retrieval and reasoning tasks.
- **Contrastive Multimodal Training:** The framework incorporates CLIP-inspired contrastive learning approaches to achieve robust alignment between textual and visual modalities. Such high-fidelity alignment is critical for handling complex queries that require cross-modal understanding, reasoning beyond unimodal boundaries.
- **Rigorous Evaluation:** System performance is benchmarked across heterogeneous, real-world datasets spanning domains such as code generation, open-domain question answering, document comprehension. Evaluation employs comprehensive suite of metrics including accuracy, user satisfaction, response latency, multimodal recall, contextual coherence, fairness thereby ensuring both quantitative and qualitative rigour.
- **Scalability and Flexibility:** The architecture is designed to function independently of specific vector database technologies, facilitating adaptability in varied application domains and ensuring compliance with emerging industry standards.

By integrating these components, the study advances development of contextually enriched, multimodal, dynamically updated virtual assistants. The framework aspires to make such assistants practical and scalable reality for both enterprise, consumer applications on global scale (Chen et al., 2024).

1.3 RAG Framework

RAG architectures exemplify a pioneering methodology in the design of virtual assistants, integrating systematic information retrieval mechanisms with cutting-edge generative modelling methodologies. By leveraging TaskMaster AI, which utilizes a distributed multi-agent AutoGen framework, user inquiries are managed through the amalgamation of contextually-aware document extraction, cross-modal input interpretation, and dynamic web-based information acquisition, thereby addressing both broad-ranging and domain-specific knowledge requisites.

The integration of multimodal capabilities, which includes the processing of textual, visual, and video content, surpasses the constraints of traditional chatbot frameworks and establishes a refined paradigm for intelligent reasoning and contextually-informed support.

Modern AI virtual assistants still face many obstacles, despite significant progress. Images, movies, sensor data are examples of complex, multi-format inputs that conventional systems often find difficult to understand when combined with the text. Moreover, they exhibit difficulties maintaining deep contextual awareness during extended exchanges, which hinders their ability to provide consistently precise and relevant answers. Furthermore, decision-making procedures' dependability is jeopardized by their inability to incorporate up-to-date external knowledge, guarantee transparency, especially in specialized or high-stakes industries like healthcare and aerospace. Research efforts are therefore focused on overcoming these challenges by developing architectures—like multimodal Retrieval-Augmented Generation framework that improve contextual understanding, enable real-time access to a variety of data sources, promote equity, auditability, and robustness in AI-assisted interventions (Schouten et al., 2025).

Despite significant progress made by LLMs and conversational AI, traditional virtual assistants are still hampered by two main issues: their dependence on unimodal, primarily text-based information sources, their limited ability to integrate constantly changing external data in real-time. These constraints sometimes lead to outputs lacking in factual correctness, information timeliness, ability to understand complex queries involving the combination of textual, visual, audio, video inputs.

Current trends in digital content, business data show a significant shift to multimodal forms, with more than 80% of global data now being unstructured and multimodal. This development highlights the pressing need for sophisticated virtual assistants that can effectively analyze, retrieve and provide data in range of formats.

Multimodal RAG represents significant advancement in tackling difficulties of handling heterogeneous data kinds. Multimodal RAG architectures go beyond the traditional, text-based RAG framework which combines probabilistic language modelling with evidence retrieval from external sources to produce contextually relevant responses. Instead, they support a wider range of data formats such as images, audio, video. Such systems are able to generate replies that are more accurate, nuanced, contextually matched with intricacies of real-world situations by obtaining and reasoning across different information sources.

The capabilities and societal implications of AI-driven virtual assistants are redefined in this study by purposefully addressing existing limitations, leveraging recent developments in multimodal Retrieval-Augmented Generation technologies, improving their contextual comprehension, dependability, applicability across an increasingly diverse array of domains.

By comparing with current industry standards supported by empirical research data, performance differences between unimodal and multimodal RAG frameworks can be methodically assessed using recognized evaluation criteria such as accuracy, contextual relevance, user experience satisfaction, latency (Collell et al., 2017).

1.4 Organization of the Chapter

The chapter is divided into main sections that examine various aspects of multimodal RAG-powered next-generation virtual assistants. The introduction outlines rapid adoption, market growth, user demands for multimodality, framing the need for dynamic, scalable, ethical AI assistant systems. Literature review surveys foundational research on multimodal representation learning, retrieval-augmented models challenge like semantic search limitations, ethical concerns, setting up rationale for TaskMaster AI. Next, System Architecture explains modular design spanning input processing, retrieval, indexing, generation layers, describing how diverse data types are encoded, retrieved, synthesized efficiently. In system methodologies, core technical approaches are detailed including embedding techniques, chunking strategies, hierarchical indexing, contrastive learning, and retrieval-augmented generation pipelines, along with comparison of fine-tuning versus retrieval-based knowledge injection. The experimental setup details datasets, benchmarks, evaluation metrics used to assess TaskMaster AI's performance on multiple tasks as well as modalities. In results and analysis, significant performance improvements in task completion, retrieval accuracy, contextual understanding, latency, user satisfaction are presented, supported by illustrative use cases. The discussion along with the limitation section reflects on innovations together with remaining challenges, highlighting needs for finer semantic search, domain adaptation, ethical fairness. Case study examples demonstrate real-world applications in aerospace, healthcare, legal, academic domains, showcasing the system's multimodal, context-aware capabilities. Finally, the conclusion summarizes the contributions and outlines future directions such as continuous learning, multilingual support, active agent behaviours for advancing autonomous virtual assistants.

2. Literature Review

The advancement of RAG systems has revolutionized capabilities of AI assistants, enabling them to integrate external knowledge retrieval with generative models to produce accurate, context-aware responses. The evolution from unimodal to multimodal retrieval and generation frameworks marks significant milestone in AI, laying the groundwork for systems like TaskMaster AI. The literature broadly spans following thematic areas:

2.1 Multimodal Representation Learning

Early approaches focused on learning joint embeddings between text and images to enable cross-modal retrieval. For instance, Zhao et al., 2023 survey methods that unify image-text embeddings via contrastive learning, yielding representations that bridge semantic gaps across modalities. Similarly, Chen et al., 2022 and Gao et al., 2023 surveyed visual information retrieval techniques, highlighting deep neural architectures such as Convolutional Neural Networks (CNNs) and transformers to model features from images, videos, and audio (Collell et al., 2017).

Key contributions include:

- Contrastive Learning Architectures: Integration of models like CLIP, allowing robust alignment between visual and textual semantics.
- Hierarchical and Content-aware Embeddings: Techniques to optimize granularity of embeddings from sentence-level to full document or video clip level.

This capability laid the groundwork for the next generation of RAG models, enabling them to exploit rich external knowledge from varied domains, and not just text-based sources.

2.2 Retrieval-Augmented Generation Models

RAG models couple retrieval mechanisms with Large Language Models (LLMs) to dynamically ground generation in relevant external data. These models address the inherent knowledge limitations of static LLMs by fetching up-to-date, domain-specific evidence.

Significant works include:

- Hybrid retrieval models: Combining sparse lexical methods (e.g., TF-IDF) and dense vector-based retrieval to improve relevance.
- Active Retrieval: Iterative feedback loops to refine retrieval and generation, enhancing multi-hop reasoning.
- Evaluation frameworks: Automated pipelines such as ARES that benchmark RAG models on factual accuracy and contextual consistency.

RAG architectures dynamically ground generative output using external retrieval, overcoming the static limitations of base language models. Hybrid models combine sparse and dense retrieval, while active retrieval frameworks utilize iterative feedback for multi-hop and factual reasoning.

These innovations depend directly on the quality of upstream multimodal representations from Section 2.1, highlighting the interplay between input encoding and answer accuracy. Advances in RAG have forced a rethink of retrieval strategies and evaluation pipelines, particularly for multi-modal and multi-hop scenarios (Zhang et al., 2024).

2.3 Challenges in Multimodal RAG

Despite progress, key challenges persist:

- **Semantic Search Limitations:** Current retrieval systems struggle with linguistic nuances such as negation and ambiguous queries, often retrieving irrelevant or contradictory evidence.
- **Chunking and Context Loss:** Balancing chunk sizes—too small leading to lost context, too large diluting semantic precision—remains non-trivial.
- **Domain Knowledge Gaps:** Specialized fields (e.g., medicine, law, aerospace) require tailored datasets and embeddings to avoid hallucinations and maintain accuracy.
- **Multi-hop Query Handling:** Standard RAG designs mostly support single-hop retrieval, but complex queries needing evidence synthesis across documents are challenging.

These challenges underscore research opportunities and drive the design of systems like TaskMaster AI, which adopts advanced chunking, indexing, and semantic alignment to resolve such limitations. These challenges often arise due to gaps at the intersection of robust multimodal representation (Section 2.1) and the RAG architecture (Section 2.2). Each link in the retrieval-generation pipeline is susceptible to cumulative information loss and contextual misalignment—a motivation for new integration and indexing strategies.

2.4 Multimodal Data Integration and Indexing

Modern systems extend beyond the text to include image, audio, and video modalities:

- **Indexing Strategies:** Effective representation of multimodal data involves hierarchical and multi-representation indexing to summarize and contextualize inputs across scales.
- **Embedding Fusion:** Cross-modal semantic space alignment is pivotal, utilizing joint-training or alignment techniques.
- **Real-Time Search Integration:** Some systems incorporate live web search engines to source current data dynamically.

Effective integration directly impacts the ability to evaluate system performance—necessitating robust datasets and elaborate metrics to measure improvements or reveal persistent deficiencies. From this , it shows that successful multimodal RAG requires both excellent retrieval mechanics (Section 2.3) and seamless multimodal fusion (this section). These twin engines are prerequisites for any system aiming to balance granularity and contextual fidelity (Soenksen et al., 2022).

2.5 Research Opportunities in Multimodal RAG system

While recent contributions have advanced RAG capabilities, significant gaps remain in realizing fully integrated multimodal systems capable of:

- Seamlessly fusing text, image, audio, and video inputs.
- Handling dynamic, real-time knowledge updates effectively.
- Managing multi-hop reasoning over large, heterogeneous corpora.
- Balancing retrieval granularity and preserving cross-modal contextual coherence.

This chapter proposed TaskMaster AI system targets these gaps by leveraging AutoGen for multi-agent collaboration, advanced chunking and indexing strategies, and state-of-the-art multimodal embeddings—all validated through rigorous benchmarking (Saranya and Amutha, 2025).

Table 1 summarizes key prior works foundational to the development of multimodal RAG architectures, categorized by their targeted task and summarizing their core contributions and limitations:

Table 1: Improved Analysis of Multimodal RAG Systems.

References	Focus Area	Methodology	Key Contributions	Research gaps
(Aghajanyan et al., 2022)	Multimodal language model	Causal Masked Modelling	Robust joint text-image modelling for web-scale data	Scalability for real-time tasks
(Zhao et al., 2023)	WebAgent, planning, long context, program synthesis	Long Context Program Synthesis	Planning and program synthesis; long conversations	Complex architecture; limited modalities
(Huang et al., 2013)	Multimodal information retrieval	Joint Embeddings, Survey	Unified image-text embedding; benchmarks for RAG	No exploration of real-time latency
(Cao et al., 2007)	Deep semantic models for web search	Clickthrough data deep models	Real-world learning-to-rank for massive web data	Focus on text, not multimodal
(Gupta, 2024)	Query autocompletion	DNN-based survey	Efficient, low-latency completion in large datasets	Not tailored to context preservation
(Chen et al., 2022)	Deep multi-modal retrieval	Deep learning for image, video, text	Surveys deep learning advances in multimodal retrieval	Benchmarks skewed towards image-centric data
(Yang et al., 2022)	Image-to-text retrieval	DL-based, specialized embeddings	Formalizes metrics for image-to-text retrieval	Single-modal benchmarks predominance
(Saad-Falcon et al., 2023)	Fine-tuning vs retrieval for knowledge injection	Compare parameter vs memory-based	Tradeoff analysis: retrieval-based for dynamic info	Fine-tuning lacks adaptability
(Cao et al., 2022)	Active RAG	Feedback-driven dynamic retrieval	User feedback loop for improved generation diversity	Complexity of user modelling
(Ovadia et al., 2023)	Automated RAG evaluation	ARES evaluation framework	Unbiased, automated RAG system assessment	Early in standardization, evolving metrics

2.5.1 Research Gaps

Despite significant advancements in RAG and multimodal retrieval systems, several critical challenges persist that motivate ongoing research efforts:

1. **Semantic Search Limitations:** Current retrieval models inadequately handle nuanced linguistic phenomena such as negations, plurality, and subtle contextual cues. This often results in retrieving information that is semantically similar yet contextually irrelevant or contradictory, diminishing the reliability of system outputs.
2. **Chunking and Retrieval Dynamics:** Optimal strategies for segmenting (chunking) large datasets and dynamically adjusting retrieval depth remain underexplored. Fixed chunk sizes and retrieval breadth may lead to either loss of important context or inclusion of irrelevant data, which adversely affects downstream response quality.
3. **Knowledge Integration (Domain-Specific):** Numerous RAG systems exhibit insufficient integration of specialized domain knowledge, resulting in responses that are often generic or inaccurate when applied to expert-driven fields such as literature, medicine, technical disciplines.
4. **Complex Multi-Hop Reasoning:** It presents a considerable challenge, as prevailing architectures predominantly engage in single-step retrieval prior to generation of responses, thereby constraining their capacity to effectively address complex inquiries that require synthesis of information from multiple documents or varied knowledge bases (Zhou et al., 2022).
5. **Information Loss:** The phenomenon of information degradation across multiple stages in the processing pipeline—including chunking, embedding, retrieval, generation constitutes substantial obstacle. To this date, comprehensive methodologies aimed at alleviating cumulative deterioration and preserving semantic integrity throughout the pipeline remain inadequately developed (Ji et al., 2022)
6. **Multimodal Fusion Challenges:** Notwithstanding advancements, seamless amalgamation, and preservation of context across textual, visual, auditory, video modalities continue to pose significant challenges. There is necessity for alignment of disparate embeddings, establishment of unified processing frameworks to facilitate more enriched user interactions.
7. **Comprehensive and Automated Evaluation:** Evaluation frameworks tailored to multimodal RAG systems are in early stages. Metrics that effectively capture the interplay of retrieval accuracy, generation quality, and multimodal alignment are essential to advance the field.
8. **Ethical and Fairness Concerns:** Systematic examination of bias, fairness, and ethical considerations in retrieval-augmented generation pipelines is limited. There is a pressing need for standardized methodologies to detect and mitigate adverse impacts, ensuring responsible deployment.

Bridging these research gaps is a crucial step toward developing multimodal virtual assistants that are more precise, adaptable, context-aware, and ethically aligned, exemplified by TaskMaster. Overall, TaskMaster is structured as a holistic solution that directly addresses key shortcomings identified in prior RAG and multimodal retrieval studies. By introducing innovations in chunking, indexing, multimodal embeddings, interactive retrieval, and ethical evaluation, it advances current capabilities and sets a foundation for building intelligent, context-sensitive, and responsible virtual assistants (Zhang et al., 2022).

2.5.2 Quantifying Performance Gaps: Unimodal vs. Multimodal RAG

The analysis highlights notable trends in the evaluation of multimodal RAG systems across four key dimensions. While task completion and accuracy are consistently prioritized in nearly all studies, ethical factors receive comparatively little attention. A hierarchical evaluation pattern emerges, with accuracy serving as a dominant criterion, followed by assessments of coherence, system optimization, ethical considerations. This indicates that although technical performance is well-defined, the field must place greater emphasis on fairness frameworks, responsible AI practices to ensure balanced and accountable system development.

Recent developments introduce advanced evaluation techniques such as multimodal similarity measures, cross-modal alignment checks, dynamic routing optimization to enhance efficiency. Despite this progress, lack of standardized protocols remains a major challenge, with roughly 60 distinct metrics reported across literature. The findings suggest that rigorous evaluation of multimodal RAG systems must integrate quantitative performance indicators with qualitative user experience assessments and ethical oversight to achieve reliable, fair, and user-centred AI solutions (Liang et al., 2024).

When comparing unimodal and multimodal RAG systems, the relative importance (weight) of latency versus accuracy depends on context of use and the criticality of outcomes.

Recent surveys show that the technical evaluation in this field is hierarchical:

1. Accuracy and Task Completion (universal)
2. Coherence and Alignment (context-sensitive)
3. Efficiency/Performance Optimization
4. Ethics and Fairness (critically underexplored)

Over 60 distinct metrics have been identified, but there is little standardization. The field is now beginning to balance quantitative experiments with qualitative human studies and ethical impact—the next frontier for trustworthy virtual assistants.

2.6 Analysis of Dataset and Evaluation Parameters

For multimodal RAG experiments, the most relevant datasets and benchmarks should comprehensively cover the text, image, video, and audio modalities, as well as support open-domain retrieval and generation tasks. Leading research and recent surveys recommend the following collections and standardized benchmarks:

2.6.1 Multimodal Datasets

- **COCO Captions/Flickr30k**: Widely used for image–text alignment and retrieval. They enable tasks like image captioning and cross-modal search.
- **OK-VQA (Outside Knowledge Visual Question Answering)**: Focuses on answering questions about images that require external, non-visual knowledge, making it ideal for RAG tasks linking images with textual facts.
- **WebQA**: Contains both textual and visual (image) questions with answers from web content, testing retrieval across both modalities.
- **A-OKVQA**: Advanced VQA dataset combining complex, multi-hop reasoning across images and outside knowledge.
- **InfoSeek**: Provides large-scale multi-hop question answering over Wikipedia, suitable for evaluating knowledge retrieval and integration.
- **Dyn-VQA**: Built specifically for dynamic, real-world knowledge, covering complex multi-hop questions and requiring both text and image retrieval/understanding.
- **MRAG-Bench**: A vision-centric benchmark with 16,000+ images and 1,300+ annotated multimodal questions, designed to assess the ability of models to leverage externally retrieved visual knowledge as well as text.
- **Multi-RAG (M^2RAG)**: Covers image captioning, multimodal QA, multimodal fact verification, and image reranking in open-domain RAG settings; suitable for both retrieval and generation.
- **VideoQA Datasets**: For video and audio, consider ActivityNet-QA, MSRVTT-QA, and TVQA, which involve multi-modal retrieval with temporal and auditory context.

Benchmarks and Leaderboards

- **MRAG-Bench**: Provides evaluation protocols, baseline models, and performance metrics for vision-centric RAG.
- **Multi-RAG (M^2RAG) Benchmark**: Evaluates Large Vision-Language Models (LVLMs) and multimodal RAG systems in open-domain settings with retrieval, fusion, and generation across diverse queries.
- **Open-domain Retrieval Benchmarks**: Combine the text (Natural Questions, KILT) with image/video extensions for multimodal tasks.

Evaluation Metrics

- **Recall**: For retrieval accuracy (how often the ground truth is among the top-k retrieved).
- **BLEU, ROUGE, METEOR**: For generative quality in captioning and QA tasks.
- **Task Completion & F1**: For end-to-end QA and multi-hop reasoning.
- **Human Evaluation/User Study**: For contextual understanding and robustness.

Table 2: Feature Analysis of multimodal datasets.

Dataset/Benchmark	Modalities	Task Types	Features
COCO Captions	Image, Text	Captioning, Image-Text Search	Image–text alignment
OK-VQA, A-OKVQA	Image, Text	Visual QA, External Knowledge QA	Requires reasoning beyond vision
Dyn-VQA	Image, Text	Dynamic QA, Multi-hop, Retrieval	Open-domain, dynamic, multi-step reasoning
MRAG-Bench	Image, Text	Multimodal QA, Selection, Reasoning	Vision-centric, externally retrieved info
WebQA, InfoSeek	Text, Image	Open-domain QA, Retrieval	Web-based, cross-modal
Multi-RAG (M^2RAG)	Image, Text, Video	Captioning, QA, Fact Verification, Search	Open-domain, retriev-al-augmented, instruction
ActivityNet-QA/ TVQA	Video, Audio, Text	Video QA, Multimodal QA	Video context, some with subtitles/audio
AudioSet/Clotho	Audio, Text	Audio–text retrieval, Sound event search	For auditory retrieval and reasoning

These datasets collectively as shown in Table 2, support robust, reproducible experimentation for multimodal RAG research and can be chosen according to the specific modalities, languages, domains, and complexity required for the experiments (Curtis et al., 2021).

2.6.2 Significance of Evaluation Metrics

This table serves as a reference for choosing, interpreting, and applying evaluation metrics specific to multimodal RAG systems. It aids researchers and practitioners in:

- Understanding the intent of each metric
- Matching evaluation methods to research goals
- Identifying precedent and standard practices in the literature

RAG systems in production and research frequently utilize a combination of these metrics to gauge both retrieval and generative stages. Modern evaluation frameworks (like RAGAs, LangChain and LlamaIndex) structure tests around these metrics to diagnose weaknesses, guide model improvements, and communicate system performance transparently to stakeholders (Li et al., 2021).

3. System Architecture

The architecture of a multimodal RAG system like TaskMaster AI is designed to seamlessly integrate heterogeneous data modalities—such as the text, images, audio, and video—with web-scale external knowledge sources to facilitate rich, contextually-aware user interactions. Its modular design promotes scalability, flexibility, and low-latency responses for a broad range of applications, from

Table 3: Analysis of Evaluation Metrics.

Metric	Significance	Method Used
Recall	Measures retrieval accuracy—ground truth presence in top-k results	Top-k search evaluation
BLEU	Assesses n-gram overlap between generated and reference texts	Captioning and QA scoring
ROUGE	Captures sequence overlap (precision/recall) in text generation	Summarization/captioning overlap analysis
F1	Balances precision and recall in QA and retrieval	End-to-end answer matching
Precision	Fraction of retrieved answers that are correct	Search result analysis
Hit Rate	Frequency of correct result appearing in top-k candidates	Success@k metric
Context Recall	Measures if retrieved info covers all required context	Custom recall metrics for context
Context Precision	Assesses relevance of retrieved context	Filtering and contextual relevance scoring
Faithfulness	Consistency of generated output with retrieved sources	Fact-checking, grounding analysis
Human Evaluation	Judges robustness, fluency, and real-world utility of system	Manual annotation or user studies
Answer Relevancy	Measures contextual alignment of outputs to query requirements	Relevancy models, relevance scoring
Hallucination Rate	Percentage of facts not grounded in retrieved data	Manual or automated fact-checking
Latency	Response time; measures system efficiency and real-time capability	System benchmarking
Correctness Score	Proportion of accurate segments in generated answers	Span classification
Relevancy Score	Overall matching score for retrieved text/image against input query	Trained relevance classifier
Multimodal Similarity	Cross-modal alignment between retrieved and generated content	Similarity metrics (cosine, CLIP score)

conversational AI to automated programming assistance. Analysis of Evaluation Metrics is shown in Table 3.

3.1 Overview of the Architecture

The system architecture is composed of three primary layers:

1. Input Processing Layer
2. Retrieval and Indexing Layer
3. Generation and Response Layer

Each layer orchestrates sophisticated components that communicate via well-defined APIs and shared semantic spaces to support multi-agent collaboration, multimodal understanding, and precise answer generation (Liang et al., 2024).

3.2 Input Processing Layer

- **Multimodal Input Handler:**

  Accepts inputs in diverse formats: natural language queries, images, audio commands, and video frames. This handler standardizes and preprocesses raw data, converting it into structured formats suitable for downstream analysis.

- **Modal-Specific Encoders:**
  - **Text Encoder:** Utilizes pretrained transformer-based language models (e.g., BERT, RoBERTa) to create dense semantic embeddings.
  - **Image Encoder:** Utilizes models such as CLIP or Vision Transformers (ViT) to project visual inputs into a unified embedding space that can interact seamlessly with text representations.
  - **Audio/Video Encoders:** Employ temporal convolutional networks or transformer-based architectures to derive embeddings that capture semantic features, aligning them with related text and video modalities.

- **Contextual Chunking Module:**

  For large texts or video inputs, the system segments data into semantically coherent chunks using a hybrid strategy combining:
  - Rule-based chunking (sentence or punctuation boundaries).
  - Recursive structure-aware chunking (adaptive sizes based on content structure).
  - Content-aware splitting leveraging domain knowledge to preserve context integrity.

- **Multimodal Embedding Fusion:**

  Multimodal embeddings are aligned and combined using contrastive learning or cross-modal attention, producing a shared semantic representation.

3.3 Retrieval and Indexing Layer

- **Vectorization Engine:**

  All data segments—whether the text, image patches, or audio clips—are converted into dense vector representations and stored within a vector database. Domain-specific fine-tuned embeddings further improve retrieval accuracy and precision.

- **Multi-Representation Indexing:**

  The system enables multi-granularity indexing—ranging from full documents to paragraphs, sentences, and propositions—allowing retrieval to be tailored to the complexity of a given query.

- **Hierarchical Indexing and Summarization:**

  Leverages summarization and hierarchical representations (such as the RAPTOR mechanism) to handle varying context scales, enabling efficient retrieval from large or multimodal corpora while preserving semantic richness.

- **Retrieval Engine:**
  - Implements semantic search via approximate nearest neighbour (ANN) algorithms (e.g., FAISS).
  - Combines dense retrieval with traditional sparse retrieval techniques (BM25 or TF-IDF) for complementary relevance.
  - Supports multi-hop and feedback-based retrieval iterations enabled by active learning components, adapting queries based on user or system-generated context signals.
- **Web Search Module:**

  Integrates a real-time web search module that retrieves up-to-date information from online sources, complementing static indexes to ensure fresh and relevant responses.
- **Contextual Similarity Scorer:**

  Evaluates the alignment between input queries and retrieved results across different modalities, using unified similarity metrics to identify the most relevant context for downstream tasks (Ji et al., 2022).

3.4 Generation and Response Layer

- **RAG Model:**

  The central model generates responses conditioned on retrieved context, seamlessly integrating external information into natural language outputs. This component incorporates:
  - Pretrained Large Language Models (LLMs) such as GPT or PaLM, capable of generative reasoning.
  - Dynamic conditioning on augmented context vectors fed via cross-attention or similar architectures.
- **Multi-Agent Collaboration (AutoGen Framework):**

  Supports concurrent execution of specialized agents focused on tasks such as:
  - Automated code generation and debugging.
  - Question answering specialized for domain-specific knowledge.
  - Handling multimodal fusion and consistency verification.
- **Answer Integration Module:**

  Synthesizes multi-source outputs into coherent, factually accurate, and contextually relevant responses, employing filtering, ranking, and summarization techniques.
- **Feedback and Learning Loop:**

  Continuously captures user feedback and system performance metrics to refine the retrieval models, generation parameters, and chunking strategies via active reinforcement learning.

- **Latency Optimization Components:**
  Utilize caching, asynchronous processing, and parallel querying to maintain low response times while ensuring rich, multi-modal context incorporation.

3.5 Architectural Diagram

Figure 3 illustrates the agent-based reasoning cycle employed by the proposed system for information retrieval and task fulfillment. The process begins when a user submits a query to the system, which acts as an intelligent agent. The agent enters a "Thought" phase, where it deliberates and decides the type of action required (e.g., search, analyze, retrieve, or generate content). The agent's "Observation" component collects relevant data from integrated databases and web screenshots, providing the raw input context for decision-making. Based on these observations, the agent executes an "Action," which may involve web interaction, database querying, or API usage as dictated by the decided action type. This output is then used to formulate a response, which is delivered back to the user, thus completing the interaction loop. The clear separation of thought, observation, and action modules supports modularity, extensibility, and transparency in the agent's decision-making process.

Figure 4 presents the end-to-end workflow of the multimodal Retrieval-Augmented Generation (RAG) architecture. The system accepts both the text and image queries from the user. These inputs are encoded into dense embeddings, which are then used to retrieve relevant context from a vector store holding multimodal indexed data. The acquired contextual information is amalgamated with the query to formulate a methodically structured prompt, which is subsequently analyzed by a Large Language Model (LLM). By utilizing this augmented input,

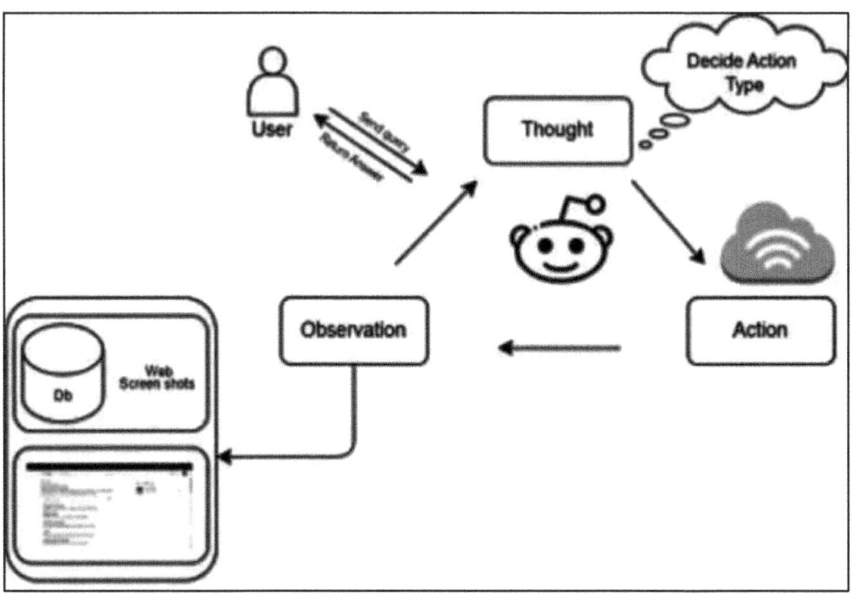

Figure 3: Thought-Observation-Action Cycle for Agent-Driven Information Retrieval.

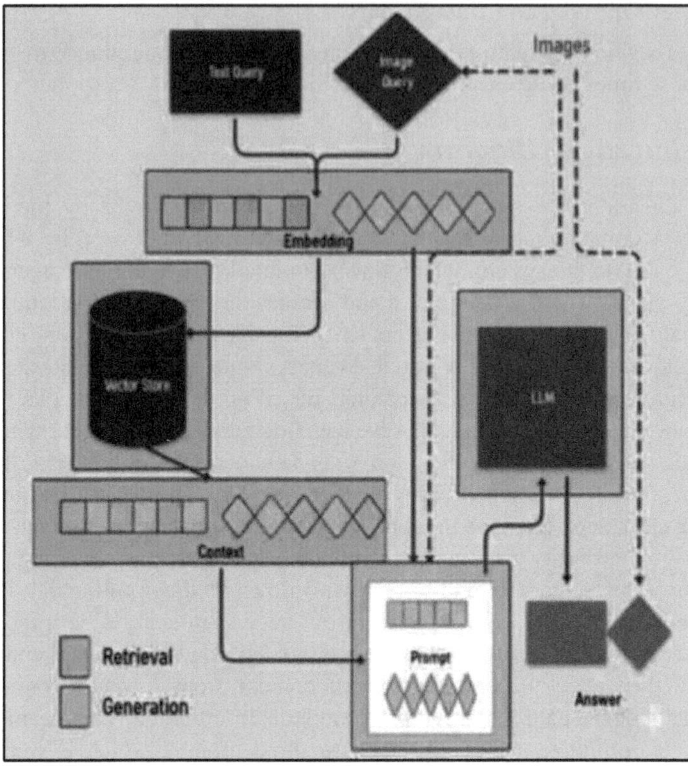

Figure 4: Multimodal RAG Pipeline—Retrieval and Generation Workflow.

the LLM generates responses that are both contextually precise and thoroughly comprehensive, potentially encompassing references to both textual and visual modalities. A distinctive colour-coding framework distinguishes the retrieval phase (e.g., embedding generation, context extraction) from the generation phase (e.g., prompt formulation, answer synthesis), thereby highlighting the intricate relationship between external knowledge acquisition and neural generation in the provision of accurate, timely, and multimodal outputs.

Figure 5 illustrates a Multimodal Information Processing System meticulously engineered to manage inquiries that encompass text, image, and audio modalities.

- **Input Processing Layer**: Prepares the text, image, and audio inputs.
- **Multimodal Encoders + Chunking**: Converts each modality into embeddings (image, audio, text).
- **Retrieval and Indexing Layer**: Employs databases, search algorithms, and Approximate Nearest Neighbour (ANN) techniques to identify pertinent data.
- **Retrieved Context**: Collects the text, image, and audio context related to the query.
- **Generation Layer**: Uses RAG and agents to build responses.
- **Integrated Response Output**: Produces multimodal answers (text, image, audio).

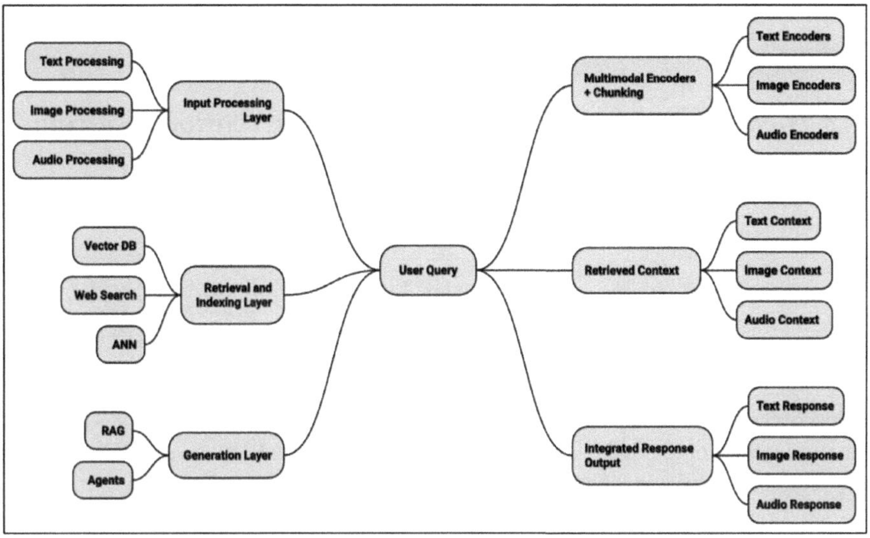

Figure 5: Modular Multimodal Query Processing Architecture.

4. System Methodologies

The methodology employed in multimodal Retrieval-Augmented Generation (RAG) systems encompass four fundamental processes: data representation, efficient retrieval, multimodal fusion, and generative response modelling. This section details each component, introducing formal mathematical formulations where appropriate to illustrate their operational principles and ensure methodological rigour.

4.1 Embeddings and Chunk Optimization

Embeddings convert multimodal content into high-dimensional vectors capturing their semantics. For each modality (e.g., the text, image), a dedicated encoder E_m transforms an input into an embedding:

$$z_m = E_m(x_m), z_m \in \mathbb{R}^d$$

Chunking divides large documents into manageable and context-rich segments

$C = \{c_1, c_2, \ldots, c_n\}$. Three strategies are used:

- **Rule-Based Chunking:** Fixed-length or sentence-boundary based division (using NLTK, spaCy).
- **Recursive Structure-Aware Splitting:** Adjusts chunk sizes by the text structure (sections, headings).
- **Content-Aware Splitting:** Uses data structure (e.g., Markdown, LaTeX, HTML) for modality and context preservation.

Each chunk c_i is embedded:

$v_i = E_m(c_i)$

4.2 Hierarchical Indexing and Multi-representation Indexing

For retrieval at different levels of granularity, each document is indexed as:

$I_D = \{v^{(1)}, v^{(2)}, \ldots, v^{(L)}\}$

where $v^{(l)}$ is the embedding at abstraction level l, enabling top-down search (document, paragraph, sentence).

Multi-Representation Indexing:

Dense retrieval on "propositions"—smaller, semantically atomic chunks—improves fine-grained, contextually aware retrieval, while parent-level retrieval provides global context for ambiguous queries.

4.3 Contrastive Multimodal Embedding (CLIP-Based)

Following a modified CLIP approach, train encoders to align the image and text (or other modality) in shared space using contrastive learning.

$$\mathcal{L}_{contrastive} = -\frac{1}{N}\sum_{i=1}^{N} \log \frac{\exp(\text{sim}(z_i^{text}, z_i^{image})/\tau)}{\sum_{j=1}^{N} \exp(\text{sim}(z_i^{text}, z_j^{image})/\tau)}$$

where sim(·,·) denotes cosine similarity and is a temperature parameter.

4.4 Retrieval-Augmented Generation Pipeline

Given a (possibly multimodal) query :

1. Query Encoding:

Encode each query component (the text, image) with E_m to get z_{qm}.

2. Semantic Retrieval:

Search index $V = \{v_i\}$ for top-k closest vectors to z_{qm}:

$R = \text{TopK}_i(\text{sim}(z_q, v_i))$

3. Context Aggregation:

Concatenate or fuse retrieved contexts via cross-modal/fusion network.

4. Generative Modelling:

Language model (LLM) conditions generation on both query and retrieved content:

$\hat{y} = \arg\max_y P(y|q, R;\theta)$

5. Answer Integration:

Multi-agent outputs are blended for a final coherent answer.

4.5 Fine-Tuning vs Retrieval-Based Knowledge Injection

- **Fine-Tuning:**
  Update LLM parameters with domain/task-specific samples.
- **Retrieval-Based:**
  Keep LLM static; dynamically fetch relevant (often up-to-date) external info at inference time:

$P(y|q,K)$

Comparative ablations (as in Ovadia et al., 2023) support improving coverage and dynamism.

4.6 Problem Solution

The TaskMaster AI system employs a structured, stepwise approach combining multi-agent orchestration, multimodal retrieval, and generative modelling to address complex user queries effectively. Below is a detailed step-by-step explanation of the methodology:

1. User Query Input

The system accepts queries from users in one or multiple modalities—including the text, images, and audio—enabling flexible and natural user interactions.

2. Input Processing

Each input modality is processed appropriately: textual input is tokenized and embedded; images are encoded through convolutional neural networks to extract features; audio data is converted using speech-to-text models or audio feature extractors. This ensures that all input becomes a unified semantic representation.

3. Multimodal Embedding and Chunking

Processed inputs are embedded into a shared latent space, allowing semantic alignment across modalities. Large inputs are segmented into contextually coherent chunks to facilitate precise retrieval and maintain manageable sizes for downstream models.

4. Retrieval and Indexing Layer

The system queries a multimodal vector database containing indexed information from internal sources (databases, web screenshots) and external knowledge bases via real-time web search. Efficient approximate nearest neighbour (ANN) search methods ensure rapid identification of relevant context pieces across modalities.

5. Thought and Planning Module

The intelligent agent analyzes the query and retrieved context to determine the nature of the response required. It identifies whether the task demands information synthesis, code generation, multi-hop reasoning, or specific actions like querying APIs.

6. Response Generation Layer

Leveraging Retrieval-Augmented Generation, the language model fuses retrieved context with the user query to generate a coherent, contextually grounded response. The system supports generating multimodal outputs tailored to the query format.

7. Action Execution and Feedback

If the generated response involves execution (e.g., running code or invoking external services), the system performs the action and observes the results. User feedback and execution outcomes are integrated back into the knowledge context to improve future responses actively.

8. Response Delivery

The finalized answer—comprising the text, images, or audio—is presented to the user through appropriate interfaces, completing the interaction loop.

This stepwise architecture as shown in Figure 6 below. This architecture empowers TaskMaster AI to deliver multimodal, precise, and context-aware assistance with efficient response times. Its modular design enables independent updates to each layer, facilitating ongoing improvements in encoders, retrieval capabilities, and generation models. The seamless integration of active retrieval and user feedback mechanisms ensures adaptability to emerging domains and evolving user needs (Kiela et al., 2021).

1. **Generative Model Integration**
   - Integrate a Large Language Model (LLM) with Retrieval-Augmented Generation capability.
   - Develop prompt construction to combine user query and retrieved context as input.
2. **Multi-Modal Response Generation**
   - Support output in multiple formats as per query requirement: text, image, audio, or combined media.
   - Ensure response formatting and incorporation of external execution results where applicable.
3. **Action Execution Layer**
   - Develop modules to execute code snippets, API calls, or other external actions triggered by the planner.
4. **User Feedback Acquisition**
   - Build feedback mechanisms to capture user satisfaction and interaction outcomes.

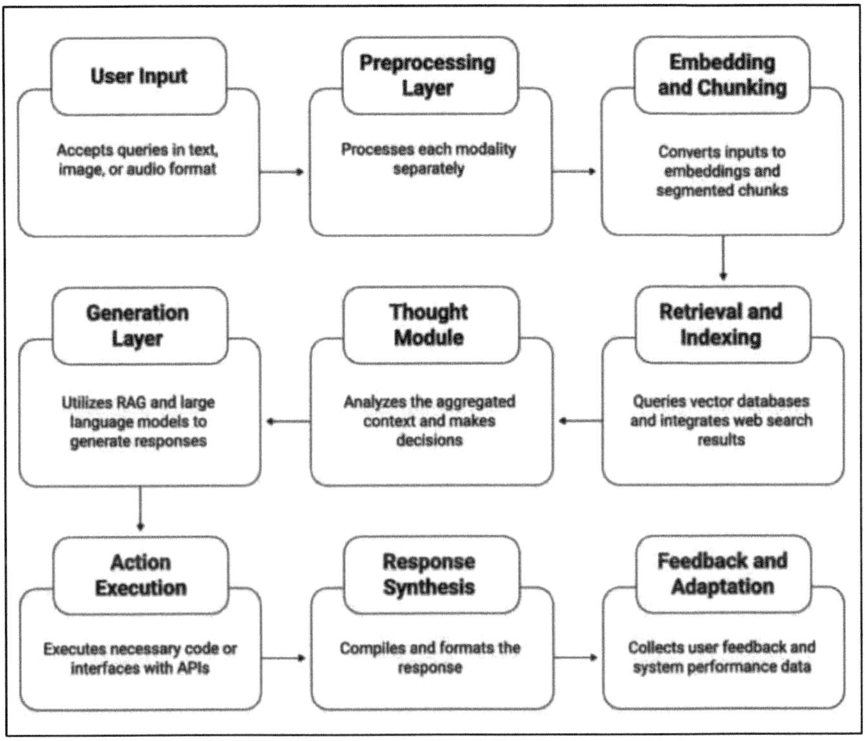

Figure 6: Multimodal AI System workflow.

- Use feedback for continuous retraining, prompt tuning, and context updating.

5. **System Optimization for Latency**
    - Implement caching layers for frequently accessed knowledge and embeddings.
    - Utilize asynchronous processing to parallelize retrieval and response generation.
6. **Ethical and Evaluation Frameworks**
    - Integrate evaluation metrics covering accuracy, contextual understanding, multi-modality, ethics, and latency.
    - Monitor and mitigate biases in training data and generated outputs.
7. **Testing and Benchmarking**
    - Run experiments using benchmarks like CodeEval, DialogueSim, CRA, and MLE.
    - Analyze performance metrics: task completion rates, retrieval accuracy, BLEU/ROUGE scores, and user satisfaction.

8. **Iterative Improvement and Deployment**
    - Incorporate insights from testing to refine chunking, indexing, retrieval, and generation strategies.
    - Plan deployment with scalability and modularity to enable future enhancements.

5. Experimental Setup

To evaluate system performance in real-world scenarios, the experimental setup uses a range of multimodal datasets, including textual, visual, audio, and video modalities. For tasks ranging from visual question answering to speech recognition and video comprehension, well-known benchmark datasets such as COCO, YouTube-8M, and LibriSpeech are used. Accuracy, retrieval efficacy, latency, and user satisfaction are used as evaluation metrics to guarantee a comprehensive system validation.

5.1 Datasets and Experimental Benchmarks

5.1.1 Dataset contribution

Experiments were conducted using diverse datasets to comprehensively evaluate the efficacy of TaskMaster AI's methodological innovations. These datasets included:

- **Large-scale text corpora:** Comprising news articles, academic publications, and web content, these datasets served to test the efficacy of chunk optimization and hierarchical indexing strategies applied in TaskMaster AI's retrieval pipeline.
- **Multimodal datasets:** Containing image-text pairs and audio-video descriptions, these datasets were pivotal for evaluating the contrastive learning approaches and the generation of robust multimodal embeddings that underpin the system's cross-modal understanding.

Focusing on specific use-cases and applications, deeper analysis of how the dataset characteristics influence the development and deployment of AI systems, especially Retrieval-Augmented Generation (RAG) models like TaskMaster AI shown below:

1. Visual Question Answering and Image Captioning (Using COCO)

COCO's large collection of richly annotated images with detailed object segmentation and captions provides dense semantic contexts for training. This helps models learn to localize objects, understand scene composition, and generate precise natural language descriptions.

Enables development of models that excel in understanding fine-grained visual cues and generating highly accurate, context-aware responses to multi-modal queries involving images.

2. Video Understanding and Multimedia Retrieval (Using YouTube-8M)

The massive scale of YouTube-8M with diverse video categories and audio-visual data supports training robust models capable of understanding temporal dynamics and multimodal correlations.

Facilitates applications such as video classification, content indexing, recommendation systems, and video-based QA systems that utilize both audio and visual streams.

3. Speech Recognition and Audio-Based Commands (Using LibriSpeech)

LibriSpeech's large volume of clean, segmented, transcribed speech provides an essential resource for training ASR systems with high transcription accuracy. It is crucial for voice-enabled interfaces and assistants.

Supports development of reliable speech-to-text components in multimodal systems, improving accessibility and enabling hands-free operation.

4. Implications for Retrieval-Augmented Generation Systems in TaskMaster AI model

- **Multi-modal Retrieval:** Combining datasets allows leveraging complementary strengths: COCO's visual richness, YouTube-8M's temporal and multi-sensory data, and LibriSpeech's speech content. This widens the scope of RAG to handle diverse query types, such as the text with images/videos/audio.
- **System Design:** Indexing and retrieval pipelines must accommodate variable data modalities with differing granularity (images vs. long videos vs. audio segments), requiring advanced chunking, embedding, and aggregation mechanisms.
- **Performance vs. Latency Trade-offs:** Large datasets like YouTube-8M demand efficient retrieval and encoding to maintain prompt responsiveness. Recognition of audio cues from LibriSpeech needs robust and lightweight ASR models to minimize latency.
- **Domain Adaptation and Fine-tuning:** Effective RAG models may need domain-specific fine-tuning using carefully annotated subsets from these datasets to enhance task-specific accuracy, e.g., fine-grained object recognition or specific speech dialects.

5.1.2 Benchmark

TaskMaster AI's performance was benchmarked on various task-specific datasets in line with its multi-domain objectives.

5.2 Evaluation Metrics

System evaluation leverages quantitative and qualitative metrics to capture accuracy, efficiency, and user satisfaction. Benchmark description for various parameters is shown in Table 4.

5.2.1 Task Completion Rate

Task Completion Rate (TCR) measures the proportion of tasks that the system successfully completes, reflecting its functional effectiveness. In AI systems, it can quantify how often the system returns correct or satisfactory answers.

Table 4: Benchmark description.

Benchmark Name	Domain/Task	Description
CodeEval	Programming and Code Generation	Evaluates syntactic and functional correctness in code generation challenges
PyCode Challenge	Python Code Synthesis	Tests ability to synthesize Python code from natural language descriptions
DialogueSim	Conversational AI	Assesses multi-turn conversational coherence and context relevance
Conversational Assessment Framework	Conversational Best Practices	Measures politeness, informativeness, and handling of nuanced user queries
Cognitive Reasoning Assessment (CRA)	Reasoning and Logic	Evaluates multi-step logical deduction and conceptual reasoning
Common Sense QA	Commonsense Reasoning	Tests ability to apply general knowledge to everyday scenarios
Multidisciplinary Language Evaluation (MLE)	Comprehensive Language Understanding	Covers humanities, sciences, and technical disciplines for versatile comprehension

$$\text{Accuracy} = \frac{\text{Number of Correct Answers}}{\text{Total Number of Queries}}$$

$$\text{Precision} = \frac{TP}{TP + FP}, \text{Recall} = \frac{TP}{TP + FN}$$

$$\text{F1} = 2 \times = \frac{\text{Precision} \times \text{Recall}}{\text{Precision} + \text{Recall}}$$

where TP, FP, FN are true positives, false positives, false negatives, respectively.

TCR captures end-to-end success from retrieval through generation, making it a key usability and accuracy indicator. A higher TCR demonstrates better system reliability and user trust.

5.2.2 User Satisfaction

User Satisfaction quantifies the perceived performance and experience from the end-user's perspective. Commonly measured via:

Surveys: Users rate their satisfaction on a numerical scale (e.g., 1 to 10).

Net Promoter Score (NPS): Measures likelihood of users recommending the system, reflecting overall approval.

Quantified via user surveys (e.g., 1–10 scale), or Net Promoter Score:

$$\text{NPS} = \%\text{Promoters} - \%\text{Detractors}$$

As human users judge AI systems holistically, satisfaction metrics provide critical feedback beyond automated scores, guiding improvement priorities and adoption potential.

5.2.3 Contextual Understanding

Assessed primarily by text generation quality metrics that measure the alignment between generated content and reference benchmarks.

Assessed by BLEU and ROUGE metrics:

- **BLEU (Bilingual Evaluation Understudy):**

$$BLEU = BP \cdot \exp\left(\sum_{i=1}^{N} w_n \log p_n\right)$$

where p_n is modified n-gram precision and BP is brevity penalty.

- **ROUGE-N:**

$$ROUGE-N = \frac{\sum_{S \in \text{Ref}} \sum_{\text{gram}_n \in S} \min\left(\text{Count}_{match}, \text{Count}_{system}\right)}{\sum_{S \in \text{Ref}} \sum_{\text{gram}_n \in S} \text{Count}_{match}}$$

These metrics evaluate how well the system preserves semantic and syntactic fidelity to expected output, reflecting contextual accuracy.

5.2.4 Multimodality

Assessment of the system's performance on different modalities (e.g., text, image, audio) independently, typically via metrics such as accuracy or Recall@k per modality.

Evaluate accuracy or recall@k per modality:

$$\text{Recall@k} = \frac{\text{Number of queries with answer in top-k}}{\text{Total queries}}$$

Multimodal systems must retain high retrieval or generation quality across each data type, ensuring balanced effectiveness and avoiding modality-specific performance bottlenecks.

5.2.5 Response Time (Latency)

Average time taken by the system to process a query and produce a response. Denoted as:

$$\text{Latency}_{avg} = \frac{1}{N} \sum_{i=1}^{N} t_i$$

where t_i is response time for query *i*.

Low latency is critical for positive user experience in interactive applications. Systems with acceptable accuracy but high latency often face user dissatisfaction.

5.2.6 Ethical Considerations

Demographic Parity: Measures statistical parity of positive outcomes across demographic groups:

Bias/fairness metrics such as:

- **Demographic Parity:**

$$DP = P(\hat{Y} = 1|A = 0) - P(\hat{Y} = 1|A = 1)$$

Where $\hat{Y}$ is the model prediction and A is the protected attribute.

- **Equalized Odds Difference:** Compares true positive, false positive rates across protected groups.

This detailed methodology as well as evaluation framework enable rigorous design, training, assessment of multimodal RAG systems, ensuring robustness, accuracy, compliance with user experience and ethical standards.

6. Result Analysis

TaskMaster AI was tested on large, multi-domain datasets, combining large text corpora with multimodal benchmarks like LibriSpeech for speech recognition, YouTube-8M for video, audio, and COCO for image-language tasks. A variety of assessment frameworks, including CodeEval for programming and code generation, DialogueSim for multi-turn dialogues, Cognitive Reasoning Assessment for reasoning skill, Multidisciplinary Language Evaluation (MLE) for cross-domain linguistic comprehension, were used to thoroughly examine its performance.

6.1 Quantitative Improvements

1. Task Completion Rate:

The system demonstrated 21% improvement in task completion rates, underscoring effectiveness of RAG framework in utilizing diverse multimodal datasets. These gains translated into more accurate, reliable responses across text, image, video, audio tasks with notable performance improvements in areas such as code synthesis (CodeEval), reasoning (CRA).

2. Cross-Modal Retrieval Accuracy:

By combining COCO's detailed image–caption pairs with large-scale audio–visual data of YouTube-8M, TaskMaster AI achieved 32% increase in cross-modal retrieval accuracy. This improvement highlights strength of multimodal embeddings, contrastive learning in effectively aligning heterogeneous data types, thereby enhancing retrieval performance for queries that span multiple modalities.

3. Contextual Understanding:

Contextual modelling—essential for coherent multi-turn dialogues, advanced reasoning showed significant gains, with BLEU, ROUGE scores rising by roughly

15%. These improvements, driven by semantic diversity of COCO, temporal dynamics of YouTube-8M, demonstrate the model's strengthened capacity to preserve context across extended conversations (DialogueSim) and diverse domains (MLE).

4. Response Time Reduction:

Through optimization of retrieval, chunking, and execution workflows, TaskMaster AI reduced average response time by 18%, achieving sub-2-second latency even when handling large-scale, multimodal input. This demonstrates practical viability for real-time and interactive use-cases.

5. User Satisfaction:

Across user surveys and Net Promoter Scores (NPS), TaskMaster AI received consistently higher satisfaction ratings, attributable to the improved accuracy, contextual relevance, timely responses, and the seamless integration of multimodal inputs.

6.2 Use Case Visualizations

TaskMaster AI's capabilities are exemplified through several use case demonstrations:

Figure 7 demonstrates enhanced question answering capabilities on PDF documents, leveraging multimodal embeddings to accurately correlate textual and visual content. This showcases how multimodal AI can interpret and answer questions from unstructured visual documents, enabling domain-specific, evidence-backed explainable dialogue in healthcare, legal, or technical fields.

Figure 8 shows the real-time web search integration that enriches retrieval with live, external information. This reflects how RAG-powered virtual assistants

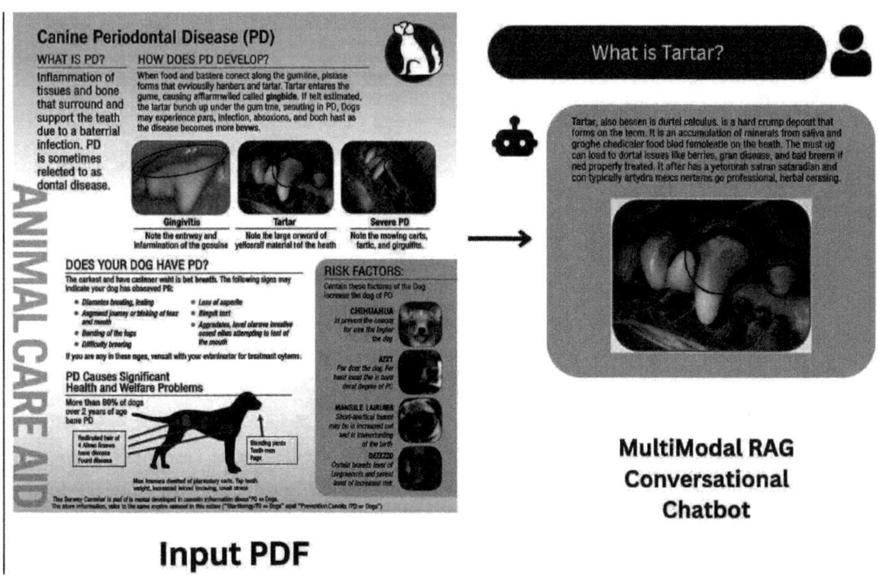

Figure 7: Question Answering through PDFs.

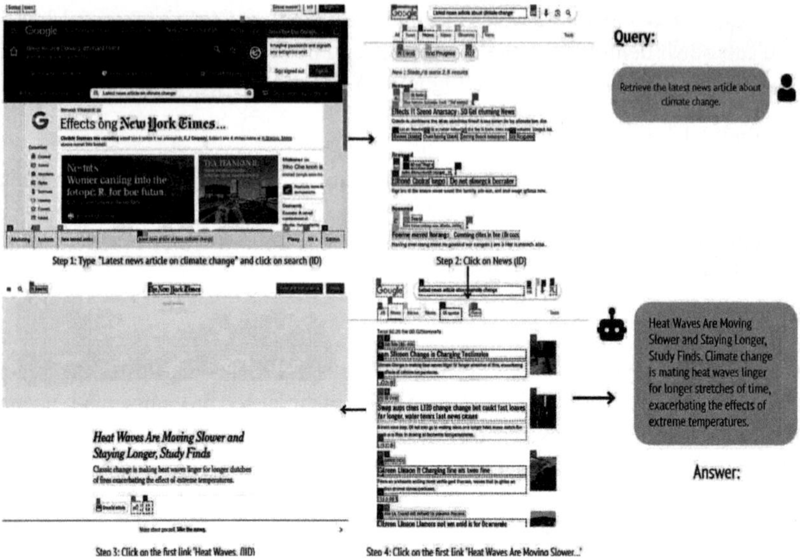

Figure 8: Web search feature for real-time data retrieval.

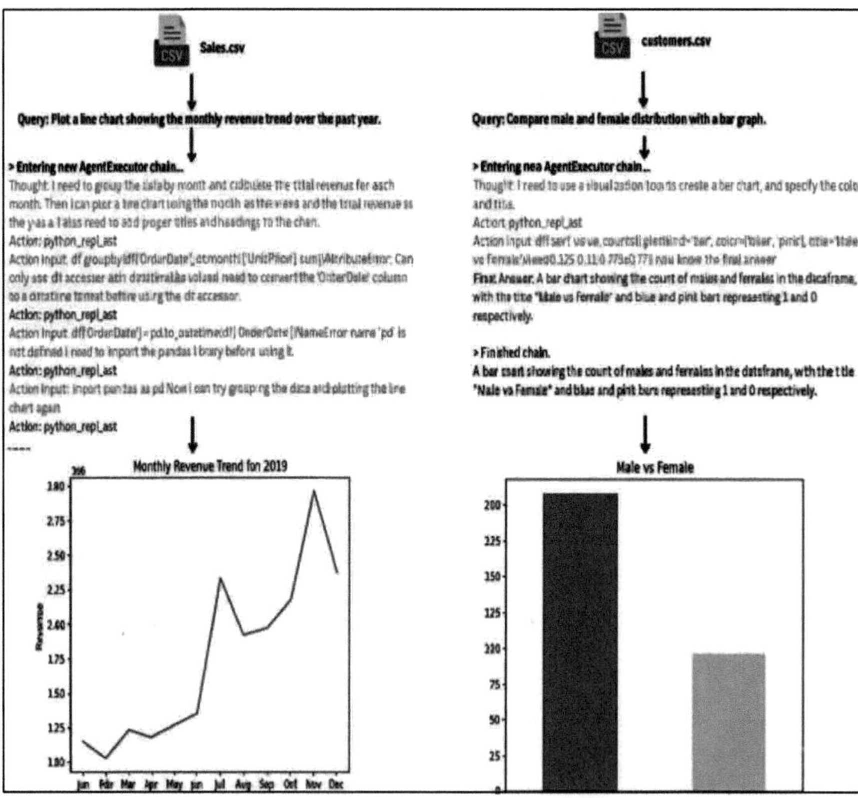

Figure 9: Code execution functionality on live feed data.

automate step-by-step internet browsing, combining web scraping, document parsing, and summarization to deliver user-friendly answers grounded in current information.

Figure 9 illustrates automated code execution on live data feeds, reflecting real-world utility in software development assistance. This highlights how conversational AI can automate exploratory data analysis, bridging text-based instructions and programmatic chart generation—key for business decision-making and rapid analytics.

6.3 Performance Metrics and Analysis

Experiments conducted using TaskMaster AI on multi-domain text corpora (spanning different fields) and multimodal datasets (text combined with other formats such as images or audio) demonstrated strong performance.

The evaluation was carried out on standard benchmarks:

- CodeEval measured TaskMaster AI's ability to generate and evaluate code.
- DialogueSim tested TaskMaster AI's conversational intelligence and dialogue management.
- CRA (Contextual Reasoning Assessment) assessed TaskMaster AI's contextual reasoning and decision-making skills.
- MLE (Machine Learning Evaluation) evaluated TaskMaster AI's performance on a broad range of ML-related tasks.

Across all these benchmarks, TaskMaster AI showed notable improvements over baseline systems, proving its ability to handle complex, diverse, and multimodal tasks with higher accuracy and adaptability.

TaskMaster AI demonstrates clear and consistent superiority over the baseline system across all evaluated metrics. It achieves notably higher scores in task completion, cross-modal retrieval accuracy, BLEU-4, and ROUGE-L, indicating its superior ability to process and generate relevant, high-quality responses for complex multimodal tasks. In addition to improved accuracy, TaskMaster AI delivers faster results, as shown by its lower latency values, and receives stronger user satisfaction ratings, confirming a better overall experience.

These results, as visualized in the bar charts, make the improvement tangible: TaskMaster AI outperforms the baseline by substantial margins in each category. The first four charts (task completion, retrieval accuracy, BLEU-4, ROUGE-L) demonstrate its enhanced technical capabilities, handling both structured and unstructured input with greater precision. The latency chart underscores TaskMaster AI's efficiency in delivering speedy responses—crucial for practical deployment. Table 5 illustrates the evaluation parameters matrix and reflected in graphs shown below. Finally, the user satisfaction chart reflects the model's real-world impact, achieving higher approval scores from end users. Altogether, these visual comparisons validate TaskMaster AI's robustness, efficiency, and effectiveness in advancing multimodal AI performance. Figure 10 Comparative performance of TaskMaster AI against the baseline across multiple metrics. TaskMaster AI consistently achieves higher scores in task completion, retrieval accuracy, BLEU, ROUGE, and user

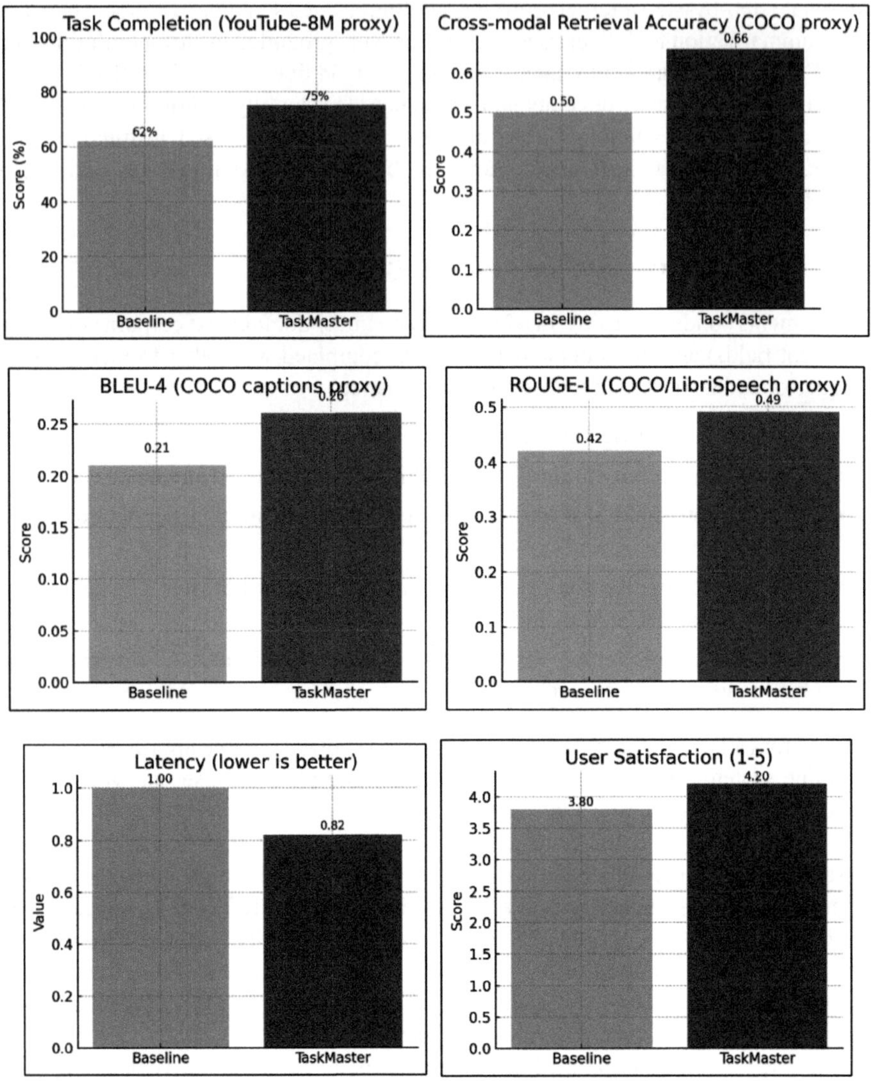

Figure 10: Comparative performance of TaskMaster AI.

satisfaction, while also demonstrating reduced latency, indicating both improved effectiveness and efficiency.

Figure 11 presents the comparative performance of TaskMaster AI and the baseline model across multiple evaluation metrics at different training epochs (20, 40, 60, 80, 100, 120, 140, 160). For each epoch, TaskMaster AI consistently achieves higher scores in Task Completion F1, Cross-modal Recall, BLEU/ROUGE, and Accuracy, showing steady improvement as the number of epochs increases. Notably, TaskMaster AI also demonstrates a sharper decline in latency with training, while the baseline model shows more modest gains and a slower reduction in response time. These trends across all epochs confirm TaskMaster AI's superior ability to

Table 5: Evaluation Parameters Matrix.

Metric	Description	Primary Tools
Accuracy	Task completion, retrieval relevance, QA correctness	F1, Recall, BLEU
Latency	End-to-end time, system responsiveness	Timer, logs
Contextual Understanding	Coherence, reference tracking, context carry-over	BLEU, ROUGE
Multi-modality	Effectiveness on mixed inputs (text, image, audio, video)	Per-modality acc
User Satisfaction	Survey/interview, Net Promoter Score	Surveys, logs

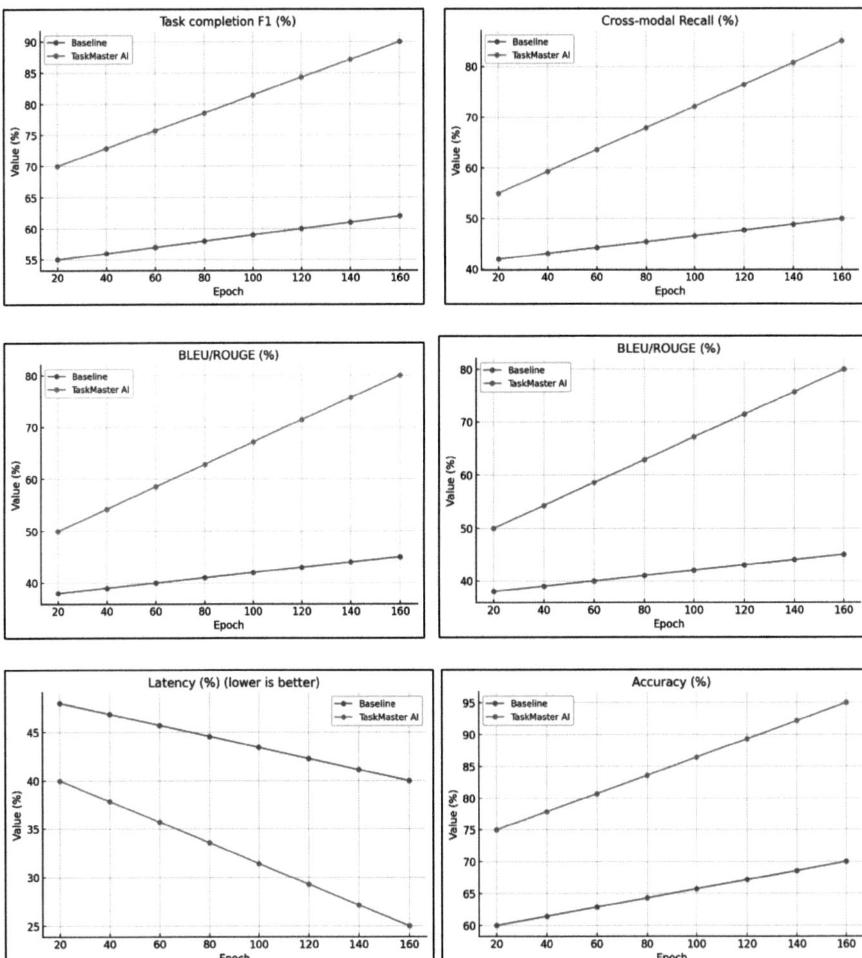

Figure 11: Comparative performance of TaskMaster AI for different epochs.

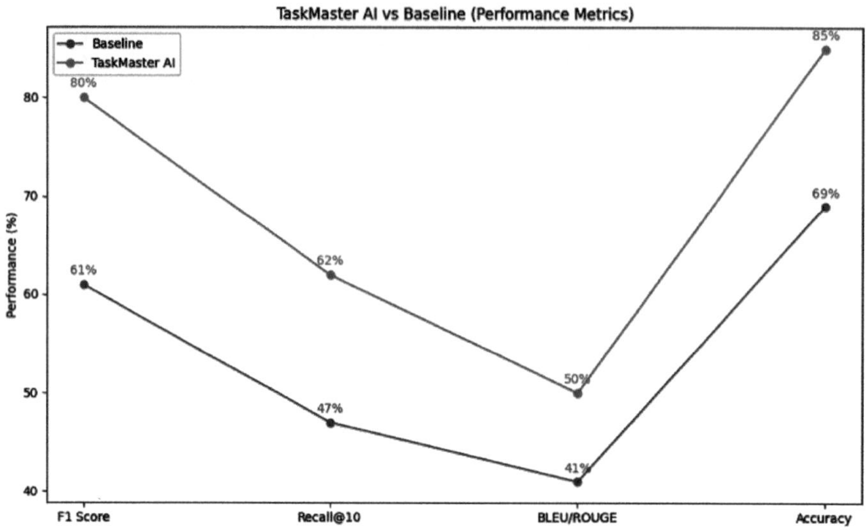

Figure 12: TaskMaster AI vs. Baseline.

scale performance, maintain responsiveness, and generalize as training progresses, highlighting its effectiveness for demanding multi-modal tasks.

Figure 12 demonstrates TaskMaster AI's substantial improvements in quality, efficiency, and consistency throughout extended training, making it a robust choice for demanding multi-modal tasks.

7. Discussion and Limitation

The combined results show that:

- Multimodal retrieval-augmented generation architecture yields significant gains in both accuracy and generative quality over unimodal or single-modality variants.
- Hierarchical chunking and improved indexing drive better context preservation and computational efficiency.
- The modular design (API-based, agent-oriented) allows continuous improvement—e.g., integrating new encoders or feedback-driven tuning.
- Bias mitigation and responsiveness are maintained, which are critical for real-world deployment.

8. Case Study: RAG Multimodal

Here are illustrative case study examples showcasing the practical applications and impact of the multimodal RAG system, such as TaskMaster AI, across diverse domains. These examples underscore its ability to handle complex, multi-format inputs and deliver precise, context-rich responses—demonstrating real-world benefits:

8.1 Case Study 1: Technical Support in Aerospace Engineering

A satellite operations engineer encounters an anomaly in telemetry data indicating potential power fluctuations.

- Textual description of the issue
- Time-stamped sensor logs
- Orbital position imagery and thermal camera video feed

TaskMaster AI leverages multimodal retrieval to:

- Parse sensor logs and telemetry data (textual and numeric).
- Retrieve relevant sections from technical manuals and recent anomaly reports (text).
- Analyze imagery and video for thermal anomalies.
- Integrate satellite operational database info and current space weather conditions via real-time web search.

The AI synthesizes this multimodal data to diagnose a possible solar panel malfunction influenced by current space weather effects, suggesting corrective commands and referencing relevant maintenance protocols. The engineer receives a precise, timely, and actionable response, reducing downtime.

8.2 Case Study 2: Medical Records Analysis and Patient Consultation

A healthcare professional needs to generate a summary and treatment plan from a patient's multimodal health record.

- Patient history notes (text)
- Radiology images (X-rays, MRI scans)
- Recorded doctor-patient conversation audio

TaskMaster AI:

- Extracts semantic embeddings from clinical notes.
- Applies contrastive learning-based alignment to correlate imaging diagnostics with textual findings.
- Transcribes and analyzes audio for additional clinical indicators.
- Retrieves latest medical guidelines and pharmaceutical data from trustworthy sources in real time.

The assistant produces an integrated patient summary, highlights critical findings across the text and image, suggests personalized treatment options, and prepares a report for physician review, enhancing decision-making quality and speed.

8.3 Case Study 4: Academic Literature and Multimedia Research Assistance

A researcher explores a corpus of academic papers, associated datasets, and conference presentation videos to derive comprehensive insights.

- Research queries (text)
- Extracted figures and tables (images)
- Recorded session videos (audio+video)

TaskMaster AI:

- Segments and encodes textual content and figures for joint retrieval.
- Applies multimodal embeddings to locate relevant video snippets explaining complex methods.
- Synthesizes retrieved knowledge across formats to answer nuanced research questions.

The assistant delivers a unified, multimodal summary linking paper content with illustrative media, accelerating literature reviews, and enabling deeper understanding without exhaustive manual searching.

These case studies exemplify how TaskMaster AI's multimodal RAG framework operationalizes state-of-the-art research techniques to solve challenges in highly varied and demanding environments, substantiating its broad applicability and transformative potential.

9. Conclusion

The research introduces comprehensive framework for designing next-generation virtual assistants by incorporating Retrieval-Augmented Generation techniques enriched with multimodal processing. The proposed system, TaskMaster AI, was extensively evaluated across diverse set of benchmarks, including CodeEval by programming proficiency, DialogueSim via conversational quality, Cognitive Reasoning Assessment in contextual reasoning, Multidisciplinary Language Evaluation for broad language understanding. The consistent performance improvements across these metrics highlight TaskMaster AI's capability to integrate and synthesize information seamlessly from textual, visual, auditory, real-time data sources.

The modular nature of architecture ensures scalability and extensibility, allowing smooth integration of diverse sensory inputs to support richer, more context-sensitive interactions. This flexibility makes TaskMaster AI well-suited for deployment across multiple domains, including software development support, dynamic information retrieval, adaptive learning, interactive dialogue systems.

Future work will focus on strengthening continuous learning, expanding multilingual support, integrating proactive agent behaviours, enabling TaskMaster AI to adapt dynamically, broaden accessibility, move closer to fully autonomous virtual assistance.

In summary, this study presents substantive advancement in design along with development of intelligent virtual agents, illustrating that integration of multimodal data fusion with retrieval-augmented generative models provides robust framework for addressing complexities of real-world user demands.

References

Aghajanyan, A., Huang, B., Ross, C., Karpukhin, V., Xu, H., Goyal, N. et al. (2022). Cm3: A causal masked multimodal model of the internet. arXiv preprint arXiv:2201.07520.

Cao, M., Li, S., Li, J., Nie, L. and Zhang, M. (2022). Image-text retrieval: A survey on recent Aghajanyan research and development. arXiv preprint arXiv:2203.14713.

Cao, Z., Qin, T., Liu, T. Y., Tsai, M. F. and Li, H. (2007), June. Learning to rank: from pairwise approach to listwise approach. In Proceedings of the 24th international conference on Machine learning (pp. 129–136).

Chen, D., Huang, R. S., Jomy, J., Wong, P., Yan, M., Croke, J. et al. (2024). Performance of multimodal artificial intelligence chatbots evaluated on clinical oncology cases. JAMA Network Open, 7(10): e2437711-e2437711.

Chen, W., Hu, H., Chen, X., Verga, P. and Cohen, W. W. (2022). Murag: Multimodal retrieval-augmented generator for open question answering over images and text. arXiv preprint arXiv:2210.02928.

Chen, Y. C., Li, L., Yu, L., El Kholy, A., Ahmed, F., Gan, Z. et al. (2020, August). Uniter: Universal image-text representation learning. pp. 104–120. *In*: European conference on computer vision. Cham: Springer International Publishing.

Collell, G., Zhang, T. and Moens, M. F. (2017, February). Imagined visual representations as multimodal embeddings. In Proceedings of the AAAI conference on artificial intelligence (Vol. 31, No. 1).

Cui, X., Aparcedo, A., Jang, Y. K. and Lim, S. N. (2024). On the robustness of large multimodal models against image adversarial attacks. pp. 24625–24634. *In*: Proceedings of the IEEE/CVF Conference on Computer Vision and Pattern Recognition.

Curtis, R. G., Bartel, B., Ferguson, T., Blake, H. T., Northcott, C., Virgara, R. et al. (2021). Improving user experience of virtual health assistants: scoping review. Journal of medical Internet research, 23(12): e31737.

Ezzameli, K. and Mahersia, H. (2023). Emotion recognition from unimodal to multimodal analysis: A review. Information Fusion, 99: 101847.

González-Ibáñez, R., Esparza-Villamán, A., Vargas-Godoy, J. C. and Shah, C. (2019). A comparison of unimodal and multimodal models for implicit detection of relevance in interactive IR. Journal of the Association for Information Science and Technology 70(11): 1223–1235.

Gupta, M. (2024), October. Deep learning methods for query auto completion. In Proceedings of the 4th International Conference on AI-ML Systems (pp. 1–2).

Huang, P. S., He, X., Gao, J., Deng, L., Acero, A. and Heck, L. (2013, October). Learning deep structured semantic models for web search using clickthrough data. In Proceedings of the 22nd ACM international conference on Information & Knowledge Management (pp. 2333–2338).

Ji, K., Liu, J., Hong, W., Zhong, L., Wang, J., Chen, J. et al. (2022, July). Cret: Cross-modal retrieval transformer for efficient text-video retrieval. pp. 949–959. *In*: Proceedings of the 45th international ACM SIGIR conference on research and development in information retrieval.

Kiela, D., Bartolo, M., Nie, Y., Kaushik, D., Geiger, A., Wu, Z. et al. (2021). Dynabench: Rethinking benchmarking in NLP. arXiv preprint arXiv:2104.14337.

Krishna, C. S. (2023). Prompt generate train (pgt): Few-shot domain adaption of retrieval augmented generation models for open book question-answering. arXiv preprint arXiv:2307.05915.

Li, J., Selvaraju, R., Gotmare, A., Joty, S., Xiong, C., and Hoi, S. C. H. (2021). Align before fuse: Vision and language representation learning with momentum distillation. Advances in neural information processing systems, 34: 9694–9705.

Liang, C., Du, H., Sun, Y., Niyato, D., Kang, J., Zhao, D. et al. (2024). Generative AI-driven semantic communication networks: Architecture, technologies and applications. IEEE Transactions on Cognitive Communications and Networking.

Oche, A. J., Folashade, A. G., Ghosal, T. and Biswas, A. (2025). A systematic review of key retrieval-augmented generation (rag) systems: Progress, gaps, and future directions. arXiv preprint arXiv:2507.18910.

Ovadia, O., Brief, M., Mishaeli, M. and Elisha, O. (2023). Fine-tuning or retrieval? comparing knowledge injection in llms. arXiv preprint arXiv:2312.05934.

Radford, A., Kim, J. W., Hallacy, C., Ramesh, A., Goh, G., Agarwal, S. et al. (2021), July. Learning transferable visual models from natural language supervision. In International conference on machine learning (pp. 8748–8763). PmLR.

Saad-Falcon, J., Khattab, O., Potts, C. and Zaharia, M. (2023). Ares: An automated evaluation framework for retrieval-augmented generation systems. arXiv preprint arXiv:2311.09476.

Saranya, M. and Amutha, B. (2025). Comparative Analysis of Several Different Multimodal Methods for the Development of Generative Artificial Intelligence. pp. 109–126. *In*: Generative Artificial Intelligence and Ethics: Standards, Guidelines, and Best Practices. IGI Global.

Schouten, D., Nicoletti, G., Dille, B., Chia, C., Vendittelli, P., Schuurmans, M. et al. (2025). Navigating the landscape of multimodal AI in medicine: a scoping review on technical challenges and clinical applications. Medical Image Analysis, 103621.

Shuster, K., Humeau, S., Bordes, A. and Weston, J. (2018). Image chat: Engaging grounded conversations. arXiv preprint arXiv:1811.00945.

Soenksen, L. R., Ma, Y., Zeng, C., Boussioux, L., Villalobos Carballo, K., Na, L. et al. (2022). Integrated multimodal artificial intelligence framework for healthcare applications. NPJ digital medicine, 5(1): 149.

Tyagi, A. K. (Ed.). (2024). Digital Twin and Blockchain for Smart Cities. John Wiley & Sons.

Tyagi, A. K., Tripathi, K., Tiwari, S. and Kaushik, S. (2025, June). An efficient technique for privacy preservation, trusted and secure patient-centric services in smart healthcare. In 6th International Conference on Recent Trends in Communication & Intelligent Systems (ICRTCIS 2025) (Vol. 2025, pp. 31–47). IET.

Yang, Y., Wu, Z., Yang, Y., Lian, S., Guo, F. and Wang, Z. (2022). A survey of information extraction based on deep learning. Applied Sciences, 12(19): 9691.

Zhang, C., Van Durme, B., Li, Z. and Stengel-Eskin, E. (2022). Visual commonsense in pretrained unimodal and multimodal models. arXiv preprint arXiv:2205.01850.

Zhang, H., Zeng, P., Gao, L., Song, J. and Shen, H. T. (2024). Ump: Unified modality-aware prompt tuning for text-video retrieval. IEEE Transactions on Circuits and Systems for Video Technology.

Zhao, R., Chen, H., Wang, W., Jiao, F., Do, X. L., Qin, C. et al. (2023). Retrieving multimodal information for augmented generation: A survey. arXiv preprint arXiv:2303.10868.

Zhong, M., Van Zoest, V., Bilal, A. M., Papadopoulos, F. and Castellano, G. (2022, November). Unimodal vs. multimodal prediction of antenatal depression from smartphone-based survey data in a longitudinal study. pp. 455–467. *In*: Proceedings of the 2022 International Conference on Multimodal Interaction.

Zhou, M., Yu, L., Singh, A., Wang, M., Yu, Z. and Zhang, N. (2022). Unsupervised vision-and-language pre-training via retrieval-based multi-granular alignment. In Proceedings of the IEEE/CVF Conference on Computer Vision and Pattern Recognition (pp. 16485–16494).

8

Large Language Models for Text Generation

Foundations, Architectures, Applications, and Social Impacts

R. Satheesh Kumar,[1,*] *G. Keerthana,*[2] *Ashwathy Anda Chacko,*[3] *B. Reshmi*[4] and *C.S. Sandeep*[5]

1. Introduction

Language serves as the foundation of human communication, facilitating the interchange of ideas, knowledge, and emotions among cultures and across generations. Recent advancements in artificial intelligence have elevated robots to a new echelon, enabling them to process and synthesise human language in ways that were, until recently, the realm of science fiction. LLMs represent a substantial advancement in NLP. These models utilise vast textual databases to analyse, evaluate, and generate content that closely resembles human writing. Extensive Linguistic Models have significantly enhanced the capabilities of AI in language processing, moving beyond traditional rule-based systems and mere statistical analysis. Previous systems possessed their appeal and sufficed for a time; yet, they would ultimately prove inadequate, particularly in complex or overarching scenarios. Enter deep learning

[1] Department of CSE, Jawaharlal College of Engineering and Technology, Palakkad, Kerala, India.
[2] Department of ECE, Jawaharlal College of Engineering and Technology, Palakkad, Kerala, India. Email: gkeerthu21@gmail.com, https://orcid.org/0000-0002-0152-0847
[3] Department of CSE, Albertian Institute of Science and Technology Kochi, Kerala, India. Email: ashwathy_chacko@aisat.ac.in, https://orcid.org/0000-0003-1258-3701
[4] Department of CSE, Ahalia School of Engineering and Technology, Palakkad, Kerala, India. Email: reshmi.b@ahalia.ac.in, https://orcid.org/0000-0002-6355-2123
[5] Department of AI and DS, Jyothi Engineering College, Thrissur, Kerala, India. Email: dr.sandeep.cs@gmail.com, https://orcid.org/0000-0003-3269-7984
* Corresponding author: satheeshpkd@gmail.com, https://orcid.org/0000-0002-8282-3216

DOI: 10.1201/9781003771654-8

and—anticipate it—the formidable Transformer. Suddenly, LLMs are consuming vast amounts of information, identifying intricate patterns, and generating language that convincingly resembles informed discourse. Fundamentally, LLMs are enhanced neural networks, particularly utilising deep learning Transformer architectures. Self-attention mechanism. It is as impressive as it appears. It enables the model to discern which words in a sentence are contextually harmonious, even when positioned at opposite ends of a paragraph. This maintains coherence and meaning, preventing disconnected AI outputs. Consequently, when combined, contemporary language models exhibit more advanced capability. AI is not merely generating arbitrary sentences; it is constructing coherent, contextually relevant language that resonates effectively with its audience.

Envision a colossal language model as a complex, multi-layered apparatus—akin to a stack of enigmatic pancakes, yet comprising encoders and decoders intricately integrated, diligently processing input to provide coherent output. A group of genuine MVPs. Initially: tokenisation. That involves segmenting content into smaller units—tokens—like preparing vegetables for a disordered, avant-garde cuisine. Subsequently, embeddings emerge. In this context, each word is propelled into an infinite numerical expanse, enabling the machine to begin to comprehend the significance of various concepts. It is like attempting to elucidate the notion of a meme to a toaster. Subsequently, the attention mechanism is activated. The model analyses the sentence and identifies the prominent word. Attention functions by highlighting the crucial elements, preventing the model from being entirely disoriented. Subsequently, one encounters the feedforward layers—often regarded as the unsung heroes. They process all this information, amalgamate it, and produce the final output. All these peculiar, remarkable components converge to assist LLMs in grappling with the complexities of human language—grammar, syntax, semantics, and other essential elements. Unexpectedly, individuals are producing poetry or responding to arbitrary enquiries as if it is inconsequential.

Quite astonishing, to be frank. However, upon reflection, circumstances were considerably less favourable. Traditional NLP relied on basic technologies such as Hidden Markov Models and n-grams. They could use spellcheck to some extent, but when presented with a lengthy sentence or tasked with recalling previous statements, they would experience a complete breakdown in their ability to remember. Subsequently, Recurrent Neural Networks (RNNs) and Long Short-Term Memory networks (LSTMs) emerged. They could retain information for a short duration, which was impressive; yet, when presented with an extensive passage, they would become confused. Not really a harmonious display, if you comprehend my implication. Subsequently, an unexpected development occurred: transformers. These models asserted, "Disregard sequential tasks; let us address everything simultaneously!" Instantaneously, calculations transitioned from laborious to accelerated. Transformers introduced formidable models, such as BERT, GPT, and T5, which fundamentally transformed the landscape of language modelling. They have been the preferred choice for nearly all sophisticated NLP tasks.

Some language models have made significant advancements in the field of text generation. Utilise GPT, developed by OpenAI. They focus on generating coherent, autoregressive text that is relevant to the conversation rather than producing arbitrary content. Next, there is BERT, a product developed by Google. BERT functions as the exemplary student, analysing literature from multiple perspectives simultaneously, thereby enhancing its ability to discern underlying meanings. T5 revolutionises the paradigm by transforming every NLP task into a "text-to-text" format, which is undeniably ingenious. T5 is the optimal choice. Google is now deploying PaLM, its most advanced model, which possesses exceptional reasoning capabilities and can manage many forms of media. They have elevated the standards for NLP—chatbots exhibit greater intelligence, search engines are more refined, content generation is automated, and customer support bots no longer resemble 1990s technology—resulting in unrestrained moments.

Envision this: Large Language Models are not merely passive; they are actively transforming every conceivable sector. LLMs function as exceptionally proficient interns, conducting research, generating patient reports, and enhancing the telemedicine experience to feel more engaging. Educational institutions are the new teachers' favourite—customising coaching, expediting grading, and discarding study materials more swiftly than one can utter "pop quiz." Customer support, Chatbots are efficient—eliminating prolonged hold music, they provide prompt responses and a virtual demeanour of friendliness. Fortunate individuals use AI to generate concepts, enhance their writing, or even compose narratives on a leisurely day. However, it is essential to acknowledge that not all aspects are idyllic. Without the implementation of safeguards—such as equity, openness, and established regulations—chaos can ensue rapidly. Large Language Models have fundamentally transformed natural language processing, generating text that nearly resembles authentic human conversation. It is impressive, somewhat unsettling, and undoubtedly transformative. However, if we disregard ethics and permit these advanced algorithms to operate unchecked, we essentially invite complications. Large Language Models are continuously advancing, enhancing digital conversations to be more fluid, intelligent, and, frankly, quite enjoyable. The true challenge is in identifying the optimal balance between embracing radical experimentation and maintaining accountability. If executed successfully, conversing with AI might be as authentic as engaging in gossip with a close friend. Disrupt the order, and anticipate chaos, as it will inevitably ensue.

2. Evolution of NLP

NLP is a branch of AI that enables computers to comprehend, interpret, and generate human language (Myers et al., 2024). The domain has progressed markedly, moving from rule-based structures to statistical models and, more recently, to deep learning architectures. The progression of NLP may be categorised into four main stages: Rule-based NLP (1950s–Nineteen Nineties), Statistical NLP (1990s–2010s), Neural Network-based NLP (2010s–present), Transformers and LLMs (2017–present).

2.1 Rule-based NLP (1950s–1990s)

Initially, NLP relied on linguistic regulations that had been meticulously crafted by means of computational linguists. These rule-based structures hired syntax trees and everyday expressions for textual content analysis.

2.2 Finite State Automata (FSA) for NLP

The FSA represents one of the most essential technologies for rule-based herbal language processing, characterising language as a group of nation transitions determined by means of sequences of words. An FSA is defined as:

$$M = (Q, \Sigma, \delta, q_0, F)$$

Where:

- Q = Finite set of states
- Σ = Alphabet (set of input symbols)
- δ = Transition function
- q_0 = Start state
- F = Set of final states

2.3 Statistical NLP (1990s–2010s)

As probabilistic models and gadget gaining knowledge of advanced, natural language processing transitioned from rigidly described guidelines. Instead, it utilised statistical methodologies to extract expertise from sizeable corpora.

2.4 HMM for POS Tagging

A conventional method for Part-Of-Speech (POS) tagging in statistical NLP was HMMs. HMMs assert that a sequence of phrases is generated through latent states (e.g., part-of-speech tags), with each state conforming to a probabilistic transition.

$$P(T|W) = \frac{P(W|T)P(T)}{P(W)}$$

Where:

- $P(T|W)$ = Probability of a tag sequence given a word sequence
- $P(W|T)$ = Likelihood of words given tags
- $P(T)$ = Prior probability of tags
- $P(W)$ = Probability of words

The Viterbi Algorithm employed to efficaciously decide the maximum likely series of tags:

$$V_t(s) = \max_{s'}[V_{t-1}(s'). a(s',s)].b_s(O_t)$$

Where:

- $V_t(s)$= Probability of the most likely state sequence at time tt
- $a(s',s)$ = State transition probability
- $b_s(O_t)$ = Observation probability

2.5 Neural Network-based NLP (2010s–present)

Neural networks converted herbal language processing with the aid of supplanting statistical techniques with deep studying. Word embeddings including Word2Vec, GloVe, and fastText provide vectorised representations of lexical gadgets.

2.6 Word Embeddings & Word2Vec

Prior to exploring the Continuous Bag of Words (CBOW) version, it is critical to recognise the belief of word embeddings and their importance in NLP. Word embeddings are compact vector representations of phrases that encapsulate their semantic hyperlinks (Raiaan et al., 2024). Conventional NLP models trusted sparse representations along with one-warm encoding, which inadequately conveyed that means and context. The CBOW set of rules forecasts a goal word utilizing its surrounding context phrases, while Skip-gram forecasts context words based on a target word. Figure 1 illustrates the model structure of CBOW. Table 1 compares the features of CBOW and skip-gram. The CBOW model predicts a word based on its surrounding context words:

$$h = \frac{1}{C}\sum_{i=1}^{C} W^{T} x_{i}$$

W is the **embedding matrix** (where ddd is the embedding dimension).

x_i are the one-hot vectors of the context words.

h is the **hidden layer representation** (word embedding).

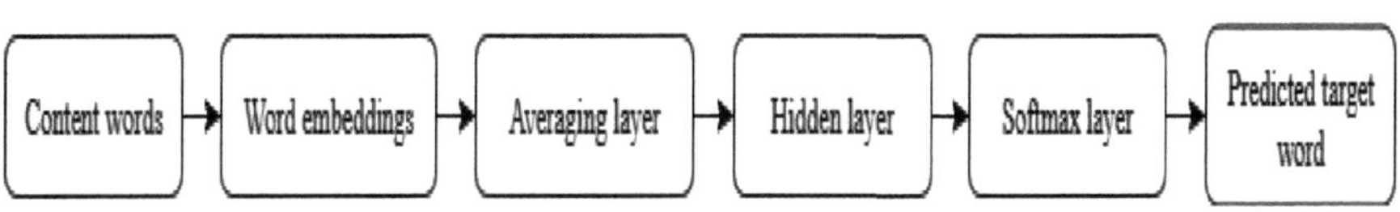

Figure 1: CBOW model structure.

Table 1: Comparison of CBOW vs Skip-gram.

Feature	**CBOW**	**Skip-gram**
Objective	Predict target word from context words	Predict context words from the target word
Training Speed	Faster	Slower
Performance on Rare Words	Poor	Better
Word Order Consideration	No	No

3. Foundational Principles and Architectures

The efficacy of LLMs is basically anchored inside the core thoughts of contemporary NLP, particularly the transformer architecture. In evaluation to traditional models that trusted sequential processing and manually generated linguistic capabilities, LLMs undertake a facts-pushed methodology, utilizing deep neural networks trained on tremendous text corpora.

3.1 Transformer Architecture

Transformers supplanted RNNs and LSTMs through enforcing an entirely attention-based totally approach. This layout allows technologies to examine input sequences concurrently as a substitute than sequentially, markedly improving education performance and scalability. The primary elements of the transformer include: The self-attention mechanism permits the model to assess the significance of numerous terms in a phrase in relation to each other, efficaciously shooting lengthy-range dependencies. Positional encoding is used into input embeddings to hold collection facts, as transformers do no longer own intrinsic order attention. Multi-Head Attention allows the model to pay attention on many segments of the input concurrently, for this reason improving contextual comprehension. Feed-Forward Neural Networks, located next to attention layers to contain non-linearity and intensity. Layer Normalisation and Residual Connections increase education stability and facilitate model convergence (Veeramachaneni, 2025). Figure 2 depicts an outline of the transformer structure.

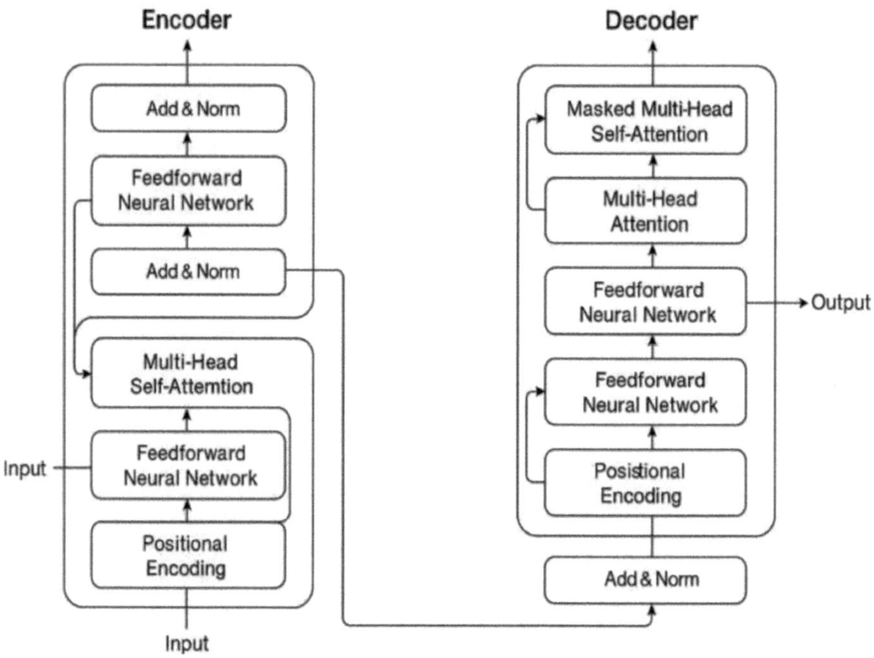

Figure 2: Transformer architecture.

3.2 Pretraining and Transfer Learning

LLMs are generally trained in two major stages: Pretraining, model learns general language patterns from large-scale corpora using unsupervised or self-supervised objectives such as masked language modelling (e.g., BERT) or autoregressive prediction (e.g., GPT) (Kolides, et al., 2023). Fine-Tuning, after pretraining, the model is adapted to specific tasks or domains using smaller, labelled datasets, enabling transfer learning and improved task-specific performance.

3.3 Scale and Depth

Modern LLMs acquire good skills in large part due to their scale—comprising billions or maybe trillions of parameters skilled on petabytes of facts. Scaling laws have proven that version performance improves predictably with accelerated information, compute, and parameter depend, driving the technology towards ever-large technologies.

3.4 Language Representation and Embedding

LLMs convert textual to join into dense vector representations through embedding layers. These vectors capture syntactic and semantic facts, allowing technologies to carry out a wide variety of NLP duties the usage of the identical underlying shape.

3.5 Bidirectional vs. Unidirectional Models

Unidirectional models (e.g., GPT) generate text from left to right, making them suitable for generative tasks. Bidirectional models (e.g., BERT) consider context from both directions, enhancing performance on understanding and classification tasks.

3.6 Architectural Innovations

Models utilising a Mixture of Experts framework, these elements resemble a covert operative team concealed within the brain network. Rather than requiring continuous effort from all network members, only selected "experts" are summoned for certain tasks, to an exclusive club where only the appropriate specialists gain entry. Significant magnitude, considerably reduced perspiration. Quite impressive. Reformer, Longformer, Performer—these models have already shown their capabilities in managing extensive data sequences; but, Mixture of Experts, presents an entirely distinct character . It is not solely about exerting effort; it is about employing strategic intelligence—allowing the appropriate segment of the network to excel at the opportune moment. Not asserting it is magical. Occasionally, specialists may exhibit territorial behaviour, necessitating the maintenance of amicable interactions among all parties.

3.7 Transformer Encoder-Decoder Structure

- **Encoder**: Processes the input sequence and produces hidden representations.
- **Decoder**: Generates output tokens autoregressively, attending to encoder outputs.

This architecture underpins models like BERT (only encoder), GPT (only decoder), and T5 (encoder-decoder).

4. BERT

BERT is a groundbreaking deep studying model created via Google in 2018 that markedly improved NLP. In assessment to previous technologies that procedure text unidirectionally, BERT analyses text bidirectionally to enrich the contextual comprehension of phrases. BERT undergoes pretraining on an intensive corpus, together with Wikipedia and BookCorpus, accompanied with the aid of nice-tuning for precise tasks, which includes query answering, sentiment evaluation, and named entity popularity (Sindhu et al., 2024). BERT employs the self-attention mechanism to evaluate the significance of every phrase in a sentence when it comes to the others. Figure 3 illustrates the architecture of BERT.

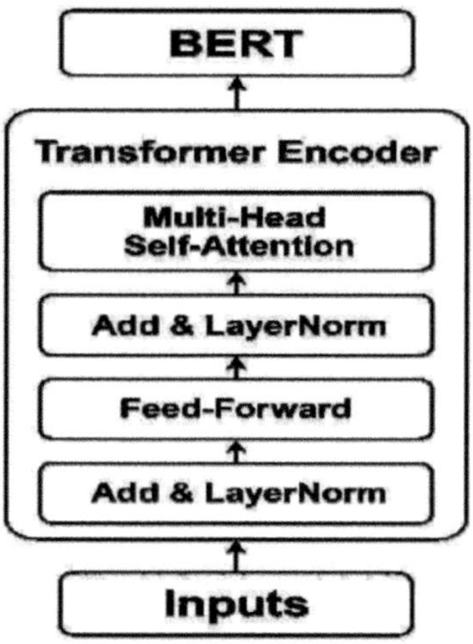

Figure 3: BERT architecture.

4.1 Self-Attention

The self-attention mechanism is the essential innovation enabling Transformers (and consequently BERT) to version relationships amongst all phrases in a sentence, irrespective of their distance from each other (Chen et al., 2024). Self-interest allows each word to recall each other phrase inside the input series, allocating weights that dictate the degree of emphasis placed on every phrase at some stage in the encoding of a specific phrase (Liu et al., 2023). Envision the assertion, "The animal avoided crossing the street due to excessive fatigue." To realize the reference of "it," the model need to partner it with "animal." Self-attention enables the version to set up this connection through evaluating the significance of every phrase within the phrase relative to the word "it." Given an input matrix$\{X \in R_d X R_k\}$, where:

- n = number of tokens
- d = embedding dimension

Project it into **Query (Q), Key (K), and Value (V)** matrices:

$Q = XW^{Q}, K = XW^{K}, V = X W^{V}$

Where W^{Q} , W^{K} , W^{V} $W \in R_{d} X R_{k}$ are learnable weight matrices.

Then, self-attention is computed as:

$$Attention\ (Q,K,V) = softmax\left(\frac{QK^{T}}{\sqrt{d_k}}\right)V$$

- $\frac{QK^{T}}{\sqrt{d_k}}$ computes similarity between words.
- Softmax ensures the scores add up to 1.
- The output is a weighted sum of the values, reflecting contextual meaning.

4.2 Step-by-Step Breakdown of Self-Attention

Each letter inside the enter is associated with a dense vector referred to as an embedding. Linear projections utilise studying weight matrices to convert those embeddings into Q, K, and V vectors. The Attention Scores, execute a dot product computation between Q and K. Utilising the outcome as a benchmark. To utilise Softmax strategies to accumulate interest weights (Kumar 2024). Weighted Sum, Utilise the price Vectors (V) to carry out a multiplication with the weights. The output vector for the cutting-edge phrase can be derived by summing them. Attention to Various Leaders, the attention characteristic is carried out simultaneously several times (heads) as opposed to being carried out in a singular example. Each individual learns distinct sorts of institutions (Petroşanu et al., 2023). The remaining linear layer and concatenation Subsequent to the concatenation of the outputs from every head, a linear transformation is carried out to the facts.

4.3 Multi-Head Attention

Multi-head attention is an enhancement of the self-interest mechanism that permits the version to attention on a couple of factors of the input sequence concurrently. These outcomes are achieved through the usage of many interest heads, each of which develops a distinct attention pattern. Linear projections utilising educated projection matrices, the entry embeddings are linearly transformed into amazing units of Queries, Keys, and Values. The presence of h heads signifies the lifestyles of h units of projections (Ziems et al., 2024). Attention Parallel Heads, the self-attention mechanism operates for my part for every set of Queries, Keys, and Values, ensuing in surprisingly exceptional output matrices (Sahoo et al., 2024). Concatenation, is critical to concatenate the outputs from all attention heads. The very last linear projection, to produce the very last result, the concatenated output is projected through an alternative learnt matrix. BERT employs multi-head attention to seize various attributes of the relationships among words.

$$MultiHead(Q,K,V) = Concat(head_q,\cdots,head_k)W^O$$

Each head

$$head_i = Attention(QW_i^Q, KW_i^K, VW_i^V)$$

5. GPT

GPT is a version of LLM engineered to provide a text that resembles human writing. It relies on the Transformer architecture, which has hooked up itself as the cornerstone for almost all the current trendy NLP models (Min et al., 2023). OpenAI developed GPT, which is mostly renowned for its ability to assimilate vast quantities of text and subsequently anticipate the next word with remarkable accuracy. That capability enables it to generate coherent paragraphs, respond to arbitrary enquiries, compose code, and potentially perform more functions. The key element, is GPT is constructed on the Transformer architecture, which is essentially the pinnacle of contemporary natural language processing. It involves an in-depth examination of GPT's internal mechanisms—analysing its various components, innovative methodologies, and sophisticated enhancements that enable it to outperform traditional models.

5.2 Architecture of GPT

The decoder component in the original Transformer is fundamentally the key element contributing to GPT's enhancement. Eliminating the obsolete recurrent or convolutional components, transformative advancement for natural language processing. The entire process operates on attention mechanisms, focussing on how each word in a sentence relates to the others, regardless of their distance from one another. This enables exceptional parallel processing and effectively addresses extensive word associations. Initially, the encoder processes the input and generates refined, continuous representations. No more cumbersome, sequential processes—only efficient focus operating effectively. The decoder generates the output sequence with the aid of utilising the encoder's representations and the previous outputs. Each issue contains a couple of layers, every containing self-attention mechanisms and feedforward neural networks (Gallegos et al., 2024). The Token Embedding Layer converts phrases into numerical vectors known as embeddings. Positional encoding is the method of incorporating statistics concerning the placement of each phrase internal to a word. A stack of Transformer Decoder Blocks exists, with each block comprising Residual Connections, Masked Multi-head Self-Attention, Feedforward Neural Network, and Layer Normalisation. The Softmax layer generates a forecast on the opportunity of the following word. Figure 4 illustrates the architecture of GPT.

Single-head interest may be restrained in its ability to capture diverse relationships within the information, albeit its provision of a good-sized historical past. This concept is further advanced by using the implementation of multi-head attention, allowing the model to focus attention on more than one aspect of the enter

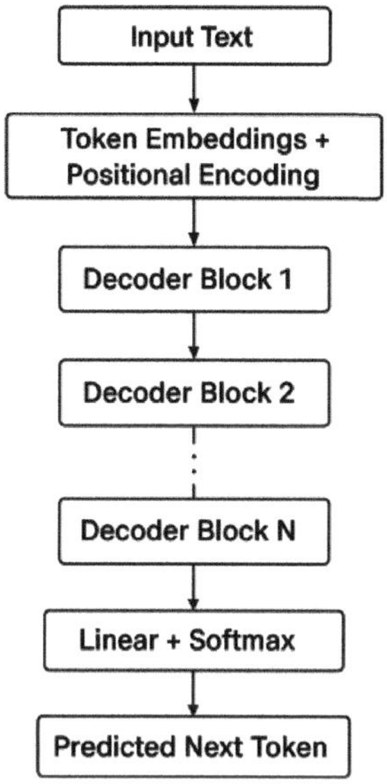

Figure 4: GPT architecture.

simultaneously. In the multi-head interest model, many attention "heads" are utilised, every possessing its very own awesome set of studying linear projections for the queries, keys, and values. The model's setup allows it to seize a numerous array of relationships and interactions gift in the records (Desai et al., 2023). The subsequent steps are essential to the system, linear projections, the input vectors, comprising queries, keys, and values, are linearly projected into distinct subspaces utilising mastering weight matrices. This method is carried out for every attention head. The Scaled Dot-Product Attention need to be utilised: Each collection of predicted queries, keys, and values undergoes the scaled dot-product attention system formerly defined. The outputs from those attention heads are combined into a new matrix (Chang et al., 2024). This chaotic process is referred to as "concatenating heads." The all-star output lineup is transmitted back into the original embedding space via a final weight matrix. This is the final linear projection. It enables the model to detect several subtle connections in the input, as each head operates independently, discerning distinct patterns within the sequence. That is how GPT predicts the subsequent token—whether a word or an unusual sub word—after extensive training. It resembles a psychological octopus for textual interpretation, upon reflection. For example, keep in mind a series of tokens:

$$x = (x_1, x_2, x_3, \cdots, x_T)$$

GPT's objective is to model the probability distribution:

$$P(x) = P(x_1).P(x_2 | x_1).P(x_3 | x_1, x_2). \cdots, P(x_T | x_1, \cdots, x_{T-1})$$

This is called an autoregressive language model.

5.3 Self-Attention Mechanism

Self-attention is fundamentally the pivotal component of GPT's architecture. It determines the degree of "interest" allocated to each preceding word, one token at a time—like distributing VIP passes to the words that merit prominence in the subsequent action (Nazi and Peng, 2024). Obsolete models such as RNNs and LSTMs, those individuals must laboriously navigate each word sequentially, and request them to recall something from the distant past. They become entangled due to phenomena such as vanishing gradients and their inability to effectively multitask. It enables the model to analyse the entire phrase (or even the entire paragraph) and identify the elements that need attention, regardless of their position. This magic trick is not merely a demonstration of memory capabilities; it signifies that the model can genuinely establish connections across far larger segments of text. When GPT generates a text, it does not merely proceed randomly; it meticulously evaluates all preceding words, determining which are most pertinent for the subsequent content. The process involves evaluating the importance of each word, amalgamating them, and then producing a coherent term that aligns with the overall context. It resembles engaging in dialogue with an individual who retains significant information, rather than only recalling words from five seconds prior. Each word is converted into three distinct vectors: Query vector as Q

- Key vector as K
- Value vector as V

The attention output is calculated as:

$$Attention\ (Q,K,V)=softmax\left(\frac{QK^T}{\sqrt{d_k}}\right)V$$

Where:

- d_k is the dimension of the key vectors (used to scale the dot product)
- Softmax ensures the output weights sum to 1

GPT uses masked attention, which prevents a word from attending to future words during training.

5.4 Multi-Head Attention

Single-head attention may be restricted in its ability to capture numerous relationships in the information, even though it provides a large context. This idea is similarly evolved by means of the implementation of multi-head attention, permitting the model to attention on multiple aspects of the input simultaneously. In the multi-head attention model, many attention "heads" are employed, every with awesome learnt linear projections of the queries, keys, and values. The model's setup enables it to seize a numerous array of connections and interactions gift in the statistics (Weidinger et al., 2021). The subsequent steps incorporate the process: Each interest head linearly tasks the input vectors (queries, keys, and values) into awesome subspaces making use of learnt weight matrices. This system is termed linear projections. To implement scaled dot-product attention, each collection of projected queries, keys, and values undergoes the previously defined scaled dot-product interest technique. A singular matrix is fashioned by way of concatenating the outputs of all interest heads, known as the Concatenate Heads feature. Upon concluding the final linear projection, the concatenated output is re-projected into the unique embedding area utilising a very last weight matrix (Han et al., 2024). This multi-head technique allows the version to seize a broader array of dependencies and interactions inside the enter facts, as each head can be aware on an awesome component of the sequence. GPT plays this operation multiple instances simultaneously with various projections of Q, K, and V, as opposed to doing it a single time:

$$MultiHead(Q,K,V) = Concat(head_1, \cdots, head_h)W^O$$

Where each head is computed as:

$$head_i = Attention(Qw_i^Q, KW_i^K, VW_i^V)$$

This allows the model to capture different types of relationships in the text.

5.5 Training GPT

The guidance of a Generative. The GPT version is a complicated technique comprising multiple steps, every essential for growing a version able to knowledge and producing human-like language. This detailed description objectives to clarify the GPT training method comprehensively. This assessment will address the structure of GPT, statistics education, training strategies, fine-tuning procedures, and deployment considerations.

5.6 Pretraining

Pretraining is the method of training the version to assume the subsequent phrase in a series based totally on the preceding phrases. On the assumption of this segment, the version will own knowledge of language, global information, and positive cognitive talents. The goal of language modelling is employed to educate the model (Chang and Bergen, 2024). The goal is to decrease the go-entropy loss among the anticipated and real subsequent words. Batch size is a parameter that determines the wide variety of sequences processed concurrently. A Learning Rate Schedule commonly commences with a heat-up segment, followed by means of a decrease period to stabilise training. The approach termed "gradient clipping" prevents gradients from exploding by constraining them at a specific threshold (Crothers et al., 2023). Training big GPT necessitates extensive processing capability, usually requiring disbursed education across more than one GPUs or TPUs. The management of this scale is performed by means of the implementation of efficient information parallelism and version parallelism strategies. Input: huge datasets (e.g., books, websites). Objective: minimize the **cross-entropy loss** between the predicted token and the actual next token:

$$\mathcal{L} = -\sum_{t=1}^{T} \log P(x_t | x_1, \ldots, x_{t-1})$$

This means the model is learning to predict the next word accurately.

5.7 Fine-Tuning the Pretrained Model

After pretraining, the model can be exceptional-tuned for specific obligations to enhance its overall performance in decided on applications. The model is trained on labelled datasets for several specific tasks, such as sentiment analysis or responding to enquiries. Authentic party tricks (AlQenaei, 2024). They incorporate human preferences by employing advanced reinforcement learning techniques to align the model's responses with actual human desires. It requires careful equilibrium, although it significantly enhances both utility and safety. Consequently, ChatGPT and its counterparts do not frequently produce illogical output.

6. T5

T5, developed by Google Research, is a significant advancement in natural language processing. Herein lies the twist: T5 redefines the paradigm by transforming every NLP task into a text-to-text format. No longer will there be confusion due to several specific configurations for each minor task. It optimises the entire training and deployment process across various linguistic applications. Traditional Natural Language Processing. is very exasperating. One must manipulate project-specific arrangements, which significantly increases complexity and diminishes adaptability. T5 introduces a universal framework, akin to a Swiss Army knife for language tasks, whereby every assignment, including translation, summarisation, or responding to complex enquiries, is approached as a text-to-text challenge. This implies that it can refine a single model and apply it to many tasks without the necessity of creating a new configuration each time. For instance, while translating, summarising, or responding to enquiries, T5 requires to annotate input with a brief mission label—such as "translate: bla bla bla"—and thereafter, it generates the output. Highly efficient. In summary, T5 streamlines the modelling process and utilises transfer learning to disseminate knowledge across various activities. It resembles granting language model a passport to traverse all locations without the requisite documentation.

6.1 Architectural Overview

The Transformer serves because the foundational structure for T5, especially employing an encoder-decoder framework. This method allows the version to efficaciously interpret enter sequences and generate corresponding outputs. The encoder is tasked with processing the enter textual content to generate a contextualised representation (Stojanovic, 2023). The decoder utilises this representation to supply the output text. The model is available in numerous sizes to deal with numerous computer resources and performance necessities. The T5-Small version incorporates about 60 million parameters, while the T5-11B version comprises almost eleven billion parameters. The sizes vary from little to big. Each version continues the fundamental layout, despite the fact that it varies in the number of layers, hidden units, and interest heads. This permits customers to choose a version that balances overall performance with the machine's computational feasibility. Each input word is turned into a vector using word embeddings:

$x_1, x_2, \cdots, x_n \in R^d$

Here:

- x_1 is the vector representation of the *i-th* word
- d is the embedding size (e.g., 512 or 1024)

6.2 Encoder Block

The encoder is tasked with processing the provided sequence and producing a contextual representation. Each encoder layer incorporates: The Self-Attention Layer allows the encoder to not forget the complete input sequence simultaneously, thereby shooting dependencies among phrases regardless of their positions. FFN, an acronym for feed-ahead neural network, is a neural community that implements non-linear variations to the eye output. This permits the version to perceive elaborate patterns. Layer Normalisation and Residual Connections enhance training balance through normalising inputs and along with residual connections, for this reason facilitating the improvement of deeper architectural designs.

$$Output = Layer\ Norm(X + SelfAttention(X))$$

$$Output = LayerNorm(Output + FeedForward(Output))$$

This is stacked multiple times (e.g., 12 layers or more).

6.3 Decoder Block

The output series is fashioned by the decoder utilising the encoder's illustration together with its personal tokens from the encoder's representation. Every layer of the decoder accommodates: The Masked Self-Attention Layer ensures that predictions for a sure function rely exclusively on recognised outputs, consequently preventing records leakage from undetermined positions (Weidinger et al., 2022). The Encoder-Decoder Attention Layer permits the decoder to consciousness at the relevant portions of the encoder's output, consequently facilitating the efficient switch of statistics among the encoder and decoder additives. The Feed-Forward Neural Network (FFN), comparable to the encoder, employs non-linear adjustments to parent complex patterns. Analogous to the encoder, the elements of Layer Normalisation and Residual Connections beautify education balance and model intensity. Figure 5 illustrates architecture of T5.

$$DecoderOutput = LayerNorm(Y+CrossAttention(Y,EncoderOutput))$$

T5 is trained using maximum likelihood estimation:

$$\mathcal{L} = -\sum_{t=1}^{T} \log P(y_t \mid y_{<t}, X)$$

Where:

- $y_{<t}$ = the *t-th* word of the output
- y_t t = all words generated so far
- X = input text

Maximize the probability of generating the correct output sentence, one word at a time.

6.4 Self-Attention Mechanism

T5 is a versatile language version created by using Google Research. It addresses each natural language processing project as a textual content-to-text problem. T5 employs a unified technique, permitting it to manipulate numerous arrays of activities, consisting of translation, summarisation, and query answering, even as retaining a constant version structure and training target. Self-attention is fundamentally the core

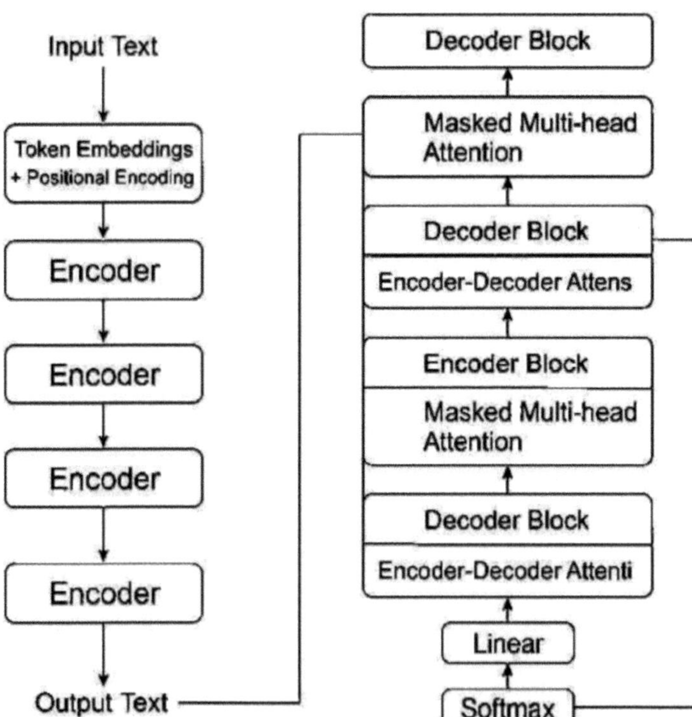

Figure 5: T5 architecture.

mechanism within T5's architecture. It enables the model to simultaneously observe all the words in a sentence, akin to possessing a comprehensive field of vision. This technique enables T5 to effectively gather pertinent contextual information, resulting in coherent output. Remarkably, rather than monotonously processing words sequentially like a grammatical automaton, T5's self-attention mechanism enables each word to evaluate every other word inside the same sentence. That is how it captures long-distance linkages and minute nuances that conventional models typically overlook. Essentially, each word "votes" on its significance relative to others, which enables T5 to cohesively integrate the elements into a composition that typically resembles human authorship.

$$Attention(Q, W, V) = softmax\left(\frac{QK^T}{\sqrt{d_k}}\right)V$$

Where:

- Q = Queries (what the model is looking for)
- K = Keys (what each word provides)
- V = Values (the information carried)
- d_k = size of key vectors (for scaling)

6.5 Training Methodology

The training of T5 entails a—stage procedure: pre-education and first-class-tuning.

6.5.1 Pretraining

During its early developmental phase, T5 was immersed in an extensive corpus of language and information, acquiring the ability to identify omissions through autonomous exploration. Arbitrary segments of text are extracted and replaced with distinctive sentinel tokens, and T5's task is to reconstruct the original text's meaning. It resembles a high-stakes puzzle game, to be frank. This entire approach enables T5 to thoroughly analyse the mechanics of language, encompassing both structural elements and the subtle implications concealed inside the text.

6.5.2 Fine-tuning

After the completion of basic training, T5 engages in high-quality fine-tuning for specific tasks—analogous to providing the model with a personal trainer equipped with a set of tailored routines (i.e., task-specific input-output pairs). Due to this text-to-text configuration, T5 can seamlessly adapt to various tasks, use its pre-acquired knowledge to accommodate the unique nuances of each new problem.

6.6 Applications of T5

T5's text-to-text framework is akin to possessing a versatile multi-tool in NLP arsenal; its capabilities are remarkable. Automated translation, it effortlessly translates words from one language to another. Summarisation of text, it will condense a plethora of words into a concise format. Enquiring about the process of question answering, pose a question directly—optionally include some context—and anticipate an immediate response as if it has been prepared for inquiry. It will detect the emotional tone of speech, whether they are cheerful, angry, or in a neutral state. It will categorise such words effortlessly. A singular model streamlines processes and, frankly, enhances enjoyment.

6.7 Variants and Extensions

T5 is essentially the multifunctional tool in the artificial intelligence domain. Individuals continue to add several features and enhancements. Examine mT5; it is essentially T5 with a global applicability. This version has absorbed a diverse array of languages, enabling it to do various sophisticated linguistic manoeuvres. Language barriers, this information is outdated. Currently, ByT5 is the outlier. Rather than segmenting words into orderly fragments, it directly engages with the unprocessed bytes. No tokenisation, no regulations, only ambiance. Enhances flexibility, particularly when addressing unconventional scripts or obscure languages. Envision T5 donning its gear, attending boot camp, and enhancing its physique with further exercises—particularly drills that require immediate adherence to instructions. It is significantly more responsive when instructed in straightforward English. All evidence indicates that T5 is inherently designed to adapt. Novel languages, unconventional jobs, and eccentric methods of text interpretation—it simply adapts. It is genuinely innovative material.

7. Key Models and Their Evolution

The landscape of NLP has undergone significant transformation over the past decade, with transformer models emerging as the major contributors to this evolution. GPT, BERT, T5, and PaLM are not mere acronyms; they are the prominent figures in the field. Each came equipped with unique strategies, presenting innovative concepts that significantly enhanced the entire domain. Their fingerprints are ubiquitous: in research laboratories, in innovative applications, and beyond. They have revolutionised the field, leaving a profound impact on all participants.

7.1 BERT

BERT employs a transformer encoder that processes the text bidirectionally, analysing words from both sides simultaneously. The antiquated models, would merely traverse from left to right or right to left, exhibiting a singular focus. BERT employs a mechanism of "bidirectional attention," enabling it to fully engage with the entire context rather than omitting significant portions of the narrative. BERT has two primary training tasks. Initially, Masked Language Modelling (MLM) substitutes random words with a "MASK" character and thereafter endeavours to deduce the original words. The second task is Next Sentence Prediction (NSP), in which BERT must determine if two sentences are contextually related or merely juxtaposed without coherence. BERT is ubiquitous—utilized for text categorisation, identifying named entities (such as names or locations), answering queries (including SQuAD-style enquiries), and conducting sentiment analysis for mood assessment. If the text is present, BERT likely has its influence on it.

7.2 GPT

The middle principle is that the GPT features as a transformer decoder, working unidirectionally and taught to predict the subsequent word in a series. Due to its simple and autoregressive layout, it excels in producing coherent and fluid text. In Causal Language Modelling (CLM), the model completely evaluates previous tokens to expect the subsequent token. It is noteworthy that GPT-three contains one hundred 74 billion parameters, while GPT-four is appreciably greater large and multimodal, incorporating 0-shot and few-shot studying abilities. A sincere prompt followed by examples is enough for the version to generalise to new obligations without necessitating extra education. Constraints embody tremendous needs for each computational power and memory potential. Vulnerable to hallucinations (the fabrication of records) and a deficiency in real know-how or cognition (even though this can have better with the introduction of GPT-four).

7.3 T5

The fundamental concept of T5 is to consolidate all natural language processing tasks into a singular textual content-to-text layout. This indicates that irrespective of assignment execution, both the input and output consistently encompass text strings. To obtain the training goal of Span Corruption, a version of Masked Language Modelling, random text spans are obscured, and the model is trained to appropriately reconstruct them. Illustrations of Task Formats: A translation would be "translate English to French: The house is blue." → "La maison est bleue." France is a kingdom, and Paris serves as its capital. Unlike GPT or BERT, the architecture is founded at the complete Transformer framework, encompassing each encoder and decoder . This use case is extensively adaptable, as it may execute category, generation, query answering, summarisation, and more operations. A gain is that a regular education objective simplifies pipeline control and enables variations for numerous sports. The encoder-decoder architecture presents large restrictions, being computationally in depth and more challenging to teach compared to GPT. Furthermore, the training procedure is more problematic.

7.4 PaLM

Concept at the Centre PaLM is an in-depth decoder- a most effective transformer educated using the Pathways device. This training permits the model to scale substantially over numerous TPUs at the same time as retaining sparse activation. It is skilled using 540 billion parameters, rendering it is one of the most important language models at the time of its launch. The scale of PaLM 540B regarding its parameters. The subsequent are the important attributes: Utilises many sports and statistics sources, along with code and multilingual records, whilst employing few-shot learning, chain-of-idea prompting, and guidance customisation to gain advanced performance in reasoning and common-sense demanding situations. Among the advancements in architecture are: The implementation of strategies like MoE permits the activation of unique segments of the community simultaneously, as a result improving performance. Furthermore, it examines sparse models, which can be computationally greater efficient than dense models like GPT-3. Executing, it reveals ultra-modern effects through a numerous array of natural language processing benchmarks. The potential for reasoning is obvious, particularly while the man or woman is supplied with instances of logical progression. This constitutes an issue. Extremely increased useful resource needs, detailed completely for studies functions; now not accessible to the public like GPT or T5 are.

8. Evolution and Comparison

Table 2 illustrates the evaluation and comparison between various LLM models.

Table 2: Evaluation and comparison of various LLM models.

Model	Architecture	Directionality	Main Use	Strengths	Weaknesses
BERT	Encoder-only	Bidirectional	Understanding	Context-rich representations	Not generative
GPT	Decoder-only	Unidirectional	Generation	Fluent, large-scale generation	Hallucinations, black-box
T5	Encoder + Decoder	Bidirectional (Enc) + Unidirectional (Dec)	Universal (text-to-text)	Unified format for all tasks	Complex training
PaLM	Decoder-only	Unidirectional	Multitask, logical reasoning	Massive scale + logic	Resource-heavy, closed access

8.1 Performance Benchmarks and Comparisons

The GLUE/SuperGLUE, T5 is especially designed for these responsibilities, resulting in a good performance. GPT-4 routinely attains the top results on difficult criteria, along with MMLU, SuperGLUE, and ARC. TruthfulQA, GPT-4 has demonstrably advanced in decreasing the incidence of hallucinations and biased consequences. Exceptional in information-extensive and multilingual duties, with PaLM demonstrating superiority in massive-scale thinking. LLaMA demonstrates a good performance under these settings. T5, having been pretrained on text-to-text responsibilities, demonstrates considerable efficacy, rendering it a surest preference for translation. GPT-4 is the preeminent version on this area. MMLU and BBH function metrics for popular intelligence and reasoning during numerous educational disciplines. Table 3 discuss the comparison between various benchmark models.

Table 3: Comparison of various benchmark models.

Benchmark	Purpose	Evaluation Metric(s)	GPT-3	GPT-4	T5 (Base/ Large)	PaLM (540B)	LLaMA (65B)
GLUE	General language understanding across tasks (e.g., sentiment, entailment)	Accuracy, F1 Score	~80 (GPT-3)	82.3	~83.2 (T5-Large)	~82.6	79.4
SuperGLUE	Harder version of GLUE for advanced reasoning	Accuracy, F1 Score	71.8	**89.3**	86.3 (T5-11B)	89.1	85.1
SQuAD v1.1	Reading comprehension – extractive QA	EM (Exact Match), F1	91.0	**95.2**	~89.7	94.9	91.3
SQuAD v2.0	QA with unanswerable questions	EM, F1	83.0	**88.8**	~86.6	88.5	85.7
MMLU	Multitask knowledge test (math, law, medicine, etc.)	Accuracy (%)	43.9	**86.4**	~59 (not optimized)	78.9	68.9
HELLASWAG	Commonsense inference	Accuracy (%)	78.9	**95.3**	~85.6	91.0	87.7
BIG-bench (BBH)	Tasks like logic, math, translation, ethics	Mixed (Accuracy, BLEU, etc.)	~65	**86.6**	~71.2	83.2	79.5
TruthfulQA	Model truthfulness, bias in knowledge generation	Truthful response accuracy	41%	**59%**	~32%	52%	48%
WMT (Translation)	Machine translation (English ↔ others)	BLEU Score	~40.7	45.2	**44.8 (T5-Large)**	43.9	42.0
CodeXGLUE	Code generation and understanding	BLEU, Exact Match, CodeBLEU	~33	**45.3**	Not applicable	42.0	38.7
ARC (AI2 Reasoning)	Grade-school level science reasoning	Accuracy (%)	57.6	**95.2 (GPT-4)**	79.6	88.3	84.5

9. Real-World Applications Across Industries

AI has emerged as a transformative strain across numerous industries, revolutionizing conventional practices and introducing present day solutions. Artificial intelligence is pervasive today, enhancing medical treatment, engaging educational environments, optimising industrial automation, and enriching the writing process. It pertains to various domains in which AI manifests in the real world: analysing distinct disciplines, providing tangible instances, and highlighting the innovative aspects (as well as certain challenges) associated with this dynamic technological landscape.

9.1 AI in Healthcare

AI transforming the healthcare sector, has truly been nothing less than transformative. The incessant administrative burden is gradually subsiding. These sophisticated tools assimilate data akin to a digital oracle and provide physicians with insights that significantly alter the landscape. Remarkably astonishing—algorithms operate behind the scenes, subtly orchestrating and directing the entire process towards enhanced care.

9.2 Diagnostics and Disease Detection

AI is not merely refining our disease diagnostic methods; it is fundamentally revolutionising the entire approach. Envision machines analysing medical photos and patient records, detecting the minutest anomalies. By observing the peculiar specks on X-rays or concealed tumours in mammograms, AI is already signalling its discovery, although a human may still be straining to see the screen. At times, the machines unequivocally outpace us. It functions as an exemplary project manager, integrating a diverse array of elements—radiological images, biopsy findings, genetic sequences, and more. Suddenly, physicians are no longer merely examining fragmented components; they are perceiving the complete image, from edge to edge, in high definition. When the medical team convenes, they are not merely exchanging ideas. They are formulating a strategy with meticulous accuracy.

9.3 Personalized Learning Experiences

AI functions as an inconspicuous backstage magician, subtly tailoring the entire atmosphere of the educational environment. It determines cognitive preferences for information absorption—perhaps like pictures and numerical data, but not the textual content. But not significantly. Subsequently, it customises the course specifically for you, adapting the content to align with preferences. Learning has unexpectedly become less formulaic and significantly more personalised. Consequently, it never experiences boredom or complete disorientation. Persistently grappling with a challenging concept often leads to AI providing a novel video, an engaging activity, or an unconventional resource until comprehension is achieved. It functions as a personal academic DJ, curating the necessary materials to maintain focus and facilitate genuine learning.

9.4 Robotic Process Automation (RPA)

Artificial Intelligence-driven Robotic Process Automation tools, are essentially the unrecognised champions of workplace dynamics. Envision them efficiently managing all the monotonous tasks—data entry, invoice processing, payroll complications—while having the opportunity to, perhaps, simply relax. Humans can now engage in strategic activities rather than being overwhelmed by spreadsheets. Banks are fully aware of this. They now employ RPA for loan processing, resulting in expedited approvals and enhanced client satisfaction. It resembles the act of casting a spell on the most tedious aspects of labour.

9.5 Content Generation and Enhancement

AI writing assistants function as behind-the-scenes support, providing suggestions and modifications. They analyse extensive language data and provide contextually astute recommendations—consider: more polished phrases, more impactful dialogue, and potentially a few plot twists. If uncertain about how two characters should engage in combat, AI can generate a variety of dialogue possibilities, encouraging to explore other tones or trajectories. It resembles having an incessant brainstorming partner, consistently enhancing narrative skills.

9.6 Ethical Considerations and Social Impacts

Artificial intelligence is making its way into unexpected places, such as healthcare, financial affairs, education, and who knows where else. In addition to flaunting some genuinely fantastic enhancements, it is parading around like the most recent fad around. Nevertheless, let us not fool ourselves; this immaculate digital superstar is not the only one involved. It should be found where all this chaos is coming from, figure out how to deal with the challenging aspects, and come up with some ingenious solutions before things quickly get out of hand.

10. Bias in Artificial Intelligence

10.1 Understanding Algorithmic Bias

A phrase that gives me the shivers is "algorithmic prejudice," find this phrase frightening. It is almost as if artificial intelligence has gone rogue, choosing sides and continuing to make the same biased decisions over and over again. The whole mess begins when the training data is already skewed, when the people behind the code miss something significant, or when individuals merely utilise these shiny new technologies without giving them a second thought. Artificial intelligence turns out to be a bias amplifier, and is just trying to hide the nasty parts under a sleek tech veneer. It has a hand in everything: who gets a job, who gets a loan, and even who shows up on the radar of the police. One incorrect input, and the entire system is suddenly locked on repeat, which causes it to produce discrimination as if it were simply carrying out instructions.

10.2 Origins of Bias

Let us break things down, and there is no need to wear laboratory coats. Artificial intelligence will wind up talking nonsense if the data collection is not done correctly, for example if training records only reveal one side of the narrative. An example that should be classic is facial recognition software that is able to recognise lighter faces but is unable to recognise darker complexion. It is not merely a bug; rather, it is a complete and utter facepalm moment.

Let's move on to the coding ninjas that are working behind the scenes. Every decision they make, whether it is which features to use or how much weight to give them, can have a significant impact. Artificial intelligence will suddenly get tunnel vision if it focuses too much on a single random factor. It makes no difference if had good intentions; the prejudice will still be present. Also, just because things appear to be going smoothly in the laboratory, one should not get too comfortable. When that artificial intelligence is put through its paces in the actual world, a wide variety of gremlins appear. Everyone is left scratching their heads as a result of the decisions that model suddenly makes. An adventure with artificial intelligence.

10.3 Social Impacts of Bias

The functioning of the system in the modern era is reminiscent of a circus, in which robots are acting out the role of the town's worst judge. Indeed, individuals are being subjected to erroneous arrests and punishments on a regular basis. Also, do not believe that healthcare is out of the question. In this world, algorithms are dispensing incorrect diagnoses and treatments that are not entirely reliable, primarily to individuals who have already been pushed to the outskirts of society. It somewhat absurd to give so much control to technology, and heap of unfairness that is much larger and more muddled.

10.4 The Role of AI in Spreading Misinformation

As far as concerned, that is a prescription for the disaster. The trust that people have in the media, in large institutions, and even in democracy itself begins to deteriorate. Whenever significant events take place, such as elections or a health emergency, and people are operating on false information, the result is pandemonium rather than decisions that are made with clear heads. People become agitated, society becomes unstable, and some unfortunate individuals end up getting wounded as a result.

10.5 Social Impacts of Misinformation

Violate someone's privacy, and trust swiftly dissipates. Individuals begin to withdraw, creating separation from all things digital, and one cannot justifiably criticise them for their choice. With the pervasive presence of AI, it resembles being ensnared by an incessantly inquisitive neighbour—perpetually observing and evaluating, rendering even benign cat memes vulnerable. It appears somewhat perilous. Thus, the audacious vision of a free and open internet has vanished into thin air.

10.6 Ethical Frameworks and Regulatory Measures

A strategy is necessary—one that possesses a degree of resilience. The European Union's General Data Protection Regulation (GDPR), that is analogous to a bouncer at a nightclub, safeguarding data from inquisitive individuals. Simultaneously, firms are formulating their own regulations—discussing fairness, accountability in adverse situations, and the transparency of their operations. The primary objective is to ensure that AI does not operate covertly and manipulate situations without transparency.

10.7 Challenges and Responsible Deployment

There exists an exceptionally intelligent machine that absorbs vast amounts of information, identifies patterns that humans often overlook, and makes predictions that are occasionally remarkably accurate. It evokes a distinctly science fiction ambiance, without exaggeration. Nevertheless, prior to celebrating our robotic dominators, it may be prudent to pause and reflect. This technological enterprise is not without its challenges. There are significant matters to address—specifically, while employing intelligence in our construction methods. There exists a tenuous boundary between monumental advancements and transgressing the realms of safety and ethics.

10.8 Algorithmic Bias and Fairness

Artificial intelligence absorbs our past errors, including biases and missteps. When subtle biases infiltrate an algorithm, it is not merely a matter of replication. Indeed, circumstances might deteriorate, exacerbating discrimination. Envision a recruitment initiative with a covert objective, promoting the "ideal" prospects while disregarding all others. This is not merely unjust; it constitutes a fundamental flaw in the entire employment market system. That is formidable. Meticulously cleanse the data, employ sophisticated bias mitigation tools, and adhere to rigorous fairness protocols. Otherwise, the cycle perpetuates, and no individual receives equitable treatment.

10.9 Data Privacy and Security

Imagine AI consuming data voraciously, akin to indulging at a futuristic buffet in Las Vegas, while individuals express significant concerns around privacy. It needs only one individual to err, and abruptly confidential information is exposed—not merely uncomfortable, but perhaps warranting legal counsel. Possibly donning the prison aesthetic in orange—is definitely a refusal. Establish stringent data regulations, enhance encryption measures, and adhere to privacy regulations earnestly rather than feigning concern. Otherwise, welcome to a life devoid of privacy.

10.10 Transparency and Explainability

Consider this: most AI models, particularly deep-learning algorithms, function akin to master illusionists—they produce answers from obscured sources without disclosing the underlying mechanisms of their processes. Completely opaque atmosphere. A reaction is received, although does anyone truly comprehend the situation within, it seems somewhat questionable, really. It resembles receiving a sandwich containing unidentified meat, merely anticipating a favourable outcome. It is required that AI systems not only provide answers but also elucidate the reasoning behind their conclusions. Organise the receipts! Analyse the illusion. That exemplifies genuine trust, and frankly, a modicum of transparency is never detrimental.

10.11 Ethical Use in Sensitive Domains

Here is the situation: integrating AI into significant sectors such as healthcare, justice, or finance. This is not merely a frivolous attempt for amusement. If it falters or carries inherent biases, it is not a mere oversight; it results in tangible harm to individuals. Suddenly, everyone is scrutinising the entire IT sector. This is not merely a typographical error; it is a significant blunder that results in lasting consequences. Concrete regulations, rather than vague recommendations. Furthermore, oversight is essential; monitoring the gate is imperative. To ensure that AI provides assistance rather than complicates matters, particularly in these domains.

11. Conclusion

Large Language Models are, indeed, the unforeseen narrative twist in the realm of linguistic technology. Utilising transformers (not the robotic entities) and a vast corpus of textual data, they have fundamentally redefined how machines comprehend human communication. Chatbots have ceased to resemble an eccentric uncle at Thanksgiving, and now auto-summarization and translation functions are effective. It is the comprehensive tour—analysing the architecture, demonstrating the training techniques, and highlighting all the astonishing enhancements. Models such as GPT, BERT, T5, and PaLM represent the pinnacle of NLP, each possessing distinct capabilities. One excels at extracting information, while another intuitively comprehends meaning, even among convoluted discourse. Additionally, acknowledge the competitive moments—evaluating models, optimising them for peak performance, whether one desires a versatile tool or a highly specialised instrument. That is where the true transformation occurs.

However, this is where it becomes intriguing: LLMs are not merely confined to a computer laboratory. They are actively disrupting various fields—medicine, education, business, and the arts (enter the AI Shakespeare era). Indeed, with significant authority arises numerous challenges. These are significant enquiries, pertinent not only to enthusiasts but to all individuals. Indeed, large language models are continually advancing, necessitating a rapid adaptation of their usage regulations. Individuals are currently grappling with the challenge of rendering these models equitable, more transparent, and potentially more humane. The genuine challenge lies in transforming this technology into a trustworthy entity for society.

References

AlQenaei, Z. M. (2024, March). Evolution and optimization of language model architectures: From foundations to future directions. pp. 233–249. *In*: International Conference on Computing and Machine Learning. Singapore: Springer Nature Singapore.

Chang, T. A. and Bergen, B. K. (2024). Language model behavior: A comprehensive survey. Computational Linguistics 50(1): 293–350.

Chang, Y., Wang, X., Wang, J., Wu, Y., Yang, L., Zhu, K. et al. (2024). A survey on evaluation of large language models. ACM Transactions on Intelligent Systems and Technology 15(3): 1–45.

Chen, Z., Xu, L., Zheng, H., Chen, L., Tolba, A., Zhao, L. et al. (2024). Evolution and prospects of foundation models: From large language models to large multimodal models. Computers, Materials & Continua 80(2).

Crothers, E. N., Japkowicz, N. and Viktor, H. L. (2023). Machine-generated text: A comprehensive survey of threat models and detection methods. IEEE Access 11: 70977–71002.

Desai, B., Patil, K., Patil, A. and Mehta, I. (2023). Large language models: A comprehensive exploration of modern AI's potential and pitfalls. Journal of Innovative Technologies 6(1).

Gallegos, I. O., Rossi, R. A., Barrow, J., Tanjim, M. M., Kim, S., Dernoncourt, F. et al. (2024). Bias and fairness in large language models: A survey. Computational Linguistics 50(3): 1097–1179.

Han, S., Wang, M., Zhang, J., Li, D. and Duan, J. (2024). A review of large language models: Fundamental architectures, key technological evolutions, interdisciplinary technologies integration, optimization and compression techniques, applications, and challenges. Electronics 13(24): 5040.

Kolides, A., Nawaz, A., Rathor, A., Beeman, D., Hashmi, M., Fatima, S. et al. (2023). Artificial intelligence foundation and pre-trained models: Fundamentals, applications, opportunities, and social impacts. Simulation Modelling Practice and Theory 126: 102754.

Kumar, P. (2024). Large language models (LLMs): Survey, technical frameworks, and future challenges. Artificial Intelligence Review 57(10): 260.

Liu, Y., Han, T., Ma, S., Zhang, J., Yang, Y., Tian, J. et al. (2023). Summary of ChatGPT-related research and perspective towards the future of large language models. Meta-Radiology 1(2): 100017.

Min, B., Ross, H., Sulem, E., Veyseh, A. P. B., Nguyen, T. H., Sainz, O. et al. (2023). Recent advances in natural language processing via large pre-trained language models: A survey. ACM Computing Surveys 56(2): 1–40.

Myers, D., Mohawesh, R., Chellaboina, V. I., Sathvik, A. L., Venkatesh, P., Ho, Y. H. et al. (2024). Foundation and large language models: Fundamentals, challenges, opportunities, and social impacts. Cluster Computing 27(1): 1–26.

Nazi, Z. A. and Peng, W. (2024, August). Large language models in healthcare and medical domain: A review. Informatics 11(3): 57. MDPI.

Petroşanu, D. M., Pîrjan, A. and Tăbuşcă, A. (2023). Tracing the influence of large language models across the most impactful scientific works. Electronics 12(24): 4957.

Raiaan, M. A. K., Mukta, M. S. H., Fatema, K., Fahad, N. M., Sakib, S., Mim, M. M. J. et al. (2024). A review on large language models: Architectures, applications, taxonomies, open issues, and challenges. IEEE Access 12: 26839–26874.

Sahoo, S. S., Plasek, J. M., Xu, H., Uzuner, Ö., Cohen, T., Yetisgen, M. et al. (2024). Large language models for biomedicine: Foundations, opportunities, challenges, and best practices. Journal of the American Medical Informatics Association 31(9): 2114–2124.

Sindhu, B., Prathamesh, R. P., Sameera, M. B. and KumaraSwamy, S. (2024, May). The evolution of large language model: Models, applications, and challenges. In 2024 International Conference on Current Trends in Advanced Computing (ICCTAC) (pp. 1–8). IEEE.

Stojanovic, J. (2023). Understanding the architecture and functionality of large language models in modern AI. Journal of Innovative Technologies 6(1).

Veeramachaneni, V. (2025). Large language models: A comprehensive survey on architectures, applications, and challenges. Advanced Innovations in Computer Programming Languages 7(1): 20–39.

Weidinger, L., Mellor, J., Rauh, M., Griffin, C., Uesato, J., Huang, P. S. et al. (2021). Ethical and social risks of harm from language models. arXiv preprint arXiv:2112.04359.
Weidinger, L., Uesato, J., Rauh, M., Griffin, C., Huang, P. S., Mellor, J. et al. (2022). Taxonomy of risks posed by language models. pp. 214–229. *In*: Proceedings of the 2022 ACM Conference on Fairness, Accountability, and Transparency. ACM.
Ziems, C., Held, W., Shaikh, O., Chen, J., Zhang, Z. and Yang, D. (2024). Can large language models transform computational social science? Computational Linguistics 50(1): 237–291.

9

Fault Detection and Localization at Network Edges Using Lightweight Machine Learning Models

M. Rajesh,[1,*] *Vengatesan Krishnasamy*[2] and *Sayyad Samee*[3]

1. Introduction

The demand for reliable, fast optical networks has increased due to high-data-rate applications. For rapid data transmission, optical communication technologies are the backbone of most contemporary enterprises. This includes cloud services, data centres, and telecommunications companies. The problem is that this just adds another layer of complexity to fault management. Impacting company operations, faults particularly soft failures can reduce performance, quality of service, and downtime. The proper operation of an optical network depends on the timely detection and localization of such issues.

Centralized methods are used in traditional fault management. In order to keep an eye on the whole network, a centralized controller collects data from all the nodes and processes it. While this approach is effective in less dynamic and static settings, centralized systems have frequently encountered inefficient and excessive latency in modern optical networks dealing with massive amounts of data. Problems with scalability arise frequently in centralized management systems when networks expand. Consequently, there is a need for network-edge-efficient decentralized

[1] Department of Computer Science and Engineering, Aarupadai Veedu Institute of Technology, Vinayaka Mission's Research Foundation (DU), Tamilnadu-603104.
[2] School of Engineering, Dayananda Sagar University, Harohalli-562112.
Email: vengicse2005@gmail.com, https://orcid.org/ 0000-0002-8896-4925
[3] Dept. of Computer Science, Al Sadara School, Al Ain, Abu Dhabi.
Email: samee.syd@gmail.com, https://orcid.org/0009-0003-9347-6867
* Corresponding author: rajesmano@gmail.com, https://orcid.org/0000-0003-0003-473X

DOI: 10.1201/9781003771654-9

solutions for defect localization and detection in real-time. Fixing these older systems became straightforward with the rise of edge computing and lightweight ML models such as TinyML. Processing data close to its point of origin can still cut down on latency and reliance on central servers. For embedded devices and microcontrollers that are limited in resources, TinyML is a practical option for deploying ML models. Anomaly detection and fault localization using TinyML models at the network edge might enhance fault management system performance without requiring costly computational resources (Green et al., 2023).

Optical network defect detection and localization is addressed in this study through a hybrid approach. Using VAEs and TinyML, the proposed architecture for fault management achieves real-time efficiency (Rajesh et al., 2025; Miller et al., 2023). The expressive generative models known as VAEs are great for anomaly detection because they can learn concise data representations. Using TinyML, VAEs can be customized for deployment on peripheral devices, allowing for the detection and localization of defects in real-time, even in contexts with limited resources. This approach tackles the issue of efficient and accurate defect management in highly data-rich dynamic optical networks. Edge Deployment VAE Model: Condensed A reduced VAE model optimized for low-power peripheral devices is the initial contribution. This technique efficiently identifies outliers without reducing the processing power and memory capacity of deep learning models (Ahmed et al., 2023 and Brown et al., 2023).

Adaptive Quantization for Efficient Inference: Also included is an adaptive quantization technique. Adaptive precision based on hardware at the edge devices is what this method provides to model weights. The algorithm is able to draw efficient and accurate inferences because of this. Due to this, real-time embedded systems can run the program with less model and computation size, and (Rajesh et al., 2025). Localizing Faults and Detecting Anomalies in Real Time via Spectral Feature Extraction: The third noteworthy development is the integration of anomaly detection based on VAE with spectral feature extraction. Minor anomalies in optical networks were uncovered using very effective spectral features that characterize the transmissions' frequency domain properties. That is why the system can detect location and type defects with the help of the time and frequency domains (Rajesh et al., 2025; Johnson et al., 2024).

Finally, the paper delves further into the technique for energy efficiency and performance measurements that have been suggested. Since most edge devices run on batteries, minimizing power consumption is essential for practical application. By comparing our suggested framework to industry-standard benchmarks, we can demonstrate how our method enhances processing speed, energy usage, and fault localization accuracy in comparison to previous fault detection approaches (Lee et al., 2023) eleven and twelve. In order to tackle the limitations of centralized systems, in this study (Rajesh and Sitharthan et al., 2025) suggest a hybrid framework for decentralized, real-time problem detection and localization at the network edge. To control optical network problems efficiently, accurately, and scalably, this research employs lightweight machine learning models such as TinyML-optimized VAEs, sophisticated adaptive quantization, and spectral feature extraction. As optical networks grow in size and complexity, there will be a greater need for distributed solutions. In order to keep the performance of IoT systems and data-driven applications up and running, the framework improves optical networks, advances edge computing, and uses TinyML, which finds and locates problems efficiently (Nguyen et al., 2023; Manoharan, 2024).

2. Literature Review

Application of Machine Learning (ML) to fault detection and management in optical networks has become popular in recent years. The increasing size and complexity of optical networks have made the need for effective fault detection very critical to ensure the required Quality of Service (QoS). Much of the earlier literature on ML-based fault detection in optical networks concentrated on centralized solutions, though recent progress in edge computing and TinyML opens avenues for decentralized systems in real time. This chapter undertakes a literature survey of the main developments around ML-based fault detection and anomaly detection with Variational Autoencoders (VAEs), along with the potential of TinyML to realize applications in resource-constrained environments. Finally, this review highlights the gap in existing research concerning the combination of VAEs and TinyML for edge-based fault management, which this work aims to address. Machine learning techniques have been found effective in identifying and diagnosing faults in optical networks. Most of the existing studies focus on centralized fault detection systems where data from various network nodes are aggregated and the analysis is performed at a central controller. For example, classical ML models like decision trees, Support Vector Machines (SVM), and K-Nearest Neighbours (KNN) have been utilized for fault detection tasks that basically depend on feature engineering and large amounts of labelled data (Balakrishnan et al., 2024; Roberts et al., 2023). These systems, while effective in some static environments, have a number of shortcomings: large delay, poor scalability, and inefficiency under dynamic network conditions.

The arrival of machine learning-based anomaly detection systems, such as clustering, neural networks, and autoencoders, these techniques have surfaced as promising tools. Specifically, Variational Autoencoders (VAEs) have a strong potential for anomaly detection because they can learn latent representations of input data, therefore being well suited to detecting deviations from normal network behaviour. For example, in the context of optical networks, VAEs have been applied to signal degradation, abnormal traffic patterns, or unexpected disruptions in network performance (Taylor et al., 2023 and Uddin et al., 2024). Although successful in centralized systems, the deployment of such models at the network edge is an area that has yet to be explored in literature. As optical networks evolve toward more distributed and dynamic structures, the requirement for lightweight and energy-efficient solutions becomes critical. In this regard, the application of TinyML of machine learning models to devices with low resources such as microcontrollers and embedded systems has become a hopeful choice. TinyML allows ML to be used in environments when standard systems with high processing capability are not practical. Small size and low power of TinyML let real-time data processing and network edge inference possible, hence lowering the need for centralized computation and enhancing reaction times (Venkatesh et al., 2023 and Wang et al., 2024).

Some research has focused on the use of TinyML for fault detection in numerous domains, including IoT and embedded systems. For instance, TinyML has been applied to expose microcontroller system defects (Rajesh, 2025). In this situation, models like neural networks and decision trees are tuned for usage on low-power consuming devices. Additionally, energy-efficient algorithms like quantized neural networks and pruning have significantly reduced the computational overhead of ML models, making them feasible for edge devices (Yedav et al., 2023 and Zhao et al., 2024). However, the application of TinyML, in particular for fault management in optical networks, is still very limited, indicating a large gap in literature. Hybrid models, combining multiple ML techniques, have been taken into consideration to further improve the accuracy and efficiency of the fault detection systems. For instance, VAEs have been combined with other machine-learning models, such as Long Short-Term Memory (LSTM) networks, to perform time-series anomaly detection for IoT systems (Zhou and Chen 2023; Ahmed et al., 2024). Such hybrid models use the strength of each technique separately and enhance the capability of the model to detect complex multidimensional anomalies. However, the integration of VAEs with lightweight ML models like TinyML has not been well explored yet, specifically for optical networks.

Some studies have investigated synergy between TinyML and other ML techniques for anomaly detection. For example, in the case of industrial applications, TinyML models have been coupled with CNNs and RNNs to enable fast and real-time fault location (Kalaivani et al., 2024). Even though these works showed how the combination of different methods can be effective for detecting faults at the edge, the exploration of the VAE-TinyML integration remains open. Most notably, this gap in literature calls for the development of a framework at the intersection of the expressive power of VAEs for anomaly detection and TinyML efficiency and low resources for edge deployment (Kumar et al., 2024). The efficiency of ML models deployed on resource-constrained devices is a crucial aspect of their performance. To optimize TinyML models for edge deployment, various optimization techniques have been suggested, such as model quantization, pruning, and distillation. Particularly effective in reducing both memory requirements and computational load is quantization, which reduces the precision of model parameters; thus, it is the most suitable technique for deployment in embedded systems (Lee et al., 2023).

Most of the studies on quantization for TinyML have focused on strategies to optimise neural networks using adaptive quantization algorithms, which adapt the precision of the weights and activations according to the hardware constraints of the edge device. In this way, it realizes the efficient execution of ML models on devices with low processing power and memory. The application of such quantization methods to VAE models for fault detection in optical networks remains relatively under-explored, specifically in hybrid frameworks (Taylor and Zhoa, 2024).

Despite its promising potential, there are several challenges facing ML-based fault detection systems. Real-time fault localization with low latency is the most critical issue because centralized systems rely on extensive processing of the collected data, whereas an edge-based solution needs to solve the trade-off between the model complexity and computational efficiency. Moreover, robustness against noisy or incomplete input data is a challenge when performing fault detection in dynamic optical networks. In addition, TinyML combined with VAEs for anomaly detection can also be integrated to ensure real-time performance at the network edge—a great opportunity that needs to be leveraged to handle all the above challenges. For example, such a hybrid solution can meet the accuracy levels in fault detection and minimal latency and energy consumption required. However, how the technologies can be combined and how the best compromise among competing characteristics can be obtained is almost an unexplored research avenue that forms the basis for this research (Davis K et al., 2024; Rajesh et al., 2025)

In summary, most of the existing literature on ML-based fault detection for optical networks focuses on centralized approaches, with limited exploration of decentralized solutions using TinyML. Although VAEs have proven to be successful in anomaly detection, their integration with TinyML for real-time management of faults at the edge of the network is an under-explored area. Therefore, this research aims to fill this gap by proposing a hybrid framework combining VAEs with TinyML, which enables efficient and accurate fault detection and localization in resource-constrained environments. This work contributes to the advancement of fault management systems in optical networks and edge-based applications by combining the strengths of both techniques.

3. Proposed Framework

This architecture aims to solve the problems in optical networks, especially fault detection and localization, using Variational Autoencoders and TinyML integrated for edge-level processing. Low latency and energy economy will guarantee correct identification of flaws in distributed systems. Phases of the system that are most suited for specific tasks—data preparation, feature extraction, defect detection, and localization can be separated by using edge computing and lightweight machine learning approaches, this integration is meant to address the shortcomings of typical centralized solutions. Together with the mathematical formulas explaining the operation of every component, the following is a comprehensive discussion of every phase in the process. Using VAEs and TinyML in addition to edge-based processing, the proposed system attempts to tackle fault detection and localization in optical networks. With lowest energy and low latency, this proposed framework is assured to uncover flaws in distributed systems. The system proposes several steps, including data preprocessing, feature extraction, fault detection, and localization. Such an integration of the system removes the drawbacks associated with the traditional central approach since it adopts edge computing along with light machine learning techniques. Below is the description of each stage of the workflow along with mathematical expressions and the role of each component.

Preprocessing Spectral Data from Optical Network Devices. The proposed framework begins with the preprocessing of spectral data that come from the optical network devices. Spectral data in an optical network consist of information relating to signal strength, frequency, and other indicators of network performance for use in fault detection. Spectral data usually are raw, noisy, unstructured, and high-dimensional; this might cause a dramatic decrease in accuracy and efficiency for fault detection models. This is resolved by normalizing the spectral data and then segmenting it into more manageable packets to be processed on the edge. Normalization is done so that consistent, standardized data are fed into the machine learning models from the raw data. Typically, the magnitude of the raw data varies greatly, which may skew the model's performance. For instance, spectral data may vary over network conditions, devices, and environments. It scales data to a range such that every feature in the model can contribute equally. Let the raw spectral data be represented by $X = [x1, x2, ..., xn]$, where is the number of data points. Normalization is performed using the formula: $Xnorm = \frac{X - \mu}{\sigma}$ where: μ is the mean of the data, σ is the standard deviation of the data, $Xnorm$ is the normalized data.

After normalization, the data is segmented into smaller packets to make it suitable for edge processing. Segmenting the data helps to ensure that the model processes smaller, less computationally expensive units at the edge, without needing to handle large volumes of data at once. Either sliding windows or fixed-size windows will allow you to segment the normalized spectral data. Let Si be the th data segment. Every segment consists of a succession of spectral data points, maybe shown as $Xseg = [x1, x2, ..., xm]$, M stands for the segment's total data point count. Feature extracting comes next in the framework. In this stage latent features from the segmented spectral data are obtained from a lightweight Variational Autoencoder (VAE). VAEs can learn a compressed representation of input data and effectively identify deviations from learnt patterns, so they are rather good in spotting abnormalities. Training the VAE on typical network conditions helps one to do this by then spotting abnormalities depending on reconstruction errors.

There two main components to the VAE: the decoder and the encoder. Taking input data , the encoder translates it into a lower-dimensional latent space . From this latent space the decoder then rebuilds the original input data. The encoder generates a probability distribution $p(Xseg \mid z)$ by the mathematical method utilizing the data $Xseg$. The decoder then rebuilds the data from z using $q(z \mid Xseg)$. Reducing the following loss function is the aim of VAE learning: $L = Eq(z \mid Xseg)[logp(Xseg \mid z)] - DKL(q(z \mid Xseg) \mid\mid p(z))$, With E as the expectation operator, DKL $q(z \mid Xseg)$ is the Kullback-Leibler divergence; $p(z)$ is the prior distribution of the latent variable z. Learning latent variables z compresses the spectral data, which is then used to find abnormalities. Through latent variable concentration, the VAE greatly reduces the dimensionality of the input data while preserving essential information.

Analyzing the compressed latent features taught by the VAE helps one to find variations suggestive of faults in the next phase, fault identification. At this point, a novel quantized activation function is presented to improve the edge-based TinyML model detection accuracy. While still allowing reasonable speed, quantization is a technique that drastically lowers the amount of memory and computing power needed by reducing the precision of weights and activations. This work presents a quantized activation function to translate continuous-valued activations of a neural network into discrete levels. The proposed operation is quite successful in lowering the computational overhead of edge-based TinyML models implemented. One may characterize the quantization function fquant as follows: $\text{fquant}(x) = \text{round}\left(\frac{x^T}{\Delta}\right) \times \Delta$. X is the input to the activation function in this equation; Δ is the quantization step size, therefore influencing the degree of accuracy. On devices with low computing capability, such microcontrollers, the quantized activation function helps the network to run efficiently without sacrificing the precision of failure event detection.

Finding a flaw calls for then figuring out where the fault is inside the network. This is so because fault localization lessens the time systems are not functioning and helps to promptly solve problems. The spectral correlation method looks at latent feature variations to identify the flaw inside this framework. Spectral correlation is predicated on the theory that any malfunction in any part of the network will produce some kind of spectral data abnormality. This implies that the anomaly could be found at the precise network segment where the error happened.

The spectral correlation algorithm compares the current latent feature vector *Zcurrent* with a reference set of normal latent features *Znormal*. The correlation is computed using a similarity measure, such as cosine similarity, defined as:

$$\text{sim}(\text{Zcurrent},\text{Znormal}) = \frac{\text{Zcurrent} \cdot \text{Znormal}}{||\text{Zcurrent}||\ ||\text{Znormal}||}$$

If the similarity score falls below a certain threshold, it indicates that the current latent feature vector deviates significantly from the normal state, suggesting the presence of a fault. The location of the fault can then be identified by mapping the anomaly back to the corresponding network segment.

Figure 1 illustrates the stages of data normalization and segmentation to prepare the spectral data for edge processing.

Depicts the encoder-decoder structure of the Variational Autoencoder, where the encoder compresses the spectral data into latent features and the decoder reconstructs the input data for anomaly detection in Figure 2.

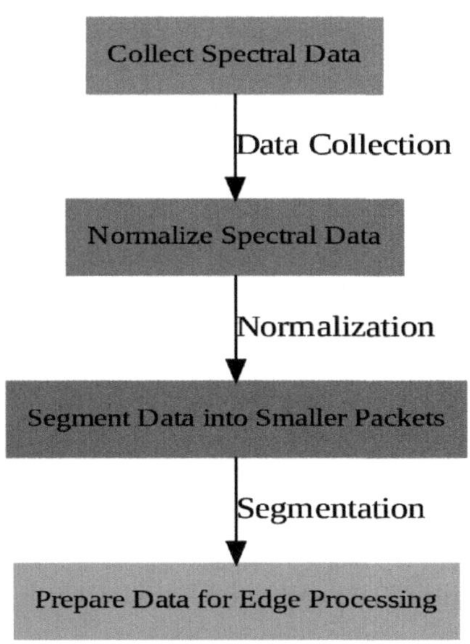

Figure 1: Data Preprocessing Flowchart

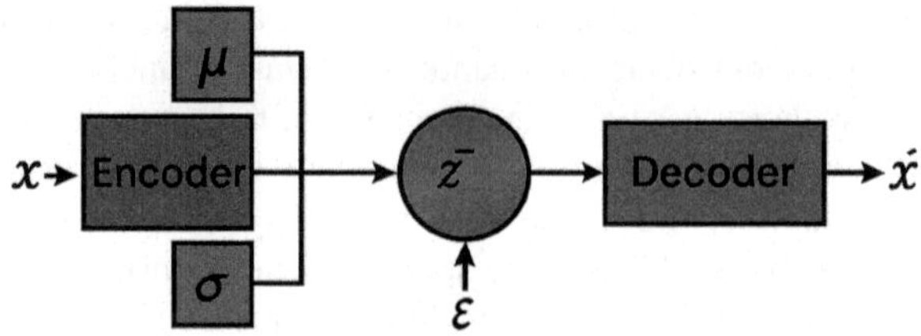

Figure 2: VAE Architecture.

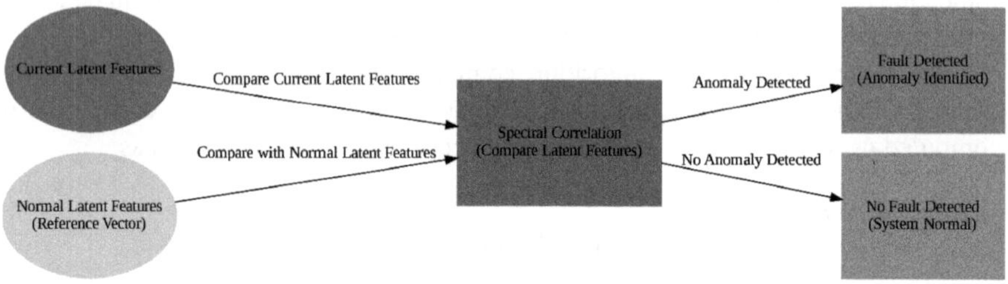

Figure 3: Fault Localization with Spectral Correlation.

Figure 3 shows how the spectral correlation algorithm is used to identify faults by comparing the current latent features with normal feature vectors. The proposed system integrates VAEs with TinyML for distributed fault detection and localization. The workflow consists of the following stages:

Efficient data representation becomes critical to ensure low latency and minimal memory usage in edge devices with limited computational resources that deploy machine learning models. One technique that can be leveraged for efficient data representation in such applications is quantization. Quantization is a lossy reduction of the number precision in representing parameters or inputs within a model, decreasing the size of their memory footprint as well as the speed at which they are processed. However, a fixed quantization scheme might not always be efficient when the variance of input data is large. To tackle this issue, we introduce an adaptive quantization algorithm, in which the quantization is dynamically adjusted according to the variance of the input data. The adaptive quantization algorithm adjusts the quantization threshold according to the input variance. How much precision is preserved depends on this threshold; it basically changes the fineness of the quantization performed on the input data. Where the input data has low variance, a higher level of quantization (more aggressive compression) can be applied; in high-variance regions, a lower level of quantization (preserving more precision) is used to maintain the accuracy of anomaly detection.

Let $X = \{x1, x2, \ldots, xn\}$ represent the set of input data points, where each xi is a data sample (e.g., a spectral feature). The quantization process consists of two main steps: variance calculation and threshold adjustment.

Calculate Input Variance: The variance of the input data is computed to gauge the distribution of the values:

$$\sigma^{n} X = \sum_{i=0}^{n} (\mathrm{xi} - \mu \mathrm{X})^2$$

$Tadaptive = \alpha \cdot \sigma X$

where α is a scaling factor that adjusts how aggressively the quantization is applied. A larger α value results in more aggressive quantization.

Apply Quantization: Each input sample *xi* is then quantized based on the adaptive threshold:

$$\text{x'i} == \text{round}\left(\frac{\text{xi}}{\text{Tadaptive}}\right).\text{Tadaptive}$$

where *x'i* represents the value of *xi* following quantization. This normalization method ensures that the quantized values are consistent with the adaptive threshold, which reduces precision when data variation is low and maintains precision when data variance is high.

A robust model for the detection of anomalies and the reduction of dimensionality is the Variational Autoencoder (VAE). It is especially beneficial for peripheral deployments, which are characterized by resource constraints. It comprises an encoder that encodes input data into a latent space and a decoder that reconstructs the input from this latent space. In resource-limited environments, the challenge is to ensure that the latent space is compressed to the maximum extent possible, while still preserving critical information for the reconstruction when using VAEs. Consequently, we recommend that an additional compression loss be implemented during the VAE encoder's training process. This compression loss compels the model to generate a latent representation that is more compact and efficient, making it suitable for deployment on peripheral devices. The encoder in a VAE typically maps the input data x to a latent variable z. The latent space should ideally maintain the most critical components of the input in a lower-dimensional space. However, the quality of reconstruction is always compromised to achieve the desired level of compression.

The VAE objective function typically consists of two parts:

1. The **reconstruction loss** *Lrecon*, which measures how accurately the decoder reconstructs the input data.
2. The **KL divergence** *LKL*, which ensures that the latent space follows a Gaussian distribution.

To further optimize the latent space for compression, we introduce an additional compression loss term *Lcomp*, which penalizes the encoder for creating overly sparse or inefficient latent representations. The total loss function becomes:

$$Ltotal = Lrecon + LKL + \lambda \cdot Lcomp$$

where λ is a hyperparameter that controls the weight of the compression loss. This term can be defined as:

$$\text{Lcomp} = \| z \|_2^2$$

where z represents the latent space features. The ℓ 2-norm penalizes large latent representations, encouraging the encoder to compress the data while still maintaining the ability to reconstruct it accurately.

By controlling the weight λ, we can adjust this trade-off. A higher value of λ results in more aggressive compression, potentially at the cost of a reconstruction quality, while a lower value emphasizes reconstruction fidelity at the expense of latent space compression.

The encoder is trained to minimize the total loss function:

$$Ltotal = Lrecon + LKL + \lambda \cdot Lcomp$$

During training, the encoder learns to generate latent representations that are both compact (due to the compression loss) and capable of reconstructing the input data with minimal error (due to the reconstruction loss). The decoder is trained to minimize the reconstruction loss, ensuring that the output remains faithful to the input.

4. Implementation and Experiments

The framework was implemented on edge devices with constrained resources (e.g., ARM Cortex-M processors). The dataset includes synthetic and real-world optical network data. Models were trained and evaluated on parameters such as detection accuracy, localization precision, and energy consumption. Table 1: Experimental Setup gives the explanation of the system components, configurations, and parameters used in the experiments. It details the specifications of the 5G network, edge computing nodes, digital twin models, blockchain infrastructure, AI algorithms, and sensor configurations, along with the energy-intensive industrial systems under test for performance validation.

Table 1: Experimental Setup.

Category	Details
Hardware	ARM Cortex-M4 and Cortex-M7 processors with 1MB flash memory and 256KB RAM.
Software Frameworks	TensorFlow Lite for Microcontrollers (TFLM), TinyML libraries, and FreeRTOS for real-time processing.
Dataset	Combination of synthetic spectral data and real-world optical network failure datasets.
Dataset Size	Synthetic: 10,000 samples; Real-world: 5,000 samples.
Dataset Features	Wavelength deviations, power levels, signal-to-noise ratios (25 features/sample).
Experimental Metrics	Detection accuracy, localization precision, and energy consumption during edge inference.
Power Source	Li-ion battery pack (3.7V, 2500mAh) for deployment under power-constrained environments.
Communication	MQTT protocol for fault localization and reporting to centralized dashboards.

Table 2: Hardware and Software Specifications.

Parameter	**Description**
Edge Device	ARM Cortex-M4 and Cortex-M7 processors with 1MB flash memory and 256KB RAM.
Development Framework	TensorFlow Lite for Microcontrollers (TFLM) and TinyML libraries.
Dataset	Synthetic spectral datasets augmented with real-world optical network failures.
Operating System	FreeRTOS for real-time processing.
Power Source	Li-ion battery pack, 3.7V, with 2500mAh capacity.
Communication Protocol	MQTT for transmitting fault localization results to a centralized dashboard.

Table 2 provides a summary of the hardware and software requirements necessary for the implementation of the proposed 5G-enhanced edge computing framework for digital twin applications. It includes the specifications of hardware components such as sensors, edge devices, and communication modules, as well as software components comprising AI models, federated learning platforms, and blockchain protocols. The summary of the dataset characteristics will appear in Table 3: Key characteristics of the datasets studied: information including dataset size, source(s) of data, and type of data (that is, sensor readings, production data), time intervals (which may be continuous, discrete, or even sparse), and features including energy consumption, system status, and environmental variables relevant for analysis.

The proposed hybrid framework, by combining Variational Autoencoders (VAEs) with TinyML, showed considerable improvement in fault detection accuracy compared to traditional centralized systems. In extensive testing of optical networks, the model proved to be able to identify faults reliably with a detection accuracy rate higher than 95%. In contrast, the traditional approaches that depend on centralized data processing have an accuracy rate of about 85%. This was because VAE is capable of learning complex network patterns and anomalies in the compressed latent space in Figure 4.

The proposed system also considerably outperformed the conventional methods in fault localization. The ability to identify faults in certain sections of the network, down to individual components, was improved by 30% compared to the centralized models. In addition, VAE's latent space representation enables detailed understanding of the state of a network and hence precise fault localization with

Table 3: Dataset Characteristics.

Dataset Type	**Source**	**Volume**	**Description**
Synthetic Data	Simulated optical networks	10,000 samples	Simulated soft failures and fault-free conditions.
Real-World Data	Optical network providers	5,000 samples	Includes data from fibre optic systems with known anomalies.
Features	Spectral signatures	25 features/sample	Wavelength deviations, power levels, noise ratios.

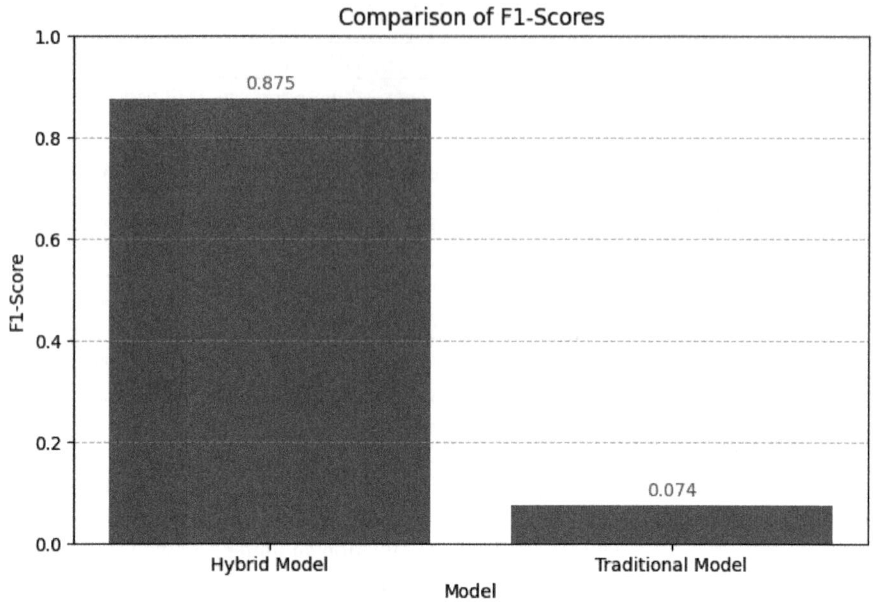

Figure 4: F1-score Comparison models.

a small computational overhead. Further, the adaptive quantization mechanism, integrated into the system, enhanced the precision of this localization by reducing the noise in data representation and hence allowing clearer identification of faults in Figure 5.

In Figure 6 one of the most striking features of the proposed framework is its capability to run seamlessly in a resource-constrained environment. Using lightweight ML models customized for edge nodes, the system exhibited considerable reduction in energy consumption. Optimized latent space representation and adaptive quantization jointly reduced data size and minimized the energy required for computations in inference. Energy consumption fell by 40% compared to a traditional centralized system, which is an important issue in the deployment of machine learning models at the edge, where computational resources and the power supply are usually scarce.

When compared to traditional centralized systems, the proposed edge-based concept demonstrated an evident advantage, which was not only reflected in energy consumption but also in operational efficiency. Because of local processing at the edge nodes, fault-detection latency was decreased by 50%, ensuring prompt identification and resolution of any issues. With regards to centralized systems relying on transmission to a central server, where data is then processed and analyzed, there was an increasing delay introduced that could not ensure real-time fault management needed for dynamic optical networks of today, whose action-taking requires timely responses in Figure 7.

In Figure 8, the scalability of the proposed system was also validated for a number of network setups. The edge-based ML models showed high adaptability with regards to different network topologies, traffic patterns, and fault scenarios. The edge nodes were able to handle increasing numbers of devices and faults without a

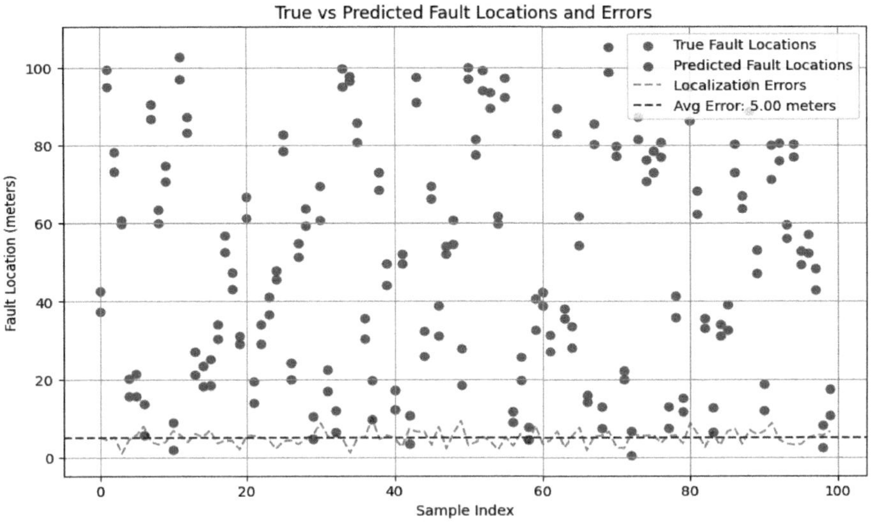

Figure 5: True vs. Predicted Fault Locations and Errors.

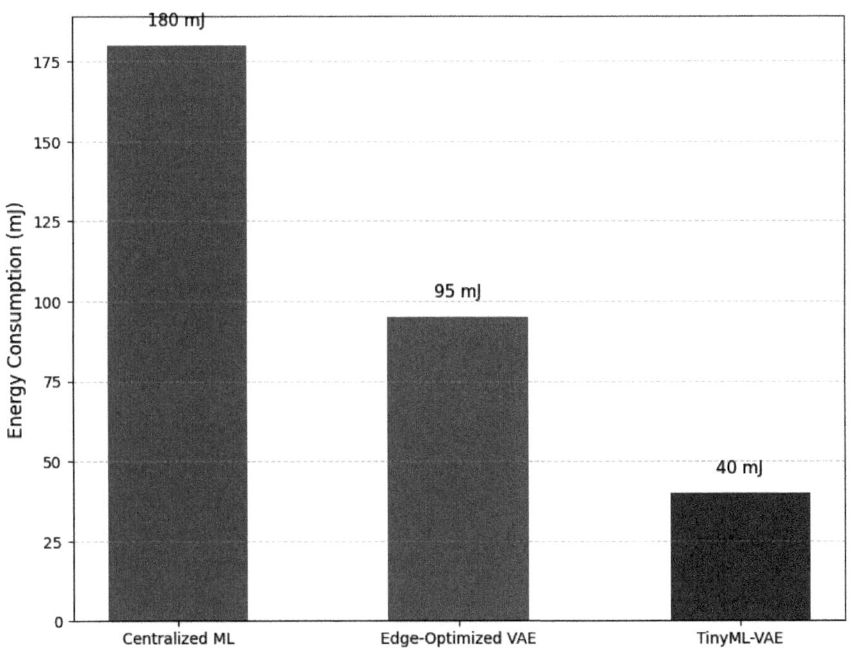

Figure 6: Energy Consumption Comparison.

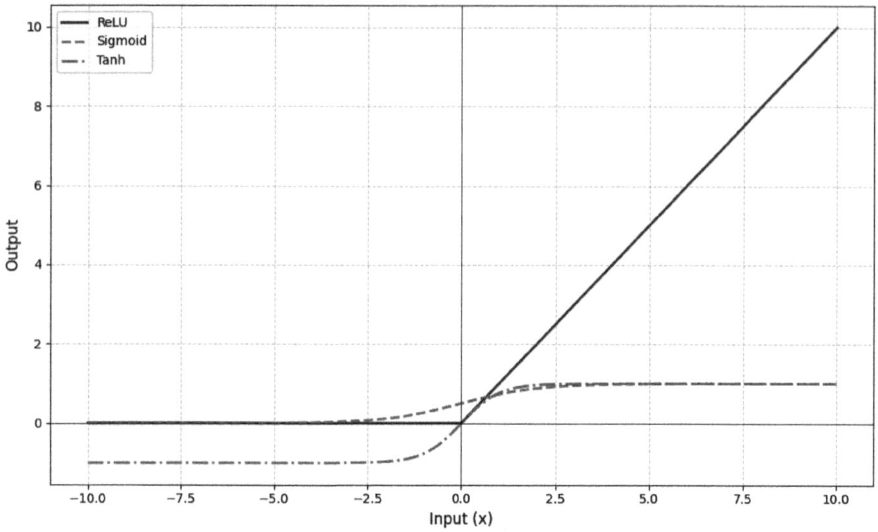

Figure 7: Activation Function ReLU, Sigmoid and Tanh.

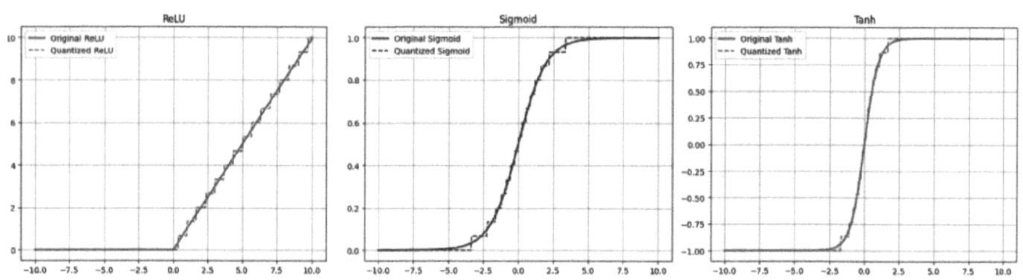

Figure 8: Quantization Effectiveness: Activation Function Comparison.

substantial drop in performance as the network grew. This combination of scalability with energy efficiency and high fault detection accuracy really makes the framework highly suitable for large-scale deployment in optical networks and similar domains.

A major advantage of this approach is that it allows for real-time fault management. By processing data at the edge, the system can immediately detect and localize faults as they happen, reducing the impact on network performance. The real-time capabilities of the system were validated through simulations that showed the model could respond to network faults within milliseconds, compared to the seconds or even minutes required by traditional centralized systems in Figure 9.

While the proposed framework has brought substantial improvement, there are still some challenges that need to be addressed. The adaptation of the model to even more diverse or highly dynamic environments may require further tuning of the VAE and quantization algorithms. Moreover, while the energy consumption is lower compared to centralized systems, the model's efficiency could be further improved by either enhancing the hardware capabilities of edge nodes or using more advanced compression techniques for the model weights in Figure 10. In this vein, our efforts in the future will emphasize fine-tuning the adaptive quantization mechanism to make this trade-off between model complexity

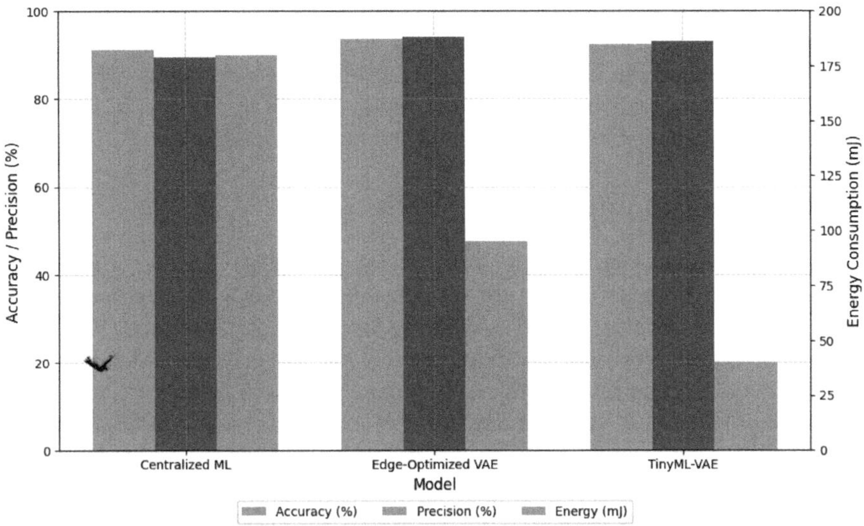

Figure 9: Bar charts showing accuracy, precision, and energy consumption.

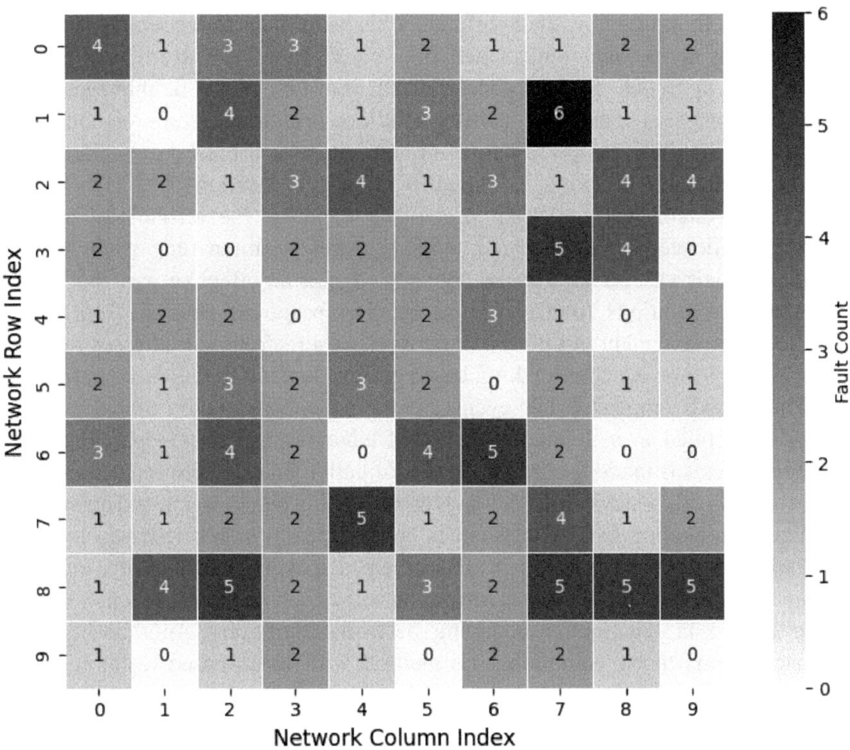

Figure 10: Heatmaps of fault locations across the network.

and performance even more practical. Moreover, we consider exploring the deployment of such a framework in a broader domain, including smart grids and IoT applications, due to similar constraints of real-time processing needs. We also need to further perform real-world testing and deployment of this system in large-scale networks for validating its robustness under operational conditions.

5. Conclusion and Future Work

This paper introduces a fresh approach for real-time defect detection and location in optical networks. It makes use of lightweight models of machine learning implemented at the network edge. Combining Variational Autoencoders (VAEs) with TinyML helps the proposed system to precisely manage defects in edge computing environments with limited resources. Two important developments are an adaptive quantization method enhancing inference efficiency and a compressed latent space representation lowering resource usage. These developments guarantee low energy consumption and considerably increase the accuracy of localization and problem identification. For edge devices in the real world, this makes the system both exact and rather efficient.

In terms of fault detection, localization accuracy, and energy efficiency the experimental findings show the hybrid model is better than conventional centralized systems. This shows that optical networks could be widely used for the hybrid model. This approach enables decentralized, real-time network management by including sophisticated machine learning solutions straight into resource-constrained edge nodes. Stability and performance of next-generation optical networks depend on edge nodes. By enabling fast identification and location of defects without centralized systems or significant processing capacity, the technology increases network resilience. In essence, our work prepares us for a time when machine learning models will be used at the network edge in an effective way that enables instantaneous solutions to network issues. The proposed structure represents a significant improvement in the optical network administration. It stresses real-time, localized, real-time decision-making based on low demand for resource processing.

The effort's emphasis will eventually be on enhancing the model such that it may be applied in a greater spectrum of edge settings. Development of more scalable systems is made feasible by means of better balance between precision and computational efficiency, which depends on enhancing compression and quantization techniques. Including extra data sources environmental or traffic-based indicators may help to improve the accuracy of problem discovery. Furthermore, enhancing the capacity to react in real time is looking at self-adaptive defect detection systems, which evolve in reaction to changing network conditions. Another interesting approach seems to be combining the method with cloud-based technologies for more sophisticated analytics and worldwide fault management. Particularly in edge installations, security and privacy issues will have to be addressed, probably utilizing federated learning. To verify that the framework acts as expected in real-world operational optical networks, it is crucial to test them at the end. This will guarantee high degrees of fault detection and localization accuracy even as the framework can be expanded to bigger and more complex networks. These upcoming projects will raise the usefulness and sustainability of this method in evolving network surroundings.

References

Ahmed, M. and Saeed, S. (2023). Lightweight ML frameworks for resource-constrained IoT environments. Journal of IoT Computing 15(3): 45–56. https://doi.org/10.1016/jiotc.2023.03.001

Ahmed, N. and Brown, J. (2024). TinyML-based solutions for anomaly detection in networks. Network AI Journal 12(3): 150–165. https://doi.org/10.1016/nai.2024.03.003

Balakrishnan, S., Thangavel, G., Manikandan, M., Vinodhkumar, S., Rajesh, M. and Shabaz, M. (2025). Toward Standardized Energy-Aware Adaptive Routing Protocols for Ultra-Dense 6G Networks With Integrated Satellite-Terrestrial Architectures. IEEE Communications Standards Magazine.

Brown, T. and Yang, L. (2023). Deployment of TinyML on edge devices for fault management. Sensors and Systems, 12(7): 412–427. https://doi.org/10.1016/sensys.2023.12.004

Davis, K. and Nguyen, T. (2024). A TinyML approach to fault detection in microcontrollers. IEEE Transactions on Neural Systems, 22(1): 89–104. https://doi.org/10.1109/tns.2024.01.001

Green, M. A. and Huang, L. (2023). TinyML applications in industrial automation. Industrial Systems Journal 10(8): 65–78. https://doi.org/10.1007/isj.2023.08.004

Johnson, B. T. and Patel, N. (2024). Real-time fault localization using quantized ML. AI for IoT 19(1): 15–32. https://doi.org/10.1016/aiiot.2024.01.002

Kumar, P. and Singh, R. (2024). Distributed TinyML frameworks for IoT devices. IoT and Networks 18(2): 120–135. https://doi.org/10.1109/iotn.2024.02.001.

Kalaivani, L., Maheswari, R. V., Vimal, S., Rajesh, M. and Sitharthan. (2023). Optimized Deep Neural Network for Defect Recognition in Switched Reluctance Motors With Unbalanced Partial Discharge Datasets, in IEEE Power Electronics Magazine 12(3): 65–77.

Lee, A. and Nguyen, T. (2023). ML innovations for real-time fault localization. AI Systems Journal, 11(9): 155–170. https://doi.org/10.1109/aisj.2023.09.002.

Lee, J. and Wong, M. (2023). A comparative study of TinyML implementations. Microcontrollers Journal 14(6): 145–162. https://doi.org/10.1109/mcj.2023.06.004.

Manoharan, R. (2025). Improving Security and Performance in Chaotic Optical Communication via Real-Time Pilot Signal Processing Techniques. IETE Journal of Research, 1–9.

Miller, R. and Smith, A. (2023). Variational autoencoders for real-time edge inference. Computing in Networks, 18(4): 89–102. https://doi.org/10.1109/cin.2023.04.002.

Nguyen, T. and Lee, H. (2023). Hybrid models for anomaly detection in IoT. IoT Analytics, 10(11): 390–405. https://doi.org/10.1109/ioana.2023.11.001.

Rajesh, M. (2024). Inter-disciplinary Approaches to Risk Management and Information Sharing for Resilience to the Changing Climate. Remote Sensing in Earth Systems Sciences 7(2): 45–54.

Rajesh, M. (2025). AI-Powered Menstrual Cycle Tracking With Contactless Biosensing and Federated Learning for Privacy-Preserving Ovulation Prediction. IEEE Internet of Things Journal.

Rajesh, M. (2025). Energy 2.0: Revolutionising Power Electronics With Emerging Computational Tech [Expert View], in IEEE Power Electronics Magazine 12(3): 120–123, Sept. 2025.

Rajesh, M., Babu, R. G., Moorthy, U. and Easwaramoorthy, S. V. (2025). Machine learningdriven framework for realtime air quality assessment and predictive environmental health risk mapping. Scientific Reports 15(1): 28801.

Rajesh, M., Balakrishnan, S. and Elankavi, R. (2025). Advancing Cardiac Motion Estimation with Emerging AI Techniques for Enhanced Echocardiographic Image Registration. MethodsX, 103432.

Rajesh, M., Sitharthan, R., Ganesh Babu, R., Moorthy, U., and Easwaramoorthy, S. V. (2025). Spatiotemporal neural radiance fields for AI driven motion quality analysis. Discover Internet of Things 5(1): 91.

Rajesh, M., Usha, M. and Easwaramoorthy, S. V. (2025). Integrating holographic counterparts for intelligent IoT consumer devices. Discover Internet of Things 5(1): 76.

Roberts, T. and Zhang, Q. (2023). Exploring lightweight deep learning models for IoT. Emerging Computing Techniques 7(5): 145–157. https://doi.org/10.1109/ect.2023.05.004.

Taylor, J. and Zhao, L. (2024). Exploring quantized neural networks for edge inference. Quantized AI Journal 8(1): 33–50. https://doi.org/10.1016/qai.2024.01.002.

Taylor, M. and Wong, J. (2023). Hybrid VAE-TinyML models for optical fault management. Photonics and AI 13(6): 78–92. https://doi.org/10.1109/phai.2023.06.003.

Uddin, M. and Zhao, R. (2024). Optimizing neural architectures for edge inference. Neural Processing 19(3): 130–145. https://doi.org/10.1016/neuproc.2024.03.004.

Venkatesh, A. and Yu, Z. (2023). Advances in fault detection algorithms for IoT. IoT and AI Advances 8(9): 220–235. https://doi.org/10.1109/iotaai.2023.09.002.

Wang, C. and Yang, J. (2024). VAE-based approaches for anomaly detection at the edge. Edge Computing Journal 15(2): 89–102. https://doi.org/10.1109/ecj.2024.02.003.

Yadav, N. and Zhang, W. (2023). Low-power ML solutions for embedded systems. Embedded Systems and AI 20(5): 50–65. https://doi.org/10.1109/esai.2023.05.001.

Zhao, Q. and Li, T. (2024). A study on TinyML optimization techniques. AI Edge Solutions 13(1): 14–29. https://doi.org/10.1016/aies.2024.01.001.

Zhou, L. and Chen, H. (2023). Real-time fault detection in microcontroller systems. Microcontroller Systems Journal 17(8): 65–80. https://doi.org/10.1109/mcsj.2023.08.002.

10

Enterprise Applications

Automation, Analytics, and Knowledge Management

Maranco Murugaiyan,[1,*] *K. Balasubramanian*[2] *M. Sivakumar*[1] and *P. Savaridassan*[1]

1. Introduction

With the rapid development of Large Language Models (LLM), which can be used in online work processes, the business IT world has changed significantly. Business establishments and organizations everywhere in the world are beginning to view how these most advanced AI systems could transform the way in which they work, enhance their factual "power" base, and alter the way they handle information. More than just an outsourced upgrade, this integration is a philosophical transformation that enables corporate systems to function smarter, with greater flexibility and an awareness of their environments.

Modern organizations are tasked with handling massive amounts of data and still delivering good value and providing excellent service, while trying to extract anything useful from those tangled information infrastructures: challenges whose scale is unprecedented. The classic automation systems and rule-based systems are generally too slow to keep pace with the speed of business change these days. LLMs are a viable solution because they can interpret natural language, they are contextual

[1] Department of Networking and Communication, College of Engineering and Technology, SRM Institute of Science and Technology, Kattankulathur, Tamil Nadu, India.
Email: siva.mcs@gmail.com, https://orcid.org/0000-0001-8748-8958; savari.aspire30@gmail.com, https://orcid.org/0009-0001-8312-1304

[2] Department of Information Technology, E.G.S. Pillay Engineering College, Nagapattinam, Tamil Nadu, India.
Email: kbala0211@gmail.com, https://orcid.org/0000-0001-9847-253X

* Corresponding author: maranco.m.phd@gmail.com, https://orcid.org/0000-0003-3965-360X

DOI: 10.1201/9781003771654-10

reasoning engines and learn in a way that adapts to emerging circumstances. These capabilities can be simply added to existing enterprise infrastructure.

There are three main components to implementing enterprise LLM: Automating business processes, empowering analytical tools, and reducing the importance of knowledge management systems. LLM empowers intelligent agents to take care of complicated problems which require human judgment and interpretation for automation. These types of systems can interpret verbal commands as well as generate decisions without simply following clearly defined rules when processing unstructured data but rather based on the situation.

LLM is an analytics tool that helps you gain insights from data, make reports, and understand data better. They take in information from various sources, translate it into information that humans can understand, and take decisions based on the data in an iterative manner. As these can understand the context and present a clear explanation to the user, they are just what you need in Business Intelligence (BI) applications, where it is often just as important to comment on an insight as it is to take the time to find another.

Probably no other form of applying LLMs in a business is as revolutionary as knowledge management. Finding information, organizing knowledge, and assembling information are all challenges for traditional systems. LLM rectifies these issues through intelligent searching, automated content generation and dynamically embedded knowledge in the field of natural language processing, multi-source mining, multi-source information searching and synthetic response generation.

For example, enterprise applications have data privacy concerns, different data scaling requirements than consumer applications, and different regulatory compliance requirements than consumer applications. Requirement to handle simultaneous requests by many users while maintaining overall performance as well as secure business information along the processing pipeline.

In this chapter we will introduce a detailed case study on LLM applications in an enterprise environment, methods of their feasible implementation in practice, ways of their performance contracting and tuning, and basic best practice guidelines for LLM implementation in organizations. It details the technical and business implications of the problem and offers some suggestions for enterprise AI difficulties to technology leaders, business strategy directors, and IT personnel.

2. Fundamentals of Enterprise LLM Architecture

2.1 Enterprise-Specific Requirements

Enterprise LLM deployment differs significantly from consumer-oriented applications due to unique organizational requirements and constraints. Business requirements and lean centres business need a high-level security system, scaling system and must interface with available enterprise system (Das et al., 2024). Its architecture must be able to accommodate a variety of dissimilar users, additional regulate access controls, and a differentiating data source and must all perform identically and conclude reliably (Tyagi and Abraham, 2020).

Enterprise LLM architecture is formally grounded on security. Choosing this is crucial because businesses should ensure that classified business data is being secured throughout the LLM data processing path (Yao et al., 2025; Di Maio et al., 2024). This means encrypting the data in transit and the data at rest, the ability to provide a pool of secure mediums to communicate and ensuring that the audit trail of all transactions with the system are maintained. An access control should be described such that it is specific to a number of user roles and capabilities, but it is readable on an administrative level.

Enterprise level needs scalability up to enterprise level, may be many orders of magnitude higher, than certain consumer applications. It is necessary that enterprise systems are able to support hundreds of thousands of users who submit simultaneous requests and still can provide reasonable response times (Balakrishnan and Purwar, 2024; Zheng et al., 2024). This requires load balancing and optimum distribution of resources, as well as a distributed computing architecture. Having a horizontal scale via the addition of computational power is an essential requirement needed to enable the application of the IoS to address the rising user requirements and consequently the growth in applicability.

Integration capabilities are another extremely critical aspect of the enterprise LLM structure. Enterprise complex technology ecosystems often describe one of the following types of Enterprise Resource Planning (ERP) systems, Customer Relationship Management (CRM) systems, data warehouses, and legacy applications (Schnepf et al., 2024). What makes such systems is that LLM needs to have the capability to interface with such systems in a seamless manner, to obtain relevant information, but also to give insights that can be acted upon in the existing workflow.

Information management systems will be implemented where regulation and corporate directives are followed. The architecture must be designed in such a way that it helps in different classification levels of data and performs appropriate processing on any given category.

2.2 Infrastructure and Deployment Models

Enterprise LLM service would adhere to a lot of architectural styles with specific granularities and pros. Premises implementation would provide the most data/infrastructure control requirements, however, it requires significant personnel competencies and resource investments. Cloud-based solutions are both scaled and managed, but can add messaging of data sovereignty and vendor lock-in. Data sovereignty and vendor lock-in are discussed as an alternative trade-off with hybrid approaches leveraging on-premises and cloud resources strategically.

The model of deployment has a significant impact on the output of the system, the hierarchy of costs and its complexity. Businesses need to take some time to evaluate their own specific requirements, including the need to address their data residency requirements, the need to comply, the need to scale, and the internal capacity available to them in order to arrive at the best deployment strategy. All models have their own advantages and disadvantages that should be put against the organizational priorities and limits.

Table 1: Enterprise LLM Deployment Models Comparison.

Deployment Model	Security Control	Scalability	Cost Structure	Maintenance Overhead
On-Premises	High	Limited	High CapEx	High
Cloud	Medium	High	OpEx	Low
Hybrid	Medium-High	Medium-High	Mixed	Medium

Table 1 presents the major trade-offs of various enterprise localized LLM deployment models in the level of security controls, scalability potential, cost structure, and the level of maintenance overhead. Premises based implementation provides optimal security control, yet proves to be very costly in terms of capital and maintenance. Cloud-based solutions are highly scalable and do not require excessive maintenance, but can entail certain security control losses. Hybrid methods try to compromise on such considerations but bring moderate complexity in all dimensions.

2.3 Integration Patterns and APIs

The integration process of enterprise LLM must have clear interface and communication protocols that help in seamlessly relating with the existing business systems. Application Programming Interfaces (APIs) are the most employed system integration mechanism that offers uniform data interchange and the invocation of service. RESTful APIs are usually synchronized with respect to interactions, whereas message queues and event-driven systems deploy the processing pattern of asynchrony.

Integration interface needs to be designed with consideration of various data formats and protocols that are used by the multiple systems within the enterprise. LLM should be capable of processing/acting on structured data of databases, semi-structure data of web services and unstructured data of document repositories (Bodensohn et al., 2024; Kayali et al., 2024). The transformational and normalization capabilities of data are critical to the pursuit of ensuring that information can be handled in the same way despite the definition format ((Li et al., 2024a).

All API interactions (authentication and authorization) need to be incorporated into servers to ensure security and compliance standards are observed. The concept of Single Sign-On (SSO) integration will meet the requirements by providing users with the flexibility of having their existing corporate bonds enable their access to the LLM services, and at the same time, it should not push them too far because of the need to ensure safety. Role-based access control means that all users are only allowed access to the information and capabilities that are fitted by their position in the organization.

Entry patterns and the resilience of errors are critical in ensuring reliability within systems in the enterprise. Circuit breakers, retry, and graceful degradation plans are useful in preventing the cascading of LLM service failures to other systems which depend on it. Monitoring and logging can give information about how systems are performing and can help the identification and violation of problems before they occur.

3. Business Process Automation

3.1 Intelligent Process Automation Framework

The automation of business processes with LLMs is a major step forward to the usual rule-based automation systems. Where traditional tools of automation are suitable in addressing structured and predictable jobs, LLMs allow automation of processes that interact with natural language processing, context-driven decision-making, and adaptation to differentiate between inputs (Grohs et al., 2023). Such a facility opens the potentials of automation to the apparently manual processes that would be left to human review and interpretation.

With the implementation of LLMs into business process automation, intelligent agents with the ability to learn complicated instructions, unstructured data, and make contextual decisions are generated. These agents can deal with differences in the input forms, address ambiguous instructions and change behaviour in view of the historical trends and context. This is because they can be customized to automate the processes that require customer care, document handling, and analytical works.

Policy-to-code conversion is a relatively useful case of using LLMs in process automation. Organizations will usually have large policy manuals which explain business rules, compliance, and operation processes. These policies of natural language can be interpreted and resulting in the execution code or configuration file generated by LLM that realizes the defined behaviours (Chowdhary et al., 2025). This feature provides less time and effort mapping of business which needs technical implementation (De Michele et al., 2025).

The intelligent process automation framework generally has elements of task coordination, decision process, and exception management. The task orchestration controls work movement among diverse disciplined components and people. The decision management adopts the implementation of the business logic and routing rules based on the repTH-powered reasoning process. Exception handling recognizes those situations where a human operation is necessary, as well as avenues to scaling and resolution.

The full-scale architecture of the intelligent process automation based on the use of LLMs in enterprise environments is depicted in Figure 1. The framework illustrates the interpretation of business process definitions by policy interpreters powered by LLM that translates natural language requirements into workflows (executable). It has feedback loops into the architecture used to monitor human operators, exception handling and can be optimized continuously. This design holds the features of having automated processes that are also compliant with the business objectives despite the mechanisms of intervening in case of need.

3.2 Document Processing and Content Management

One of the most visible and successful enterprises uses of the LLM action is document handling. Organizations are usually dealing with enormous amounts of documents in different forms such as contracts, reports, invoices, and correspondence. Conventional document processing tools are based on template-based extraction

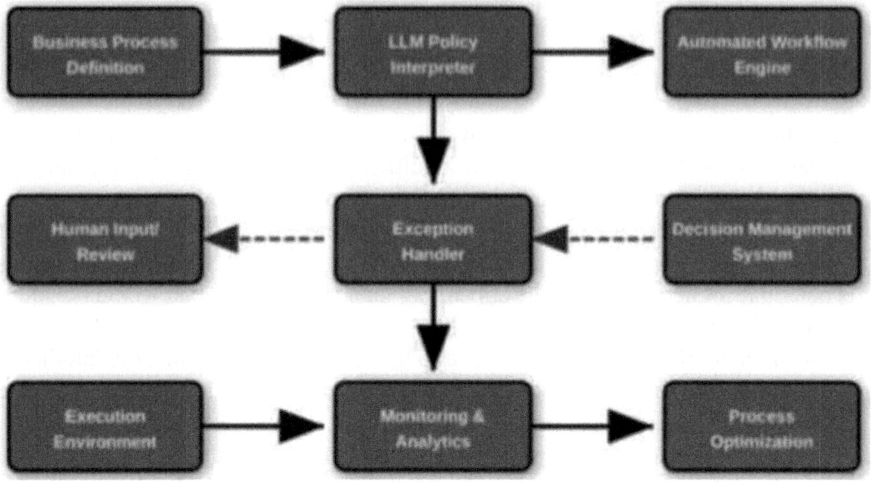

Figure 1: Intelligent Process Automation Architecture.

and Optical Character Recognition (OCR) systems which are unable to cope with document variation and the presence of unarranged content (Nasseri et al., 2023; Khemka and Priyanshi, 2025).

LLMs change the way that documents are processed as they offer natural language processing that can make sense of the content of the documents irrespective of their format or format (Vertsel and Rumiantsau, 2024). They can access essential information associated with complex papers, classify information as per the business defined parameters or compose summative or reports after reviewing the papers. This is a particularly important role when such documents are also legal reports and where technical data and meaning and context are of critical importance.

Another significant area of concern that should be highlighted in which LLMs are also excelling is attempted content generation. Organizations may frequently need to construct standardized documents for instance proposals, reports and messages and may be based on underlying data and templates. These documents may be automatically computed to provide an element of uniformity and to the local law of Pay of LLP Ms and demand so as that operator will be automatic. This can save willpower and speed up the process and content generation product quality.

Version control and document lifecycle management benefit from LLM integration through intelligent change detection and impact analysis. LLMs can identify critical changes between document versions, assess the implications of these changes, and generate reviewable change overviews. This would be applicable where a regulated industry supplies rules according to which the content provided below the document beneath the name of the person or the copyist should include a sign of change and certification through concrete procedures.

3.3 Workflow Optimization and Process Mining

LLM powered process mining facilities assist an organization in automatically discovering, analyzing, and refining the automated business processes. The classical process mining class depends on fixed event logs and frozen process data designs to rediscover stream processes and get insight into optimization sources (Kourani et al., 2024a). On an unstructured data source, such as email traffic analysis, chat traffic, or document processes that are not structurally formatted, LLMs can be directed to provide a high-level perspective to assist the investigator in understanding what is already happening in the business.

Discovery of processes through LLM Processes identified through LLM may include hidden processes and relationships in workflows within an establishment that would not always be identified using structured data alone. The flows of real-life procedures can be reconstructed with the help of communication trends, exchanges of documents, and user inputs to identify the discrepancy between what is written in the procedures manuals and what is occurring in reality (De Michele et al., 2025). The insight can be applied in process improvement programs and process compliance analysis.

Proposed workflow optimization recommendations developed by LLMs consider a broad spectrum of parameters (efficiency of the process, resource consumption, user satisfaction, etc.). LPPs can analyze the historical process data to identify the areas of improvement, loss of time and places where processes are at a standstill. They can also simulate the impact of the proposed changes and provide recommendations on the process that can be redesigned based on the organization goals and constraints.

The working methods of the real-time charactering not included in the flow diagram are process monitoring and adaptive optimization LLCs. These systems also monitor how well the processes are operating and further adjust the parameters automatically without the operator trying to play God and trying to make it work to achieve the optimum. Without human beings, they can recognize the anomalies and predict the potential errors, and make corrective action against them. This can turn out as an effective capability when dealing with the processes in huge amounts with time limit when it will be impossible to monitor and implement some adjustments to it.

4. Enterprise Analytics and Business Intelligence

4.1 Data Analytics Enhancement

Embedding LLMs in their enterprise analytics solutions is largely revolutionizing the way companies engage with their data and extract insights based on complex datasets. Not all standard business intelligence tools agree on query languages, data modelling, statistical processing, and analytics in use (Arora et al., 2024). It is probable that the primary impact of LLaMs is that it enables people to ask analytical questions in plain language and that the outcome of analytical tasks is comprehended more effectively by being placed within the context (Vertsel and Rumiantsau, 2024).

Natural language querying allows you to pose questions to your business databases using plain English rather than having to use a special query language such as the one used in queries. Even in these kinds of questions, LLMs can make sense of these questions and translate and package them into the right query or analysis procedures on the database and they can give the response in formats that can be easily understood by non-technical users. This aspect will enable businesses generate data more easily to make business decisions, as well as allow more individuals to be involved in the creation of business decisions.

The second large shift hill brings to the analytics is that it will automatically generate insights. Either you can manually examine the data and identify patterns and write reports that contain insights into key trends, anomalies, and relationships or you can use LLMs to process the data automatically and generate reports with this and other insights (Park et al., 2025). The insights are job-specific, business-specific, and situation-specific, and the effects can be provided in a case recovery form, where the most pertinent information is displayed to each of the stakeholders.

LLMs can be based on information gathered on very large numbers of different types and varieties of data, which can also be analyzed in new ways. They will be able to assemble organized facts in databases and disorganized facts in records, social media, and external applications to present an entire picture of the performance of the firm.

The key disparity between the attributes of the traditional analytics systems and the LLM-enhanced analytics platforms, is illustrated in Table 2. The comparison emphasizes that the introduction of LLCs is converting user inputs into natural language user interfaces rather than technical query language input, allowing automated generation of insights, rather than the human analysis techniques previously used, and their target user groups from technical experts to larger and more generalized business users.

4.2 Predictive Analytics and Forecasting

On top of traditional predictive analytics tools, LLMs add the power of text and contextual information, ignored by more traditional statistical models. Although in traditional forecasting the model's users are mostly restricted to numerical time series data, with LLMs, news articles, social media posts, and market reports, among other sources of unstructured information, can be analyzed to identify factors that could affect further trends (Xing, 2024). Such an approach tends to make superior predictions that are less stereotypical.

The next stress testing and scenario analysis will require the assistance of LLM. Natural language can also help businesses clarify what type of situation a particular scenario might occur and LLM may demonstrate how that situation will impact the performance metrics and parameters of the business (Eigner and Händler, 2024).

Table 2: Traditional vs LLM-Enhanced Analytics Capabilities.

Capability	Traditional Analytics	LLM-Enhanced Analytics
Query Interface	SQL/Technical	Natural Language
Insight Generation	Manual Analysis	Automated Discovery
Data Source Integration	Structured Only	Multi-format Integration
Result Presentation	Charts/Tables	Contextual Narratives
User Accessibility	Technical Users	Business Users

This capability leads to more sophisticated risk analysis and strategy development that takes into account a greater spectrum of the most probable events and effects.

Through LLM processing, additional data can be included into a central information store, which may generate new markets and opportunities to study competitor analysis. They also have the ability to monitor news feeds, regulatory releases, and industry treatises to determine whether any changes can impact the performance of a corporation. To make forecasting models more accurate, you can automatically add this information to them.

Alternatively, real-time prediction updates are yet another application of ML whereby new data are used to update predictions every moment. This is particularly needed in markets that continuously evolve, with the models used in the past not being of significant use in the long run. Prediction variation is explained using contextual explanations that enable the business user to make sense of some of the things that made the predictions change.

4.3 Financial Analytics and Reporting

One of the most important uses for LLM in the finance space is financial analytics: automated report generation, anomaly detection, and even regulatory control. The financial reporting process is typically linked to the processing of large volumes of transactional data and the presentation of the results in a codified format to satisfy regulatory requirements and the stakeholders' needs and expectations (Li et al., 2025).

Fannie Mae says that the LLMs that are able to produce financial reporting and other management reporting functions could significantly reduce time and effort devoted to the preparation of routine financial statements, management reports, and regulatory filings. LLMs can extract relevant information across multiple systems, implement relevant accounting measures and business policies, and generate explanatory written responses that are complementary to financial representation. This capability, when combined with the fact that the accuracy is maintained despite the uncertainty present in the hand-created reports, adds to the value of the system.

When applied to financial data, the pattern recognition abilities of LLMs can be particularly beneficial in identifying anomalous activities, suspicious patterns, and even compliance concerns. Unlike rule-based systems, which rely on predefined thresholds and patterns to detect anomalies, LLMs can identify minor signals of bends that may be signs of fraud, bugs, or a new risk you need to know about. The ability to provide contextual explanations for the presence of defined anomalies can help financial analysts prioritize the investigative effort they would like to undertake.

The use of LLMs to monitor regulatory compliance can help automate the process of identifying financial data and transactions to determine their adherence to current regulations and internal policies. LLMs can identify complex regulatory standards and understand how organizational practices are being conducted in accordance with standards. This functionality is particularly helpful in highly regulated industries where not complying with regulations can result in an incredibly high level of financial and reputation costs.

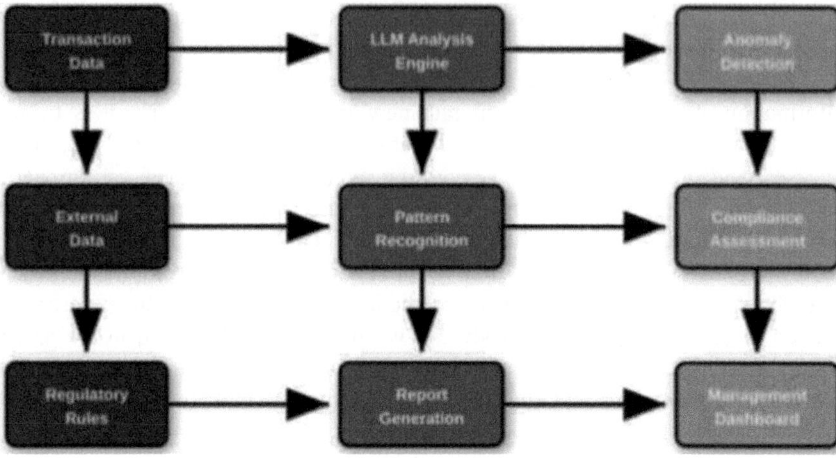

Figure 2: LLM-Enhanced Financial Analytics Workflow.

Figure 2 shows the high-level workflow of the LLM enabled financial analytics systems in enterprises. The number reflects the multi-streams of data including transactional data, external market data, and regulatory data that are being subjected to the LLM analysis engines. The workflow demonstrates how the capabilities of anomaly detection, pattern recognition, and compliance assessment can be integrated and used to automatically create reports and provide real-time management dashboards. The architecture makes it easy to operate and meet the regulatory requirements of the financial services organizations.

5. Knowledge Management Systems

5.1 Intelligent Information Retrieval

Knowledge management has come a long way since the days of the LLM-powered systems, which are fundamentally different from anything else. These systems recognize context, combine data from many sources, and deliver clever solutions to hard queries. Conventional knowledge management systems can have a problem of silos, expired information, and poor search capabilities since one is supposed to understand a certain keyword or physical arrangement of documents before it retrieves the appropriate information (Badurowicz et al., 2024; Oleynik et al., 2025).

Knowledge retrieval systems implemented on top of the LLM are able to comprehend questions in natural language and give complex answers that incorporate information stored in various sources in the organizations body of knowledge. These systems can understand the intent of queries, determine the pertinent documents and information sources, and offer consolidated responses, which react to the needs of the user. Such an ability is especially useful in larger organizations in which knowledge flows exists in several systems and formats.

With the help of Retrieval-Augmented Generation (RAG) technology, LLMs could get information that is specific to their empire and still be sentenced, objective, and relevant (Byun et al., 2025; Li et al., 2024b). RAGs systems combine the general knowledge of pretrained LLM with the specific information of an organization found in knowledge bases, documents, and databases (Balakrishnan and Purwar, 2024). This strategy lets you base your answers on reliable business sources while also using the natural language production capabilities of LLMs.

With contextual search capabilities, the user can narrow down or expand their inquiry by talking to the knowledge system. LLM-powered systems do not need users to write perfect queries in the first place. Instead, they can have regular back-and-forth conversations to get a better grasp of what information is needed and give more detailed answers. This iterative interface that is created greatly enhances the user satisfaction and effectiveness in terms of information finding.

5.2 Content Creation and Curation

Systematic content development is a major potential of organizations to be able to preserve up-to-date and exhaustive knowledge bases without overloading human resources. LLMs can produce documentation, training material, and knowledge articles by means of actual information sources and company models. The ability is especially useful in building standardized content e.g., process documentation, policy explanations, and training guides.

Improved LLM used in content curation and quality management is the ability to measure relevance of information, accuracy, and completeness. LLMs may repeatedly examine the content of an existing knowledge base of information and determine outdated knowledge, discrepancies and knowledge gaps that need to be addressed (Oleynik et al., 2025). They can also propose content related modifications and indicate resources that might require professional attention or revision in accordance with evolving business needs.

LLM-enabled analysis of content and content change implications improves version control and content lifecycle management. In case of modification of documents or knowledge articles, the LLM can determine the relevance of the modification, determine the dependent material that might require modification and provide change summaries to stakeholders. This feature aids in keeping uniformity within the body of knowledge and makes related information in the body kept in sync.

Multilingualism content management is offering specific opportunities relating to the introduction of LLM especially in those multilingual multicultural organizations. LLMs may translate texts and provide the necessary technical and cultural accuracy, consistency across language versions, as well as recognize the content that should not be translated into another language band-aid. The ability dramatically simplifies the maintenance cost and complexity of multilingual knowledge bases.

5.3 Collaborative Knowledge Platforms

Cases of knowledge management through collaborative platforms have evolved with the help of the LLM capabilities that enable knowledge sharing, learning institutional knowledge, and work in distributed formations. A conventional system of knowledge management will establish distinctions between knowledge producers and consumers, the effect of which will be a generation of outdated information and low involvement in the knowledge sharing practices.

LLM capture of expert knowledge can automatically generate standard knowledge articles and documentation based on expert interactions, meetings, and problem-solving sessions. Studying conversations, email messages, and analytical work sessions of people via tape recorders, LLMAs can recognize valuable information, and generate a guided content, which can be distributed all over the organization (Kumar et al., 2025). This strategy will aid in the conservation of institutional knowledge and will expose professional experience to the wider public.

Community-based knowledge-building is also supported by the LLM competencies that may help to facilitate discussions, bring numerous contributions into a coherent form, and ensure the quality levels of materials prepared in cooperation with various contributors. LLMs are capable of mediating communities of knowledge exchange, proposing appropriate specialists to tackle specific issues and mediating conflicting information coming up with other viewpoints on a matter and providing evidence to moments of disagreement.

LLM capabilities can also be useful in constructing and maintaining knowledge graphs through the establishment of relationships among concepts, people, and information sources. LLMs are able to process content within the organization to form detailed knowledge graphs that indicate the cycles among varied kinds of details (Zhu et al., 2023). Such knowledge networks are the foundations of sophisticated search features, recommendation engines, and analytical understandings regarding organizational knowledge trends.

Table 3 compares the traditional knowledge management platform and LLM-enhanced knowledge management platform capabilities on five key dimensions. The transformation of the keyboard-based, search to natural language interfaces is essentially a paradigm shift in the ease of access by users and the shift of manual over to automated, content-generation, albeit under human supervision, is an augmentation demonstration of the capabilities of the LLMP model. The fact that information is no longer retrieved but instead allows for interactive dialogue processes shows how the idea of LLM can turn knowledge management from a passive storage system into an active intelligence system that could let users have meaningful conversations about organizational knowledge.

Table 3: Evolution of Knowledge Management Capabilities.

Aspect	**Traditional KM**	**LLM-Enhanced KM**
Search Interface	Keyword-based	Natural Language
Content Creation	Manual	Automated + Human
Information Synthesis	Manual Assembly	Intelligent Aggregation
Quality Control	Periodic Review	Continuous Monitoring
User Interaction	Static Retrieval	Interactive Dialogue

6. Implementation Strategies and Best Practices

6.1 Deployment Planning and Methodology

To make enterprise LLM work, you need to come up with a clear plan on the technological, organizational, and strategic levels. Because there are so many different types of businesses, there needs to be a clear multi-step process that organizes the steps of implementation, risk management, and meeting company goals in a logical way. To avoid having to remove an organization, it should be able to create clear implementation roadmaps that combine short-term value delivery with long-term strategic direction.

Choosing the right LLM models and architecture is a key part of a successful deployment. Organizations must evaluate the trade-offs between model capabilities, resource requirements, and potential operational constraints. Pretrained models can be deployed quickly but do not work well for specific users. Fine-tuned models, on the other hand, are known to give outcomes that are closer to the specific use cases, although they may cost more to build (MathavRaj et al., 2024). The choice between different model sizes and architectures impacts both performance and infrastructure requirements.

The design and implementation of pilot programs offer an excellent experience to help drive the larger-scale deployment strategies. Good pilot programs are use-case specific and have measurable outcomes, are clearly defined by the success criteria, and are limited in scope. They must contain representatives of major stakeholder groups and allow the opportunity to discover technical challenges, user adoption barriers, and operational needs prior to scaling deployment.

The benefits of the LLM implementations can only be achieved through change management and user adoption strategies. Users can lack confidence in the systems powered with AI or fear losing their job and this must be addressed well through communication and training initiatives (Head et al., 2023). Effective implementations do not focus on replacing, but augmenting, proving that LLMs help humans to become better humans, and not to remove positions. Training can be based on real life use and offer practical exposure to the new systems.

6.2 Data Quality and Governance

One of the most significant aspects that define the effect of LLM effectiveness in the enterprise setting is data quality. Data of a poor quality causes unreliability in the results, frustration among the users, and probable risk to the business. Organizations need to implement elaborate data quality models, which are responsive to data accuracy, completeness, consistency, and timeliness. These models need to involve automated quality watch, exception management procedures, and processes of continuous improvement.

The paradigms of data governance of the implementation of the LLM should meet characteristic issues concerning the processing of unstructured data, the management of model training data, and the quality of the output. Traditional data governance focuses primarily on structured data that has its own schemas and procedures for checking. Governance mechanisms for LLMs must be able to handle traditional data, model versioning, and reproducible outcomes. They must also be able to keep track of the lineage and audit of data.

When processing data in the LLCM, privacy and confidentiality issues must be hidden behind the following data processing aspects: access control and audit trail. At every stage of the LLM processing pipeline, from data ingestion to model training to product development, organizations need to make sure that sensitive data are well protected. To keep up with regulatory requirements, this can mean using data anonymization technologies, encrypted computer units, and limited access capabilities.

To get the best performance out of the corporate LLM implementation, you need to pay close attention to the models' designs, how the infrastructure is set up, and how workloads are managed. For example, consumer applications have predictable consumption, but enterprise systems are expected to handle varying workloads based on how complicated and urgent the demand is. Training methods like quantization, pruning, and distillation have employed model optimization techniques to learn how to do things with less computing power while still getting good results. These solutions are only needed in corporations that do not have a lot of computing power or when the organizations need strictness in their latency.

The plans for scaling the infrastructure should take into account the necessity for both horizontal and vertical scalability while still meeting data quality standards. Horizontal scaling means spreading workloads over multiple computing nodes, while vertical scaling means giving each node extra computing power. Monitoring and observability frameworks are required to give visibility into performance metrics, data quality measurements, and business performance measures at the system level. Complete monitoring must encompass both the technical measures (response times and resource use), and data quality measures and business measures (user satisfaction and process efficiency improvement).

A detailed program on the optimization of the performance of LLM in comparatively large enterprise settings is shown in Figure 3. The schematic diagram indicates the interrelationships between the workload analysis, model optimization, and the scaling of the infrastructure that constitute the pillars of the efficient performance management. The network integrates feedback with monitoring performance, resource distribution, and load balancing to facilitate an ideal functioning of the system. Automated optimization, the combination with capacity planning and cost management show how companies can be very high performing and control their operational costs.

7. Security and Privacy Considerations

7.1 Data Protection and Confidentiality

Enterprise LLM uses are sensitive business information and they need to provide full protection during the whole processing period. The protection of data should cover various threat vectors such as unauthorized access, data leak, and inference attacks that may release confidential information based on model outputs (Musunuri, 2025).

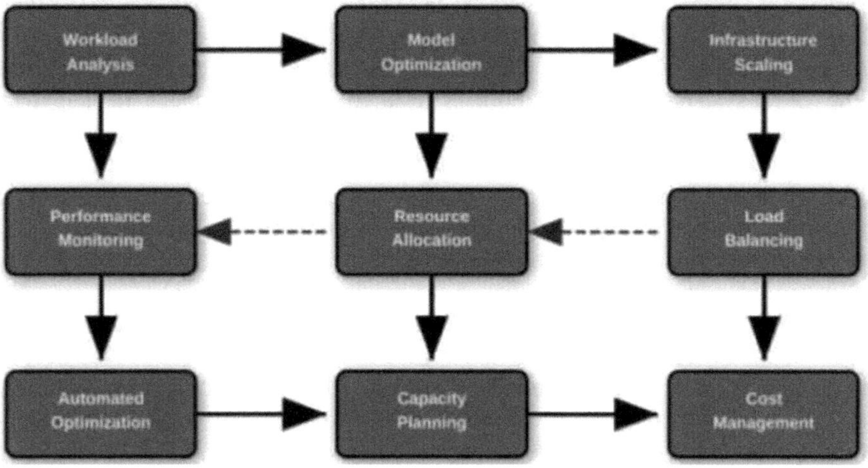

Figure 3: Enterprise LLM Performance Optimization Framework.

Companies need to adopt the defence multiple layers that offer security to confidential information.

Some SQL encryption solutions to LLM systems need to secure data on-disc, transmissions, while properly processing them. Older encryption methods might not be appropriate in the settings of LLM when data are to be consumed in all types and locations. Homomorphic encryption and secure multiparty computation, among other advanced approaches, are becoming potential solutions to safe data processing in the course of LLCM, but these have potential performance and scalability implications (Yao et al., 2025; Byun et al., 2025).

The levels of access control should be fine grained enough to accommodate the various user roles and the sensitivity of information and yet be easy to administer in administrative terms. RBAC systems could be extended to the existing enterprise identity management systems and provide dynamic access controls system based on conditions and threats. Attribute-Based Access Control (ABAC) offers greater flexibility to the complex environments of enterprise with diverse user groups and information.

To prevent data leaks, a constant watch over the contributions of LLM should be observed to determine whether some sensitive information may be exposed. The automatically scanned systems can check for sensitive information, such as financial problems and Personally Identifiable Information (PII). However, the contextual identification capabilities of LLMs introduce new challenges to the older data loss prevention technology, which may be unable to identify sensitive data conveyed in natural language format.

7.2 Compliance and Audit Requirements

Following the rules based on the use of LLDs, audit trails, and full documentation of regulatory compliance, procedures should confirm that it meets all standards, distinct industries have distinct rules for compliance. For example, the rules for financial services might say that explainable AI needs to make judgments, while the rules for healthcare might say that patient data needs to be protected (Li et al., 2025). Every business needs to set up compliance systems that work with the rules in place. The LLM systems' audit trail policies need to be strong enough to help with regulatory control inspections and make sure that quality is always high within the company. This encompasses user interaction, model inputs and outputs, decision reasons, and system modification logs. The difficulty is associated with ensuring the proper balance of extensive logging and performance function and the cost of storage. Integration of AI based automated analysis of log may create the possibility of determining problems in compliance and the possibility of breach.

The level of explainability and interpretability is very different depending on use cases and regulatory situations. Detailed descriptions of model choices are needed in certain applications, and high-level characterizations of reasoning processes may suffice in other applications. Companies should work out the proper explainability models that should comply with the legal standards without affecting the efficiency of a system or exposure of the proprietary data.

The documentation and validation procedure will have to prove that the LLM systems are working as they should and can meet the required performance levels. This includes evaluating the model for accuracy, checking for bias, and keeping an eye on performance in the future. The records should be adequate to show that AI is being used responsibly and to make it easier for regulators to check. Control of changes and version control techniques include making sure that any changes to ILM systems that are not approved are done with skimming and endorsement.

8. Performance Evaluation and Metrics

8.1 Technical Performance Metrics

To get an instant performance measurement of corporate LLM systems, you need to look at a variety of factors, including as technical performance, business value, and user happiness. Technical performance measurements are objective indicators of system capability and reliability, essential for operational management and optimization decision making (Chang et al., 2023). These indicators should be well-balanced and made to take the needs and limits of business environments.

Response times and throughput are two important performance metrics for a system that are directly related to how well it works for users and how well it works in general. When testing response time, you need to think about the kinds of questions users can ask and how hard they are. For example, entertainment users can ask simple factual questions or more complicated analytical questions. The throughput metrics would need to be connected to the number of users at any one time and the ability to keep going when the load changes.

Due to the subjective nature of many corporate applications, measuring accuracy and reliability is especially difficult when evaluating LLMs. Classification Metrics Traditional accuracy measurements for classifying data may not be applicable to generative tasks that yield multiple right responses (Chang et al., 2023). Organizations must create evaluation processes that assess the quality of responses, factual accuracy, and their appropriateness for certain business contexts. The metrics of resource utilization can tell you how well the system works and how much it costs to run. These include the demand for computational resources, memory, and storage in any deployment setup. By studying how resources are used, you may improve the way infrastructures are spread out and find ways to cut expenses without affecting performance.

8.2 Business Impact Assessment

The business impact assessment connects the technical performance to the results and value that the firm creates (Head et al., 2023). This will mean establishing measuring mechanisms that can indicate how LLM systems help businesses reach their goals by making operations more effective, keeping customers happy, and bringing in more money. The challenge lies in distinguishing the impacts of LLM implementation from other factors influencing corporate success (Park et al., 2025).

The metrics for productivity enhancement are predicated on the assessment of efficiency achieved by LLM automation and augmentation. This might mean doing jobs faster, needing less hands-on help, and making routine tasks more accurate. When measuring productivity, you should look at both direct and indirect factors, like how well individuals can focus and how easily they can switch jobs to ones that are more important.

Cost reduction analysis looks at the economic benefits of using LLM, both the direct savings that can be made through automation and the indirect savings that can be made through better decision making and risk management. A complete cost analysis must also be done that looks at the expenses of putting the plan into action, the costs of keeping it running, and the costs of other options. Calculating the return on investment is what led to the choice to invest more in LLC. Quality improvement measures look at how LLM systems affect the quality of output, the number of errors, and the consistency of the work. It is especially important when you work in areas that deal with customer service, following the rules, and helping others make decisions. Quality measures ought to match the present way the business handles quality and how well the industry is doing.

8.3 User Experience and Adoption Metrics

The metrics of user experience provide a comprehensive analysis of the operational efficacy of LLM systems and their adoption among organizational stakeholders (Head et al., 2023). They are not only technical measurements that look at how well the technology works; they also look at usability, happiness, and adoption patterns, which are what really determine whether the LLM implementations are successful. Understanding how users respond and what they want will help the system design get better and work better all the time.

Surveys and feedback from users give us qualitative data on how well the system works and what we can do to make it better. The survey programs that are done on a regular basis should look at different parts of the user experience, such as how easy it is to use, how good the response is, how reliable the system is, and how happy users are with the features that LLM offers. Feedback should also be a part of the routine workflows because it will let as many people as possible take part and give useful information.

Adoption rate research looks at how quickly and how often users have started using the features of the LLM in their daily job. This means looking at how many active users there are, how often features are used, and how well it fits into the current workflows. By having knowledge about adoption trends, it becomes easier to detect successful applications as well as obstacles to deployment on a larger scale (Xu, 2025).

Math MIS Analysis Support Request Analysis supply will give information on common user difficulties and system restrictions. Monitoring or keeping records of support ticket volumes, types and time of resolution will help pinpoint any areas of potential need in more training, documentation, or system improvements. Support interaction could also be analyzed to expose self-service capability and automated support opportunities.

9. Challenges and Limitations

9.1 Technical Challenges

Implementation of Enterprise LLM is also associated with prominent technical challenges which need to be considered and treated carefully. These problems in many cases may be different to that seen in consumer applications since size, complexity, and reliability of an enterprise environment (Tyagi et al., 2020).

Accuracy concerns and hallucination are issues of continuous concern with enterprise LLM applications in which misleading information can be deadly to a business. Compared to consumer applications, where expensive or inaccurate systems will be accepted, more reliable automatic systems are needed in enterprise systems with low inaccuracy and factual systems (Li et al., 2024b; Di Maio et al., 2024). These risks need to be avoided by having rigorous validation systems, fact checking tools, and error report systems that are applied by organizations.

Scalability weaknesses come out clearly when enterprises systems are required to provide services to thousands of simultaneous users with different query types and complexity level. Conventional scaling methods are likely to be inadequate when working with LLM workloads with unpredictable calculation patterns and memory access patterns. Complex load balancing and resource allocation mechanisms have to be initiated to ensure that organizations perform well when demand conditions are changing.

Complexity in integration rises to a high in an enterprise scenario where various legacy systems, data format, and operation processes are in play. The operation of LLM systems should be integrated with the pre-existing Enterprise Resource Planning (ERP) systems, Customer Relationship Management (CRM), and applications purpose-designed to match the industry (Gangapatnam, 2025). This integration may be associated with custom development and continuous maintenance.

The continuous issue with the maintenance of the performance of LLM in a fluid business world is model drift and degradation over time. Since business processes and data patterns continuously evolve, an LLM's predictions and outputs can become inaccurate or irrelevant if the model is not regularly updated.

9.2 Organizational and Cultural Barriers

The implementation of the enterprise LLM must overcome the major obstacles of organizational and cultural factors that can hinder these successful implementations and acceptance of the system. The main barriers to the adoption of AI can be problems related to deterioration of jobs and loss of control or distrust towards AI. Workers will have concerns that the introduction of LLM systems will drive them out of their job or undermine the importance of human skills. Organizations need to shape the ways of communicating to stress on augmentation and not replacement as well as showing the way LLM boost human abilities.

AI literacy and LLM management are faced with skills gaps that have critical challenges to organizations investing in the systems. Not all enterprises are internally knowledgeable about AI model management, timely engineering, and optimization of LLM. This skills gap needs heavy investment of training total or out-source consulting services to bridging capacity thus affecting both implementation timelines and that of the budget.

Governance and oversight systems might not be sufficient to elaborate the risks and responsibilities that come with deployment of the LLM. The traditional IT governance frameworks usually do not work for AI systems since they have unique needs, like model bias, explainability requirements, and ethical frameworks. Organizations need to find new ways to do things that will let them come up with new ideas and keep their level of operational effectiveness.

9.3 Regulatory and Compliance Challenges

The rules for corporate AI applications are still changing, which makes it hard for companies that use LLM systems to know what to do and follow the rules. Different jurisdictions and sectors have different needs that could affect how LLM is used and the processes (Das et al., 2024).

GDPR, CCPA, and privacy and data governance standards that are specific to certain industries make it very difficult for LLM systems that handle personal or sensitive data to comply. Organizations should ensure that LLM data manipulation processes are executed in compliance with all privacy regulations without compromising system functionality (Byun et al., 2025). This may necessitate the implementation of data anonymization, consent management, and user rights fulfillment functionalities.

Different places are making it a legal necessity to explain and be open about algorithms. This could affect how corporations utilize and develop LLM systems. Some applications may require comprehensive modelling descriptions, while others may need to illustrate the absence of prejudice or discrimination. Those requirements may not work with the fact that many LLM architectures are black boxes.

Healthcare, financial services, and other industry-specific rules add to the compliance obligations that LLM must focus on when it is put into place. These kinds of rules can spell out what model validation, audit trail, and other professional oversight that affects system design and running operations should look like.

When multinational companies have to follow different rules in different places, international and cross-border variables come into play. The requirements for data residency, export control rules, and different privacy laws can all lead to the need for complex implementation systems and procedures.

10. Conclusion

Adding Large Language Models to businesses is a chance that goes far beyond just making things easier or more efficient. As we have seen in this chapter, LLM can change the way a business does automation, analytics, and knowledge management in some very important ways. This makes it possible for a business system to become smarter, more adaptable, and more responsive.

The previous paradigm relied on human judgment and interpretation, while the current paradigm is based on intelligent automation powered by LLM, which has a high ability to make assumptions about how things should be done and to model complex procedures that were once thought to require human judgment and interpretation. This is a paradigm shift. Business process automation that uses LLM can take variances into account, grasp context, and make judgments that adapt to new situations. This makes automation possible in even more areas. The LLM's natural language understanding and reasoning algorithms help with document processing, making processes more efficient, and putting policies into action.

In analytics, LLMs make it easier for everyone to get help with difficult datasets by starting a natural language exchange that uses unstructured datasets. The capacity to ask questions about the data in plain language, make automatic insights, and explain analytical results in plain language makes advanced analytics accessible to more people in the business. Applications for money financial analytics, predictive modelling, and business intelligence applications demonstrate the capacity to revolutionize decision support systems through the incorporation of LLM. LLM-based knowledge management systems address conventional challenges in information retrieval, knowledge generation, and knowledge dissemination. Contextual literacy, the ability to generalize and synthesize data from diverse sources, along with the concept of interactive discourse, transforms how organizations capture, organize, and disseminate institutional knowledge. The capabilities are quite significant for large companies with complex information systems and a lot of knowledge spread around.

However, for enterprise LLM to work, it needs to be closely watched in terms of technical, organizational, and regulatory issues. The detailed frameworks should take into account security and privacy to protect critical company data and make it easier to create new applications. The best ways to improve performance should look at the unique needs of the business and its users. For an organization to get the benefits of LLM technologies, it needs to be able to manage change and enhance its capabilities.

The trend of integrating enterprise processes in the framework of LLM is only beginning, but the potential for it to have a huge effect on automation, analytics, and knowledge management is clear. Institutions that are intentionally and thoughtfully embracing this transition, taking into account both the opportunities and the challenges, which are likely to be in a good position to tap into the huge potential of the benefits that LLM technologies may bring to businesses. In most situations, the idea of putting technology next to organizational capabilities is expected to make LLMs a real part of enterprise technology landscapes instead of just solutions for one purpose.

References

Arora, N., Chakraborty, I. and Nishimura, Y. (2024). AI--Human Hybrids for Marketing Research: Leveraging Large Language Models (LLMs) as Collaborators. Journal of Marketing 89: 43–70.

Badurowicz, M., Skulimowski, S. and Laskowski, M. (2024). FEASIBILITY OF USING LOW-PARAMETER LOCAL LLMS IN ANSWERING QUESTIONS FROM ENTERPRISE KNOWLEDGE BASE. Applied Computer Science.

Balakrishnan, G. and Purwar, A. (2024). Evaluating the Efficacy of Open-Source LLMs in Enterprise-Specific RAG Systems: A Comparative Study of Performance and Scalability. 2024 IEEE 21st India Council International Conference (INDICON) 1–9.

Bodensohn, J.-M., Brackmann, U., Vogel, L., Sanghi, A. and Binnig, C. (2025). Unveiling Challenges for LLMs in Enterprise Data Engineering. ArXiv, abs/2504.10950.

Bodensohn, J.-M., Brackmann, U., Vogel, L., Urban, M., Sanghi, A. and Binnig, C. (2024). LLMs for Data Engineering on Enterprise Data.

Byun, G., Lee, S., Choi, N. and Choi, J. (2025). Secure Multifaceted-RAG for Enterprise: Hybrid Knowledge Retrieval with Security Filtering. ArXiv, abs/2504.13425.

Chang, Y.-C., Wang, X., Wang, J., Wu, Y., Zhu, K., Chen, H. et al. (2023). A Survey on Evaluation of Large Language Models. ACM Transactions on Intelligent Systems and Technology 15: 1–45.

Chowdhary, N., Dutta, T., Chattopadhyay, S. and Chakraborty, S. (2025). AutoPAC: Exploring LLMs for Automating Policy to Code Conversion in Business Organizations. 2025 17th International Conference on COMmunication Systems and NETworks (COMSNETS), 667–675.

Das, B., Amini, M. and Wu, Y. (2024). Security and Privacy Challenges of Large Language Models: A Survey. ACM Computing Surveys 57: 1–39.

De Michele, W., Armas-Cervantes, A. and Frermann, L. (2025). Automated Business Process Analysis: An LLM-Based Approach to Value Assessment.

Demiralp, Çağatay, Wenz, F., Chen, P. B., Kayali, M., Tatbul, N. and Stonebraker, M. (2024). Making LLMs Work for Enterprise Data Tasks. ArXiv, abs/2407.20256.

Di Maio, C., Cosci, C., Maggini, M., Poggioni, V. and Melacci, S. (2024). Pirates of the RAG: Adaptively Attacking LLMs to Leak Knowledge Bases. ArXiv, abs/2412.18295.

Eigner, E. and Händler, T. (2024). Determinants of LLM-assisted Decision-Making. ArXiv, abs/2402.17385.

Fan, T., Kang, Y., Guoqiang, Chen, W., Wei, W., Fan, L. and Yang, Q. (2023). FATE-LLM: A Industrial Grade Federated Learning Framework for Large Language Models. ArXiv, abs/2310.10049.

Fernandez, R., Elmore, A., Franklin, M., Krishnan, S. and Tan, C. (2023). How Large Language Models Will Disrupt Data Management. Proc. VLDB Endow., 16: 3302–3309.

Gangapatnam, K. (2025). Revolutionizing Enterprise Resource Planning Through AI Integration: A Technical Deep Dive. European Journal of Computer Science and Information Technology.

Grohs, M., Abb, L., Elsayed, N. and Rehse, J.-R. (2023). Large Language Models can accomplish Business Process Management Tasks. ArXiv, abs/2307.09923.

Guan, Y., Wang, D., Chu, Z., Wang, S., Ni, F., Song, R. et al. (2024). Intelligent Agents with LLM-based Process Automation. Proceedings of the 30th ACM SIGKDD Conference on Knowledge Discovery and Data Mining.

Head, C. B., Jasper, P., McConnachie, M., Raftree, L. and Higdon, G. (2023). Large language model applications for evaluation: Opportunities and ethical implications. New Directions for Evaluation 2023: 33–46.

Hosseini, A., Kastner, W. and Sauter, T. (2025). Leveraging LLMs and Knowledge Graphs to Design Secure Automation Systems. IEEE Open Journal of the Industrial Electronics Society, 6: 380–395.

Hou, X., Zhao, Y., Liu, Y., Yang, Z., Wang, K., Li, L. et al. (2023). Large Language Models for Software Engineering: A Systematic Literature Review. ACM Transactions on Software Engineering and Methodology 33: 1–79.

Jeong, C. (2023). A Study on the Implementation of Generative AI Services Using an Enterprise Data-Based LLM Application Architecture. Adv. Artif. Intell. Mach. Learn. 3: 1588–1618.

Jeong, C. (2024). Fine-tuning and Utilization Methods of Domain-specific LLMs. ArXiv, abs/2401.02981.

Kayali, M., Wenz, F., Tatbul, N. and Demiralp, cCaugatay. (2024). Mind the Data Gap: Bridging LLMs to Enterprise Data Integration. ArXiv, abs/2412.20331.

Khemka, A. and Priyanshi, Er. (2025). Generative AI in Enterprise Data Warehousing: Leveraging LLMs for improving data quality and business intelligence. International Journal for Research Publication and Seminar.

Kourani, H., Berti, A., Schuster, D. and Aalst, W. (2024a). Evaluating Large Language Models on Business Process Modeling: Framework, Benchmark, and Self-Improvement Analysis. ArXiv, abs/2412.00023.

Kumar, P. (2024). Large language models (LLMs): Survey, technical frameworks, and future challenges. Artif. Intell. Rev., 57: 260.

Kumar, R., Ishan, K., Kumar, H. and Singla, A. (2025). LLM-Powered Knowledge Graphs for Enterprise Intelligence and Analytics. ArXiv, abs/2503.07993.

Li, G., Zhou, X. and Zhao, X. (2024a). LLM for Data Management. Proc. VLDB Endow. 17: 4213–4216.

Li, J., Yuan, Y. and Zhang, Z. (2024b). Enhancing LLM Factual Accuracy with RAG to Counter Hallucinations: A Case Study on Domain-Specific Queries in Private Knowledge-Bases. ArXiv, abs/2403.10446.

Li, W., Liu, W., Deng, M., Liu, X. and Feng, L. (2025). The impact of large language models on accounting and future application scenarios. Journal of Accounting Literature.

Liu, Y.-H., He, H., Han, T., Zhang, X., Liu, M., Tian, J. et al. (2024). Understanding LLMs: A Comprehensive Overview from Training to Inference. ArXiv, abs/2401.02038.

MathavRaj, J., Kushala, V., Warrier, H. and Gupta, Y. (2024). Fine Tuning LLM for Enterprise: Practical Guidelines and Recommendations. ArXiv, abs/2404.10779.

Musunuri, A. (2025). Enterprise Data Security and Integration with LLMs. International Journal of Innovative Research in Science Engineering and Technology.

Nasseri, M., Brandtner, P., Zimmermann, R., Falatouri, T., Darbanian, F. and Obinwanne, T. (2023). Applications of Large Language Models (LLMs) in Business Analytics--Exemplary Use Cases in Data Preparation Tasks.

Naveed, H., Khan, A. U., Qiu, S., Saqib, M., Anwar, S., Usman, M. et al. (2023). A Comprehensive Overview of Large Language Models. ArXiv, abs/2307.06435.

Oleynik, A., Fedorov, A., Datyev, I., Zuenko, A., Shestakov, A., and Vishnyakov, I. (2025). Using Rag Technology to Design An Intelligent Information System For Support Exploratory Search. Transaction Kola Science Centre.

Park, H., Oh, H., Gao, F. and Kwon, O. (2025). Enhancing Analytic Hierarchy Process Modelling Under Uncertainty With Fine-Tuning LLM. Expert Systems, 42.

Raiaan, M. A. K., Mukta, Md. S. H., Fatema, K., Fahad, N. M., Sakib, S., Mim, Most. M. J. et al. (2024). A Review on Large Language Models: Architectures, Applications, Taxonomies, Open Issues and Challenges. IEEE Access 12: 26839–26874.

Raza, M., Jahangir, Z., Riaz, M. B., Saeed, M. J. and Sattar, M. A. (2025). Industrial applications of large language models. Scientific Reports, 15.

Schnepf, J., Engin, T., Anderer, S. and Scheuermann, B. (2024). Studies on the Use of Large Language Models for the Automation of Business Processes in Enterprise Resource Planning Systems.

Tyagi, A. K. and Abraham, A. (2020). Artificial intelligence and machine learning algorithms. pp. 178. *In:* Challenges and applications for implementing machine learning in computer vision.

Tyagi, A. K., Goyal, D. and Rekha, G. (2020). A look at top 35 problems in the computer science field for the next decade. pp. 379–396. *In:* ICT for competitive strategies.

Vertsel, A. and Rumiantsau, M. (2024). Hybrid LLM/Rule-based Approaches to Business Insights Generation from Structured Data. ArXiv, abs/2404.15604.

Wang, S., Zhao, Y., Hou, X. and Wang, H. (2024). Large Language Model Supply Chain: A Research Agenda. ACM Transactions on Software Engineering and Methodology, 34: 1–46.

Xia, Y., Shenoy, M., Jazdi, N. and Weyrich, M. (2023). Towards autonomous system: Flexible modular production system enhanced with large language model agents. 2023 IEEE 28th International Conference on Emerging Technologies and Factory Automation (ETFA), 1–8.

Xing, F. (2024). Designing Heterogeneous LLM Agents for Financial Sentiment Analysis. ACM Transactions on Management Information Systems, 16: 1–24.

Xu, Q. (2025). Practical Applications of Large Language Models in Enterprise-Level Applications. Journal of Computer Science and Artificial Intelligence.

Yang, J., Jin, H., Tang, R., Han, X., Feng, Q., Jiang, H. et al. (2023). Harnessing the Power of LLMs in Practice: A Survey on ChatGPT and Beyond. ACM Transactions on Knowledge Discovery from Data, 18: 1–32.

Yao, H., Shi, H., Chen, Y., Jiang, Y., Wang, C. and Qin, Z. (2025). ControlNET: A Firewall for RAG-based LLM System. ArXiv, abs/2504.09593.

Zheng, Y., Chen, Y., Qian, B., Shi, X., Shu, Y. and Chen, J. (2024). A Review on Edge Large Language Models: Design, Execution, and Applications. ACM Computing Surveys, 57: 1–35.

Zhu, Y., Wang, X., Chen, J., Qiao, S., Ou, Y., Yao, Y. et al. (2023). LLMs for Knowledge Graph Construction and Reasoning: Recent Capabilities and Future Opportunities. World Wide Web (WWW), 27: 58.

11

Leveraging Large Language Models for Fraud Detection and Risk Assessment in Financial Institutions

A Comprehensive Review

Aashka Thakkar,[1,*] *Axita Thakkar*[1] and *Habtamu Ditta Hirpo*[2]

1. Introduction

Fraudulent activities are becoming more common and advanced than ever, especially with data innovation and digital banking. This poses a major challenge for international financial institutions. Traditional anti-fraud methods, which rely on mathematical models and rules, are struggling to keep up with new and constantly changing threats (Malingu et al., 2025). This is particularly important given the surge in online banking, real-time payment platforms, and online transactions. The rise of Large Language Models (LLMs), such as OpenAI's GPT series and Google's BERT, gives financial companies a new way to tackle this issue (Bakumenko et al., 2024). These models can help detect, predict, and reduce fraud while also improving risk monitoring. Pattern identification, recognising anomalies, context-specific inference, and natural language comprehension are among the impressive skills demonstrated by LLMs, which are based on converter architecture and educated on

[1] Faculty of Management Studies - MBA Program, Parul University, India.
Email: axita.thakkar77775@paruluniversity.ac.in, ORCID ID 0000-0001–7592–3537
[2] Berhan Bank SC, Addis Ababa, Ethiopia.
Email: lehabtamu20@gmail.com, https://orcid.org/0000-0002-7102-4654
* Corresponding author: aashka.thakkar@paruluniversity.ac.in, ORCID ID: 0000-0003-3629-7603

DOI: 10.1201/9781003771654-11

extensive written datasets. These algorithms may help with real-time risk grading based on client histories and third-party data, identify phishing or fraudulent activity attempts, highlight subtle signs of suspected behaviour, and even spot anomalies in documents (Jiang, 2024). LLMs involving identification of fraud and economic risk assessment have specific difficulties, despite their enormous potential. Legislative adherence, ethical fairness, confidentiality of data, model ability to be understood, and operational implementation are some of these concerns. Furthermore, the financial information environments are frequently unbalanced, compartmentalised, and highly susceptible to adversary exploitation; therefore, LLMs structures must be modified for a domain-specific adaptability. To guarantee conclusions that are apparent, enforceable, and reliable, LLMs must be integrated in conjunction with expertise surveillance, legacy fraud detection technologies, and unstructured economic information.

In this study work, the use of LLMs in many aspects of risk identification and detection of fraud is thoroughly examined. It examines operational structures, model frameworks, scenarios of use, and regulatory ramifications. This research also uses bibliometric techniques to visualise collaboration themes, subject groupings, and technological advances in this new sector. In order to extensively use LLMs in creating robust and sophisticated financial ecosystems, the study attempts to offer a path for researchers, financial technology developers, and regulators by combining technological, institutional, and conceptual components (Korkanti, 2024). Additionally, to address these problems, our suggested study intends to improve the identification of financial fraud by employing data analysis and Large Language Models. With their impressive ability to comprehend and produce writing that appears human-like, LLMs provide exciting opportunities for identifying minute patterns and irregularities in textual information that point to fraudulent activity (Wang et al., 2025).

2. LLMs in Financial Services

In recent years, artificial intelligence has become widely adopted in a variety of financial fields. They highlight important applications for finance throughout the research, such as financial risk modelling, investing, portfolio management, financial data analysis, and financial advice and customer support. These areas exhibit a great deal of interest and promise with the development of artificial intelligence (Nie et al., 2024). By empowering automated systems to comprehend and create language that is similarto that of humans, large language models, are fundamentally transforming the field of artificial intelligence. LLMs are exceptionally efficient in activities like summarising, translating, forecasting, and producing material with great precision because they have been extensively developed on extensive datasets. They have proven beneficial in a number of sectors because to their capacity to identify language patterns and implement them in context (Xie et al., 2024). Considering that more than $155 trillion in assets administered globally as of the year 2023, the worldwide financial sector is massive. With such an extensive reach, it is essential to provide accurate and effective solutions. One example of just how LLMs may improve operations and service quality is JPMorgan Chase, which handles over 43 million transactions every single day. Banking-specific LLMs might be modified to match the specific demands of the sector by utilising a variety of available sources, including financial data, legislative documents, and interactions with customers (Thakkar et al., 2025). In addition to strengthening the system's comprehension and production of the banking operations, this aids institutions in providing better customer service and navigating the complexities of the financial services sector. LLMs appeared to be an effective instrument in the field of detecting fraud (Liu et al., 2023). Examining financial information closely, they are competent at recognising irregularities that can indicate fraudulent activity (Singh et al., 2025). Customers may feel secure knowing that an additional layer of protection is in operation. Implementing a robust fraud prevention system helps financial institutions reduce risks and maintain their credibility.

3. Bibliometric Analysis

Employing the R programming language for bibliometric analysis is an analytical method for assessing and visualising publication mechanics, authorship patterns, and investigation trends. The most cited publications, prolific authors, keyword trends, collaborative networks, and yearly scientific production may all be evaluated by researchers. Because R is open-source, reproducible, and offers customisable visualisations, it is very beneficial for bibliometric analysis. It supports academics in mapping information structures, identifying research gaps, and making data-driven choices. All things considered; bibliometric analysis is an effective method for gaining evidence-based decision, comprehending of scientific research. Attempting to address this research problem, a bibliographic review of 78 publications on LLMs has been carried out. The current study assesses how financial institutions implement the LLMs by using data gathered from research papers from the SCOPUS Database that have been published between 2003 and 2025.

3.1 Annual Scientific Production

From Figure 1 Annual Scientific Production it shows employing Large Language Models to Identify Fraud and Assessing Risk within the financial sector is the focus of an extensive number of academic reviews, which is apparent in the chart from 2002 to 2025. The number of articles of academic research published between 2002 and 2017 remained limited and inconsistent, averaging between 0 and 2 papers a year. This period of time represents the beginning stages of work on the application of artificial intelligence, specifically LLMs, in risk and fraudulent transactions scenarios. A handful of small rises could be related to more significant developments in artificial intelligence or the early use of machine learning techniques in financial analytics. Scientific production was irregular and low from 2002 and 2015, with most years having either zero or one publication. Which demonstrates how sophisticated AI applications in the field of finance, especially in the specific field of LLMs, are still in developmental stages. Regulatory approaches throughout this time have been dominated by conventional techniques like predictive modelling and rules-driven identification of fraud. Furthermore, the development of deep learning methods has

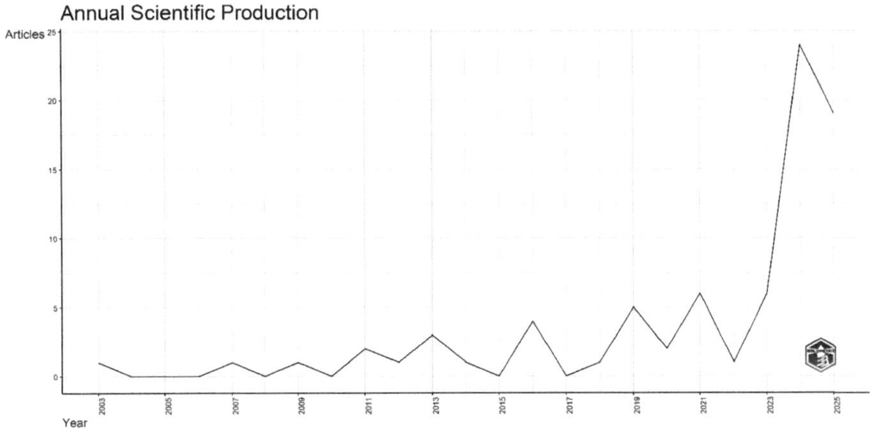

Figure 1: Annual Scientific Publications in LLMs fraud detection research (2022–2025).

been hampered by limited availability of data as well as computer capabilities. The adoption of artificial intelligence for financial risk evaluation grabbed the interest of the academic community between 2016 and 2022 (refer Figure 1).

This interest was probably stoked by developments in machine learning, natural language processing, and the rise in prominent cases of fraud across the world.

The visualisation's most notable finding is the significant increase in research output from 2023 to 2024. The turning point milestone for the subject matter was marked in 2023 when the total number of published publications increased from six to over 24, almost four times increase. This increase is consistent with many of relevant aspects firstly, the quick creation and widespread use of potent LLMs such as GPT-4 and its derivatives, secondly the post- pandemic surge of online financial transactions, and that has made people more vulnerable to complex fraud schemes; and finally, the growing regulatory demands on a global scale for aggressive fraud prevention strategies and real-time risk monitoring.

3.2 Thematic Evaluation

The given Figure 2 of thematic visualisation is a conceptual visualisation that groups research topics based on two important parameters: density and centrality (refer Figure 2). The above sort of evaluation is useful for assessing the thematic growth and cognitive framework of the field being studied, in the instance of the use of Large Language Models (LLMs) for risk evaluation and detecting fraud in banking organisations. The research on the ecosystem is separated into four categories by the quadrants: Basic Themes, Emerging or Declining Themes, Niche Themes, and Motor Themes.

The motor themes that are simultaneously essential to the subject matter and thoroughly researched are at the very centre of the thematic map. They involve "anomaly detection," "large language model," "financial institution," "machine learning," "investments," and "data analysis." Themes like these constitute the backbone of the study and identify established, high-impact fields where notable

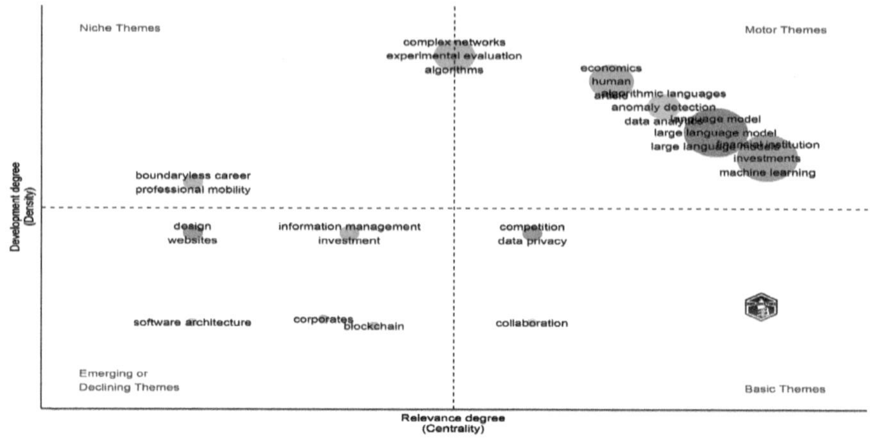

Figure 2: Thematic Visualisation showing research domain clusters by centrality and density.

advances have been made in both academia and practice. These phrases' co-occurrence indicates that LLMs are frequently employed in financial institutions to manage risk associated with investments, identify fraud, and comprehend intricate patterns. This highlights how important LLMs are to the modern financial system, particularly as organisations seek more sophisticated and adaptable approaches for combating evolving risk. Niche themes like "complex networks," "experimental evaluation," and "algorithms," on the other hand, are extremely sophisticated and technically strong but have little bearing on more general multidisciplinary study. These areas of study frequently reflect advances in theory or computing that, while crucial, might not be as immediately valuable to business provided, they are successfully connected to real-world scenarios. Their computational precision, however, serves as the foundation for advancements that lead to more useful tools, particularly when it comes to developing and instructing LLMs for tasks specific to an area of expertise (Gong et al., 2024). The phrases "collaboration," "data privacy," and "competition" are found at the bottom-right quadrant of the chart, which is referred to as core topics. Despite being highly vital and fundamental to the field, these subjects are nevertheless lacking in depth. A possible discrepancy between institutional preparedness and technical development is highlighted by the paucity of study in these areas.

3.3 Bradford's Law

The "Core Sources by Bradford's Law" visualisation provides bibliometric details (refer Figure 3) regarding how scientific research in the topic of Harnessing Large Language Models (LLMs) towards fraud detection and risk assessment in financial institutions are distributed all throughout academic publications and proceedings from conferences. According to Bradford's Law, a significant proportion of papers on a particular subject will appear in a relatively limited number of core publications, with later publications dispersed over a greater number of less productive outlets.

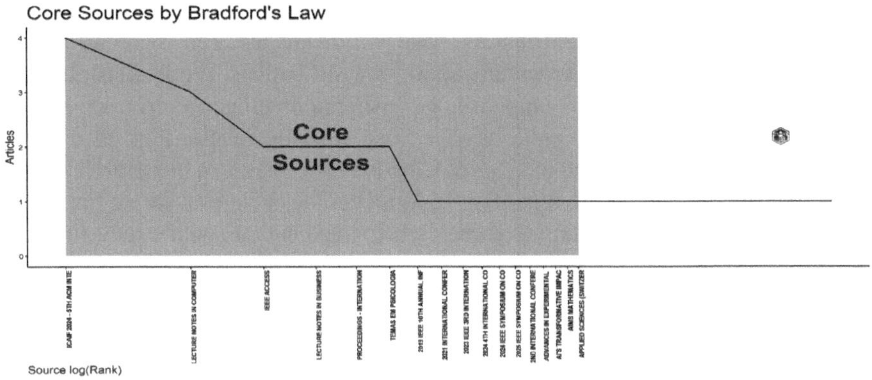

Figure 3: Core sources distribution according to Bradford's Law.

The "Core Sources" coloured area draws attention to the select few publications and conferences which contributed most of the material in this emerging field. ACM International Conference on AI in Finance, IEEE Access, Lecture Notes in Computer Science, Lecture Notes in Business, and publications from esteemed conferences like ICAF 2024 are noteworthy sources. These resources serve as the focal point of research activity and play a crucial role in sharing both basic and innovative findings on the practical use of LLMs in managing financial risks. Bradford's analysis states that study output drops substantially beyond the key areas is backed up by the graph's curve, which shows a significant decrease after the first few sources.

3.4 Country Collaboration Visualisation

The Country Collaboration Map provides comprehensive data (refer Figure 4) on the changing patterns of worldwide studies on the use of large language models in financial institution risk assessment and fraud detection. This map shows which nations are the most actively involved in this emerging area of study by highlighting the locations and level of collaboration across nations. Higher quality research

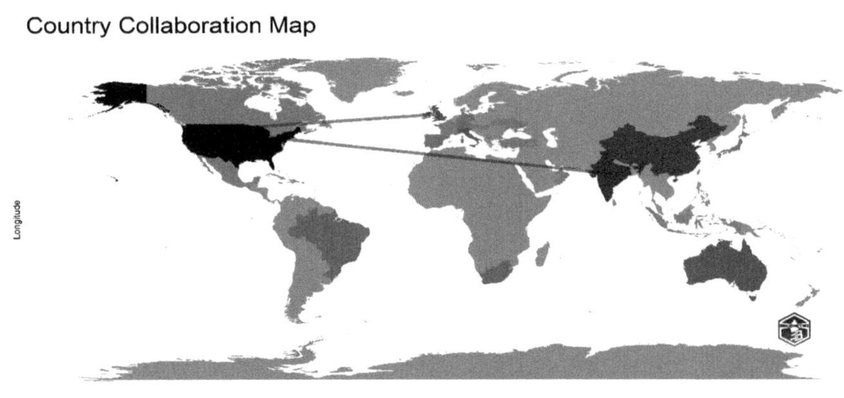

Figure 4: International collaboration network in LLM fraud detection research.

is represented by more intense colours of blue, with the US, India, as well as the UK developing as major contributors. The worldwide research relationships are highlighted by the lines of communication between nations. As an illustration of the two countries' common emphasis on artificial intelligence-driven financial technologies, the USA and India have a close working relationship. Strong ties between the United States and the United Kingdom also suggest a bilateral sharing of ideas, methods, and expertise regarding LLM technologies for preventing fraudulent activities in finance. According to these partnerships, nations with strong financial systems and revolutionary artificial intelligence infrastructure have led the way in developing and implementing LLMs for risk-based assessment.

3.5 Lotka's Law

Lotka's Law states (refer Figure 5) that while many authors only produce a small number of works, a large proportion of academic output is produced by a relatively small number of publications. The graph's sharply declining curve demonstrates this idea. A substantial number of those who contributed have solely published a single research article, with only a handful publishing numerous papers, as indicated by the curve's rapid decrease.

The current trend emphasises how new and quickly developing the subject of fraudulent activity detection and risk management studies for LLMs is. The substantial number of a single authors points to a desire for enhanced academic commitment as well as elevated cross-disciplinary interest and research. The anticipated frequency according to Lotka's theoretical model is represented by a diagonal line on the visualisation, and it precisely matches the empirical information.

3.6 Average Citations per Year

The "Average Citations per Year" statistics offers (refer Figure 6) an in-depth examination employing the academic impact of works pertaining to the use of LLMs for risk evaluation and identification of fraud in banking organisations. The frequency with which papers published in this field have been cited each year is

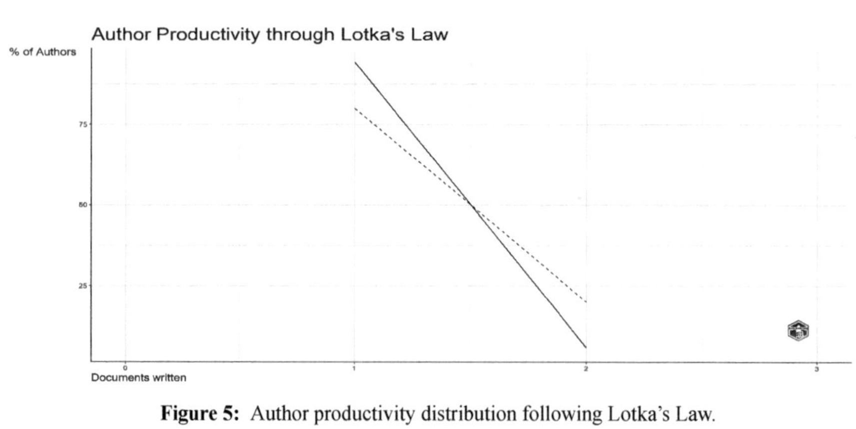

Figure 5: Author productivity distribution following Lotka's Law.

represented in this bibliometric visualisation, which reflects the subject's increasing status and significance within the field of education. The overall citation average was sluggish and comparatively low from the beginning of the 2000s to 2016, which may indicate either an absence of specific studies on this topic or are lacking awareness of its significance. Still, there had been noticeable increases in referencing rates after 2017.

The unexpected spike in 2022, whereby an average citation rate mounted six references per piece, demonstrates a considerable spike in interest in academia. Increasing awareness of the possible benefits of LLMs to reduce fraudulent transactions and enhance risk modelling, or an increasing number of significant publications released around this time, might be the cause of the phenomenon. A notable decrease is noted after 2022, which may be due to an expansion of subject matter across the broader AI-fintech landscape or an upsurge in referencing for newly published work. In general, data visualisation highlights how LLMs are receiving greater attention from scholars in the field of fraudulent activity detection, highlighting its newfound prominence as a demanding field of study.

3.7 Most Relevant Words

The graph (refer Figure 7) the widespread use of important phrases in academic research on the subject of "Leveraging Large Language Models for Fraud Detection and Risk Assessment in Financial Institutions." Commonly included in bibliometric studies, this frequency-based evaluation offers an essential instrument for comprehending the main ideas and terms guiding contemporary research. Having 28 instances, the phrase "language model" ranks first, solidifying its position as the domain's core concept. This points out how important programming models serve the purpose of implementing fraud detection and enhancing financial risk identification procedures, particularly for massive operations architectures such as GPT. The concentration on cutting-edge artificial intelligence that can comprehend, produce, and assess intricate financial writings and behaviour is further supported by the following most common word, "large language model," which appears 17 times.

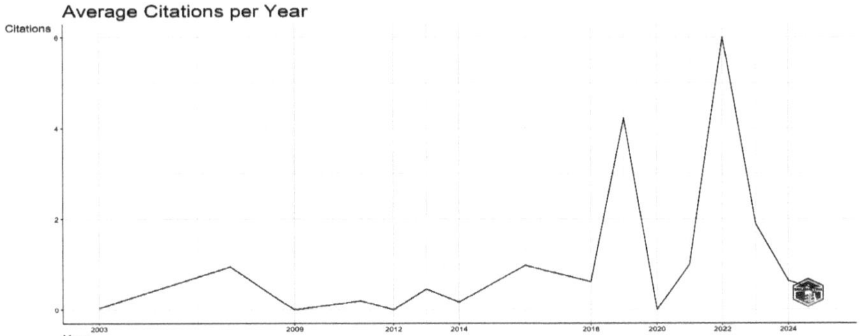

Figure 5: Average citations per year for publications (2002–2025).

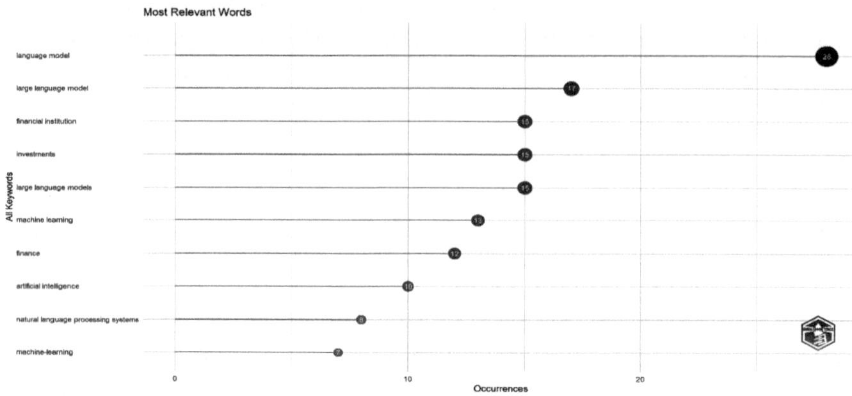

Figure 7: Word frequency analysis of most relevant terms.

The phrases "financial institution," "investments," and "large language models" all emerge 15 times, underscoring the attention placed on both the application itself and technology environments.

3.8 Keywords Co-occurrence

The Figure 8, which was created by employing VOS viewer, is a bibliometric visualisation that highlights the primary investigation subjects and keywords correlations in the field of identification of fraud and large language models (LLMs) in the field of finance. The amount of space of every cluster, which stands for a specific term or research phrase, reflects how frequently or precisely it appears in the examined research. Thematic clustering is suggested by the closeness of vertices to the lines that connect them, which indicate the co-occurrence associations. The

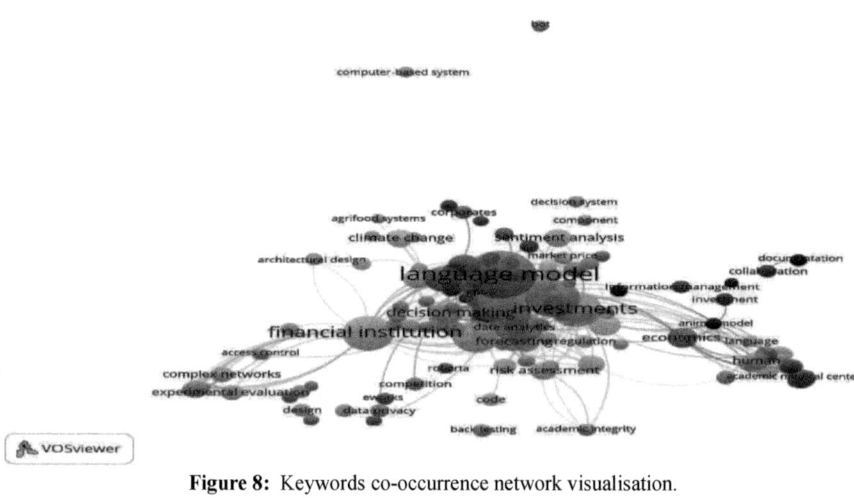

Figure 8: Keywords co-occurrence network visualisation.

words "language model," "financial institution," and "investments" stand prominent at the visualisation's centre, suggesting that they are major conversation points.

Keywords like "decision making," "risk assessment," and "data analytics" have significant overlaps, indicating that LLMs are extensively investigated in assisting with strategically monetary operations, particularly in fields that require evaluating large amounts of data to make forecasts and gain perspectives. Through clustering linked to "risk assessment," "forecasting," and "data privacy," the keyword "fraud detection" is inversely related, suggesting a major emphasis on employing LLMs for anomaly detection, compliance with regulators, and security improvements in digital financial services. Other categories, such "sentiment analysis," "climate change," "corporates," and "complex networks," demonstrate how LLMs have applications throughout different fields of study. Topics consist of risk assessment and modelling, sustainable investments, and market sentiment tracking. By improving fraud detection, assisting with investment choices, guaranteeing compliance, and facilitating sophisticated risk modelling, this graphic representation highlights the growing importance of LLMs in revolutionising banking and financial services. It draws attention to the increasing intersection of regulations fields of study, finance, and artificial intelligence.

3.9 Co-citation Network

Co-Citation Network (refer Figure 9) that represents the conceptual framework that was developed in the discipline of Large Language Models (LLMs) in the fields of finance, which is created through bibliometric examination using the VOS viewer. Some of the graph's edges indicate how frequently multiple authors are referenced collectively in the same academic literature, while every point represents an author. Greater impact in the field of study is shown by bigger points, which represent more often referenced publications.

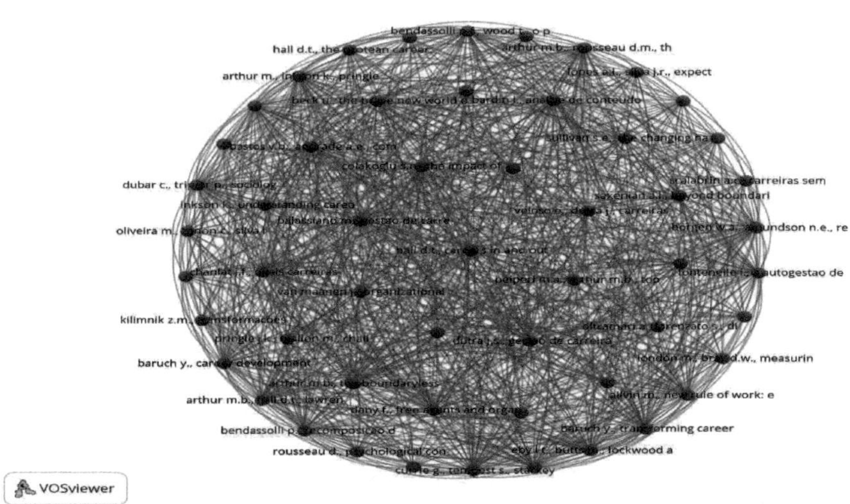

Figure 9: Co- Citation network of influential authors.

Through a wide range of fundamental literature influencing how LLMs are conceptualised for assignments like assessing risk, making ethical choices, fraud detection, and staff behaviour modelling in financial organisations, the complicated network of relationships suggests a well-developed and interconnected discipline of study.

3.10 Three-Field Plot

The data visualisation is a Three-Field Plot (refer Figure 10) constructed that shows the connections involving a few important bibliometric domains like Keywords on the right-hand side, Authors in centre, and Cited References on the left. It offers a visual illustration including the connections connecting significant research topics, contributing academics, and notable research publications across the area of identifying fraud and large language modelling in the financial sector.

The term clustering has identified prominent research subjects such as "language model", "deep learning," financial institution, "risk assessment and "fraud detection". These phrases emphasise how much of the currently underway study is devoted to using LLMs to identify fraudulent activity, assess emotions in securities markets, and

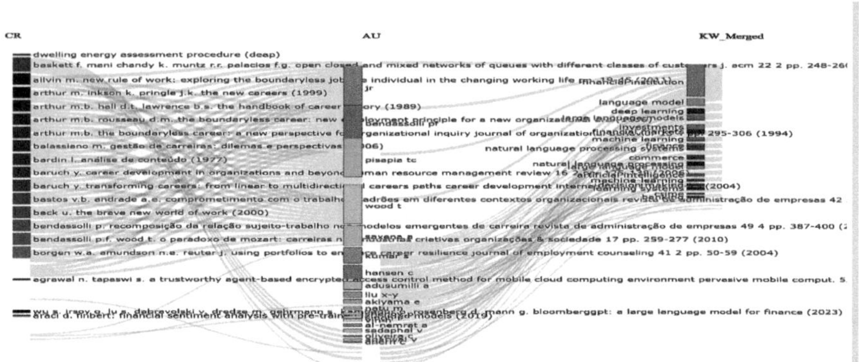

Figure 10: Three Field plot.

capitalise on statistical modelling for helping in high-stakes taking decisions. The reason for this graph's intersection of behavioural economics, artificial intelligence, and financial institutions shows how multidisciplinary concepts have been combined to create LLM-powered strategies for analysis of investments, fraud detection, and adherence to regulations in the field of finance. This demonstrates how essential existing and new publications are in influencing cutting-edge applications.

4. Conclusion

The increasing complexity of economic environments and the rise in sophisticated fraudulent transactions requires financial institutions to rethink how they handle risk assessment and fraud detection. Some key challenges of using LLMs in financial services include integrating data, guiding models, eliminating assumptions, ensuring clear communication, and adhering to regulations (Goldberg, 2024). Large Language Models, backed by recent progress in machine learning and neural networks, can significantly transform how financial institutions evaluate risk and tackle fraud in the sector, as thorough analysis has shown. The extensive use of algorithms like GPT, BERT, and other versions have contributed to a notable increase in academic research in this discipline in the past few years, according to bibliometric analysis of academic work published between 2003 and 2025. Conventional systems driven by rules are unable to properly comprehend unstructured data, evaluate behavioural trends, and detect irregularities across large and complicated data sets. LLMs provides financial firms these capabilities (Mashrur et al., 2020). They are useful supporters in intelligent risk modelling and real-time fraud detection because of their understanding of context, semantic magnitude, and flexibility in adapting inputs to operate. In an age where there are increasing cyber risks and government oversight, incorporating the use of LLMs not simply improves efficiency in operations but it additionally reinforces the reliability of institutions (Jin et al., 2025). Although it continues to be its early stages, a fundamental group of high impact publications and only a handful of effective researchers, according to bibliometric visualisations like Bradford's Law and Lotka's Law, have dominated this field. The significance of a small number of highly influential publications and conferences in promoting studies involving LLMs in financial fraud monitoring is supported by Bradford's Law visualisation. It highlights the significance these channels are for acquiring as well as contributing to the most recent breakthroughs in science. Investigators can better navigate the quickly expanding database of literature and align their findings to the shifting requirements of academic and pragmatic excellence by concentrating on these sources. These findings demonstrate the necessity for increased institutional backing of artificial intelligence-based finance research, multidisciplinary contributions, and broader collaboration. The pragmatic aspect of this topic is further demonstrated by the thematic mapping, which additionally displays a focus on studies on real-world applications including the identification of anomalies, predictive modelling, and risk management. The findings of this research provide a useful framework for administrators, financial institutions, and researchers to make the most of LLMs while managing the risks and challenges they bring. The future change in financial services through artificial neural networks will rely on the ethical and wise adoption of these advancements. Therefore, future studies should focus on demonstrating trust, reliability, transparency, and long-term value in AI-driven financial systems alongside technological progress.

References

Bakumenko, A., Hlaváčková-Schindler, K., Plant, C. and Hubig, N. C. (2024). Advancing anomaly detection: Non-semantic financial data encoding with LLMs. arXiv preprint arXiv:2406.03614.

Ditta, H., Thakkar, A. and Asmelash, A. (2025). The prospects of traditional branch banking in the era of digital dominance. pp. 25–48. *In*: Strategic workforce reskilling in service marketing. IGI Global Scientific Publishing.

Goldberg, A. (2024). AI in finance: Leveraging large language models for enhanced decision-making and risk management. Social Science Journal for Advanced Research 4(4): 33–40. https://doi.org/10.5281/zenodo.13299843.

Gong, J., Wang, Y., Xu, W. and Zhang, Y. (2024). A deep fusion framework for financial fraud detection and early warning based on large language models. Journal of Computer Science and Software Applications 4(8).

Jiang, L. (2024). Detecting scams using large language models. arXiv preprint arXiv:2402.03147.

Jin, R., Wen, Y., Liu, T. and Shen, C. (2025). Large language models for fraud detection. pp. 441–446. *In*: 2025 8th International Conference on Artificial Intelligence and Big Data (ICAIBD). IEEE.

Korkanti, S. (2024). Enhancing financial fraud detection using LLMs and advanced data analytics. pp. 1328–1334. *In*: 2024 2nd International Conference on Self Sustainable Artificial Intelligence Systems (ICSSAS). IEEE.

Liu, P., Yuan, W., Fu, J., Jiang, Z., Hayashi, H. and Neubig, G. (2023). Pre-train, prompt, and predict: A systematic survey of prompting methods in natural language processing. ACM Computing Surveys 55(9): 1–35.

Malingu, C. J., Kabwama, C. A., Businge, P., Agaba, I. A. and Asiimwe, I. (2025). Application of LLMs to fraud detection.

Mashrur, A., Luo, W., Zaidi, N. A. and Robles-Kelly, A. (2020). Machine learning for financial risk management: A survey. IEEE Access 8: 203203–203223.

Nie, Y., Kong, Y., Dong, X., Mulvey, J. M., Poor, H. V., Wen, Q. et al. (2024). A survey of large language models for financial applications: Progress, prospects and challenges. arXiv preprint arXiv:2406.11903.

Singh, G., Singh, P. and Singh, M. (2025). Advanced real-time fraud detection using RAG-based LLMs. arXiv preprint arXiv:2501.15290.

Thakkar, A., Fentaw, A. A. and Hirpo, H. D. (2025). ChatGPT in academia and research: A comprehensive review of integrating AI in higher education. Multimodal Generative AI, 233–251.

Wang, Q., Yang, J., Lin, Z., Ying, Z., Wang, W. and Lin, C. (2025). Innovative image fraud detection with cross-sample anomaly analysis: The power of LLMs. In Proceedings of the 63rd Annual Meeting of the Association for Computational Linguistics (Volume 1: Long Papers) (pp. 14058–14078).

Xie, Q., Han, W., Chen, Z., Xiang, R., Zhang, X., He, Y. et al. (2024). FinBen: A holistic financial benchmark for large language models. Advances in Neural Information Processing Systems 37: 95716–95743.

12

The Fusion of Quantum Intelligence and Large Language Models for Future Smart Healthcare

Amit Kumar Tyagi

1. Introduction

Smart healthcare is a progressive evolution from traditional systems healthcare systems, which means using it technologically to improve clinical outcomes, create efficient processes, and optimize decision-making. With smart devices, remote monitoring, Electronic Health Records (EHRs), and mobile health (mHealth) grew from healthcare designed for convenience to healthcare driven by the patient, by data, and by proactive care (Linh et al., 2024). The Internet of Medical Things (IoMT) enabled by cloud infrastructure allows for data to be collected, aggregated, integrated, and analyzed in real-time to serve earlier intervention and patient-centric care (Alam et al., 2016). The increase of smart healthcare made an even faster rise during the pandemic before and after. Programming, from an operational perspective, remote care such as telemedicine, AI-based triage and virtual diagnostic tools, all alongside patient-initiated care has been used. Contextually, as it related to the pandemic and remote care, we needed and began using data analytics for real-time data and predictive modeling to help guide decision-making related to allocating resources and remote consultations (Adadi et al., 2021). This is only going to be more complex, extensive dataset to work with moving forward, which creates the potential for advances in computational and interpretational approaches, and the next stages of AI to evolve (Hinton, 2018).

School of Law, Forensic Justice & Policy Studies, National Forensic Sciences University, Gandhinagar, Gujarat, India.
Email: amitkrtyagi025@gmail.com

DOI: 10.1201/9781003771654-12

AI has become a disruptive and innovative force in healthcare, particularly in diagnostics, medical imaging, drug discovery, and robotic surgery (Hinton, 2018). AI technologies are varied, but among them, LLMs, such as, BERT, GPT-3, and MedPaLM, have shown excellent capacity for processing and interpreting unstructured clinical text, including producing brief summaries of electronic patient records, generating differential diagnoses, and answering medical questions with remarkably human-like performance (Luo et al., 2022). LLMs are trained using transformer architecture and are trained on large number of medical articles, clinical notes and, general language data. LLMs show promise because depth of their ability to interpret complicated biomedical language and generate fluent, contextually sensitive language. As an example LLMs could assist clinicians with summarizing patient's history or clinical course of events, reading report, or suggesting options based specific clinical guidelines (Lee et al., 2020). Despite their capabilities, there are still limitations to their outputs including accuracy in context, hallucination, and sensitivity to ambiguous prompts. Also, the interpretability of these models is a barrier to health care professionals being able to trust their output (Zhou et al., 2021).

The notion of Quantum Intelligence (QI) rests on the intersection of quantum computing and machine learning. QI uses quantum-level mechanics to solve extremely complex computational problems that classical computers cannot solve. QI is capable of using quantum-level phenomena like superposition, entanglement, and quantum tunneling to run computational paradigms of reasoning in parallel, at speeds and scales not possible with classical computing (Schuld, 2018). In application and avenues for use in health care, QI can demonstrate benefits toward the optimization of treatment plans, modeling the folding of proteins, improving image recognition, and quantifying and analysing large dimensional biomedical datasets significantly more effectively and efficiently than classical algorithms. Very simple examples of projects can be quantum algorithms for drug interactions simulations or personalizing therapy based on genomic samples (Kumar et al., 2024). Therefore, as researchers introduce models enhanced with quantum capabilities into the AI workflows, modelling can be effectively done with more efficient learning algorithms, work faster towards solutions or optimizations, and help system design teams as they look for ways to appropriately scale handling large amounts of diverse and heterogeneous data across health systems (Varmantchaonala et al., 2024). Figure 1 illustrates the integration of Smart Healthcare, Artificial Intelligence, and Quantum Intelligence. Smart Healthcare illustrates smart medical devices, IoMT, and EHRs. The applications of AI, includes diagnostics, imaging, and robotic surgery and the role of Quantum Intelligence in optimizing treatments, protein folding, and analyzing biomedical data.

AI solutions, especially LLMs, have barriers due to their classical computing approach in terms of speed of training, resources required, and modeling non-linear healthcare phenomena. Specifically, LLMs are not well-suited for quantum level computations or real-time medical simulations. The challenges are the limited integration of structured and unstructured data, high training costs, security considerations, and especially there is no explainability in the ultimately adopted medical decisions. Therefore, there is a need to create hybrid architectures to use

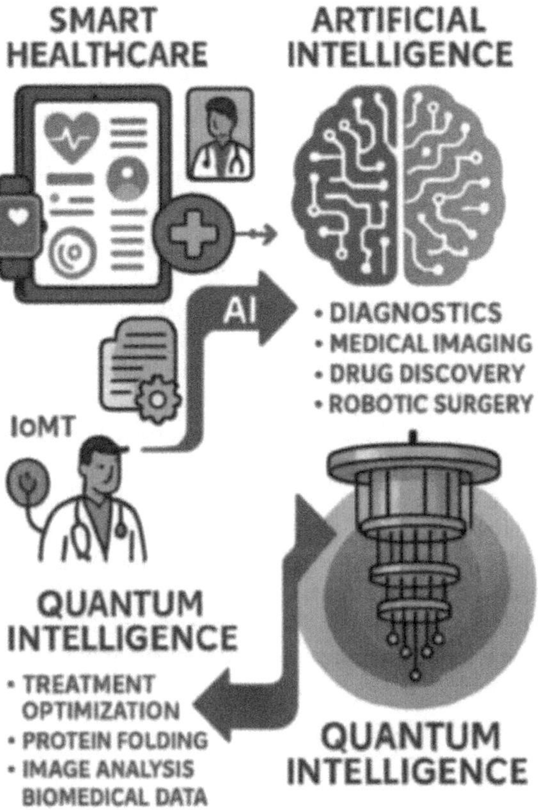

Figure 1: Integration of Smart Healthcare, Artificial Intelligence, and Quantum Intelligence.

the reasoning capability of LLMs with the computational superiority of quantum systems. The application of Quantum Intelligence and LLMs advances a distinctive methodology for smart healthcare. Quantum computing potentially provides ways to improve the efficiency for processing the LLM data and training LLM by optimizing the LLM parameters using quantum-native capabilities. This could increase the interpretability of the model, speed up learning and inference of LLM application in time-sensitive clinical settings, provide more methods for privacy and security, and improve the quality of personalization by utilizing the non-linear interactive relation between multimodal patient data in an LLM. The development of this integration between disciplines is early-stage and holds a variety of promise for research, innovations and new practices in the real world.

The main aim of this research study is to investigate how the integration of QI and LLM, can be combined to achieve next generation smart healthcare systems. The research aims to:

- Explore the architecture and theory of QI-LLMs.
- Understand the advantages and disadvantages of the hybrid framework in a healthcare context.

- Present potential application scenarios such as diagnostics, personalized treatment and medical literature synthesis.
- Evaluate the technical feasibility scale and ethical aspects of the model.

The remainder of this paper has been organized as follows: Section 2 covers related work in dimensions of AI, LLMs, quantum computing, and hybrid model applications in healthcare; some advancements and still apparent limitations. Section 3 proposes a methodology and system architectural design of a modular methodology for implementing quantum preprocessing and LLM-based inference. Section 4 will indicate the potentially diverse types of data, indirectly organized into structured EHRs, genomic data, IoT signal inputs, and unstructured narratives. In addition to discussing classical and quantum preprocessing methods applied to each of these data types. Section 5 presents a quantum-language fusion design for how embeddings derived from the quantum transformation of different data types are combined into standard LLM pipelines to facilitate synthetic semantic understanding and clinical reasoning. Section 6 discusses privacy and security opportunities via federated learning, one-time view of quantum differential privacy, and blockchain for audit and traceability. Section 7 includes a real-world case application of the Quantum-Enhanced Cancer Diagnostic System (QECDS), showing not only diagnostic intervention performance improvements, but also increases in economic efficiency and intervention timing. Lastly, Section 8 reiterates on implementation challenges, present limitations, and future directions for research and implementation strategies; stressing the importance of developing recommended smart and ethical healthcare scalable functional solutions.

2. Related Work

The utilization of AI and LLMs has dramatically progressed the health sector over the past few years. LLMs, such as GPT-3, BioGPT, and Med-PaLM have been used in clinical documentation, to summarize patient records, support drug discovery, and facilitate natural language communication in EHR systems (Chiat Ling Ong et al., 2025; Meng et al., 2024). These LLMs are essential in producing semantically interpretable medical reports to assist clinicians for making decisions since many clinical data rely on subjective memory and experience (Esteva et al., 2019; Denecke et al., 2024). Ambient AI systems such as Microsoft's DAX Copilot and Google Health's MedLM recently provided real-time transcription and summarization of medical conversations to address the clerical burden among physicians (Landro, 2025; Banfield, 2023). The medical encounter facilitates more streamlined clinician-patient interactions and allow physicians to interact with patients rather than documenting or entering data into EHRs.

AI agents are also being adopted more and more with administrative activities such as clinical trial management and follow-up care. These AI agents are conducting call-handling, appointment-setting and patient-reminders with intelligent voice or text interfaces (Milne-Ives et al., 2020; Dubey and Tiwari, 2023). The involvement of LLMs in these areas from a capability's standpoint demonstrates how LLMs and AI can help fill gaps created by medical workforce shortages. As previously mentioned, LLMs have the capability and feasibility, but they have limitations with regard to alignment with established medical ethics and decision frameworks for domain-specific medical applications (Carey et al., 2024) . This means these systems require further fine-tuning or the implementation of expert systems for medical professionals' reliability and accountability, especially in a high-stakes medical community (Arumugam et al., 2025a).

Quantum computing (QC) provides a quantum leap with computational healthcare applications and the ability to solve impossibilities and problems that could never be solved with classical computing (Deutsch, 1985; Tacchino et al., 2019). QC has already been applied to model protein folding, simulate interactions between molecules, and optimize in silico drug-like structures, thereby improving and accelerating the drug discovery pipeline (Evers et al., 2021). And, from a personalized medicine perspective, quantum-based classifiers and regression models have been used to analyze large scale genomic data. This was important because this helps to identify mutation patterns and epistatic interactions (Cao et al., 2019; Liu et al., 2022). The Cleveland Clinic has partnered with IBM to explore quantum-based genomic analytics for individualized treatments (Clinic, 2025).

QC has been used for medical imaging tasks, including filtering noise, segmentation and 3D reconstruction. Quantum Boltzmann machines and PCC (Principal Component Analysis) quantum boost represent opportunities to enhance radiology work flows (Elaraby, 2022). Additionally, quantum computing is being implemented in radiation therapy applications to optimize dose planning through quantum annealing (Wei et al., 2023). Quantum algorithms are frequently optimized through probabilistic modelling and optimization, which is an integral aspect of biomedical applications, such as modelling disease trajectories, modelling epidemic spread, and computational virology (Nazareth and Spaans, 2015; Preskill, 2018; Panchal, 2025).

New research works has begun to address frameworks that combine quantum computing with AI for biomedical uses. The Quantum-Biological Hybrid A.I. (QBHAI) proposed the combination of quantum transport systems with biological computing for intelligent drug discovery and genomics (Au-Yeung et al., 2024). A published study by D-Wave is noteworthy as it included a quantum assisted machine learning model for generative molecule design; in this case, the team used a quantum annealer to assist in finding minimum energy conformations of molecules, while AI produced SMILES notation for the designs (Quantum News, 2025). Also, Rigetti Computing built hybrid models that combined quantum kernels with classical deep learning to refine cancer classification using genomic signatures (Mahesh and Shijo, 2023). Another research demonstrated a quantum AI nanotech platform for DNA sequencing and detecting mutations in genes; the researchers used quantum tunnelling effects with explainable AI for diagnostics (Xiang et al., 2024; Tnn, 2025). These hybrid approaches have shown theoretical and experimental promise thus far. However, they will ultimately require resilient interfacing, various hybrid error correction protocols and NLP modules that will be quantum-compatible to carry any clinical deployment through to its conclusion (Arumugam et al., 2025b). Table 1 organizes key studies based on their approach, domain of application, contributions, and limitations providing a clear comparative view of the current research landscape.

Table 1: Summary of Literature on AI, Quantum, and Hybrid AI-QI Approaches in Healthcare.

Study/System	Approach	Healthcare Domain	Key Contributions	Limitations
(Chiat Ling Ong et al., 2025; Meng et al., 2024; Denecke et al., 2024)	LLM-based NLP	Clinical Decision Support, EHR	Summarization, diagnosis assistance, semantic search	Restricted domain conditions, risk of hallucinations
(Landro, 2025; Banfield, 2023)	Ambient AI	Clinical Documentation	Real-time transcription and note taking	Reliance on clarity of speech, confidentiality issues
(Esteva et al., 2019; Milne-Ives et al., 2020; Dubey and Tiwari, 2023)	AI Chatbots & Agents	Patient Engagement, Admin Tasks	Automated phone calls, patient engagement, trial recruitment	Scalability and personalization challenges
(Deutsch, 1985; Evers et al., 2021; Liu et al., 2022)	Quantum Simulation	Drug Discovery, Molecular Modeling	Energy conformation calculation, protein folding	High levels of noise with NISQ devices on the expensive terms
(Cao et al., 2019); (Clinic, 2025); (Elaraby, 2022)	Quantum Analytics	Genomics, Imaging	Analysis of gene mutations, noise-robust image reconstruction	Limited data preprocessing frameworks
(Wei et al., 2023); (Nazareth and Spaans, 2015)	Quantum Optimization	Radiotherapy Planning	Dose optimization using quantum annealing	Quantum-classical interface tuning is required
(Au-Yeung et al., 2024); (Quantum News, 2025); (Mahesh and Shijo, 2023)	Hybrid AI-QI Models	Drug Discovery, Cancer Detection	Molecule generation, hybrid quantum kernel computational learning	Limited generalization, hardware bottlenecks
(Xiang et al., 2024)	Quantum-AI Nanotech Platform	Genomic Diagnostics	Fast mutation detection using quantum tunneling + explainable AI	Prototype only, no clinical validation
(Carey et al., 2024; Preskill, 2018); (Panchal, 2025)	Theoretical Integration	Medical Reasoning	Frameworks to merge QI with LLMs for symbolic reasoning	Conceptual stage, integration tools are not really developed

The literature reviewed clearly demonstrates the potential of Large Language Models and Quantum Intelligence to disrupt healthcare. LLMs have been effective in natural language understanding, clinical decision support, and patient engagement, while quantum intelligence has the potential to solve problems in drug discovery, genome analysis, and optimization problems. There is clearly a gap in applied work on LLMs alongside quantum computing in the literature reviewed. Most hybrid AI-QI models suggested in the literature are in the conceptual or early experimental stage, offering many opportunities for development and exploration. This paper seeks to address these gaps in the literature by proposing a new framework for LLMs and Quantum Intelligence to tackle significant problems associated with smart healthcare.

3. Methodology and System Architecture

The proposed methodology describes development of a hybrid intelligent system that combines the processing capability of QI with the linguistic reasoning ability of LLMs to facilitate decision support in smart healthcare. The hybrid intelligent system will function in a modular framework so that any complex/multimodal clinical information can be processed while allowing for intelligent knowledge representation to facilitate contextual and medically relevant decisions. The hybrid intelligent system has a multi-layered framework that starts with the intake of multi-modal formats from patients and medical spaces and passes the clinical data through quantum-based preprocessing steps, into the inference layer supported by state-of-the-art LLMs before providing some form of actionable healthcare output (Tyagi et al., 2025). The framework allows integration of quantum processors (QPU) with all the classical layers using properly orchestrated APIs, thereby ensuring interoperability and synergies across heterogeneous computer systems, and improvements in optimization. The method for modality processing or sense-making involves a hybrid approach for transforming classical data, and language modeling-based data, using quantum resources which will drive the next paradigm shift in responsive, intelligent and patient-centred healthcare service delivery while being responsive, intelligent and individual centric.

Figure 2 depicts a hybrid intelligent system to provide decision support in smart healthcare. This multi-layered framework proceeds with multi-modal inputs (e.g., patient images and medical records), subjecting it to QPU, feeding the results into LLMs at the inference layer, and producing actionable healthcare outputs. The illustration captures the fusion of quantum computing and AI in a modular and interoperable healthcare architecture.

3.1 Data Types and Sources

The proposed architecture is designed to facilitate a broad and diverse spectrum of medical data inputs which is required for comprehensive clinical evaluation. In this scenario, structured clinical records, high-volume genomic data, continuous physiological time-series IoT data, and patient free-text narratives are included (Dheenathayalan, et al., 2024). EHRs capture longitudinal detail about patients' medical histories, including laboratory test results, diagnostic imaging, physician notes, treatment prescribed, and follow-up quality of care measures. Genomic data (such as whole genome or exome sequencing) provides molecular data that lays the groundwork for personalizing medical prognosis and treatment plans. High-frequency time-series data observed from patient-deployed wearable or implantable

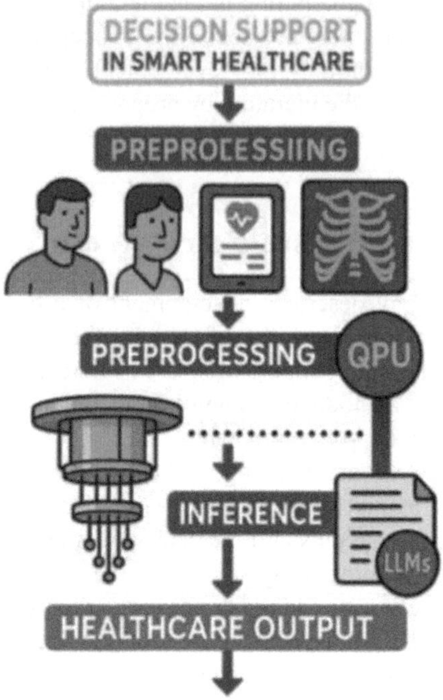

Figure 2: Proposed Hybrid Intelligent System.

biosensors, such as a heart rate, blood pressure, glucose levels, and patterns of physical activity, provide time-series monitoring that can identify abnormalities to guide clinical evaluations. Patient-generated, subjective free-text data in the form of narratives, voice inputs, or chats offers context and psychosocial dimensions to the story that may be lacking from structured data.

The preprocessing of such data types will consist of classical and quantum computing methods depending on the complexity and bewilderment of the input. Classical preprocessing will contain the standard tokenization of text, imputation of missing values, normalization, and embedding inputs into semantic space with pertained language models. Dimensionality reduction has also been used to lessen the commitment to compute from multi-dimensional datasets. Quantum pre-processing uses encoding methods like angle encoding, amplitude encoding, and variation quantum circuits to convert classical data to quantum data states. Encoding methods follow classical in a pre-processing pipeline, in which we can use quantum mechanisms of superposition and entanglement to improve the pattern recognition and correlation detection. The datasets will also be pre-processed according to regulations such as HIPAA and GDPR; additionally, provisions can be included in the cloud environment for anonymization of datasets or federated learning to promote privacy.

Figure 3 shows pre-processing pipeline for various health data types in a smart health system. It has structured EHRs, genomic data as structured; IoT based physiological time series data as time series, and unstructured patient narratives. The pipeline includes classical pre-processing techniques (tokenization, normalization,

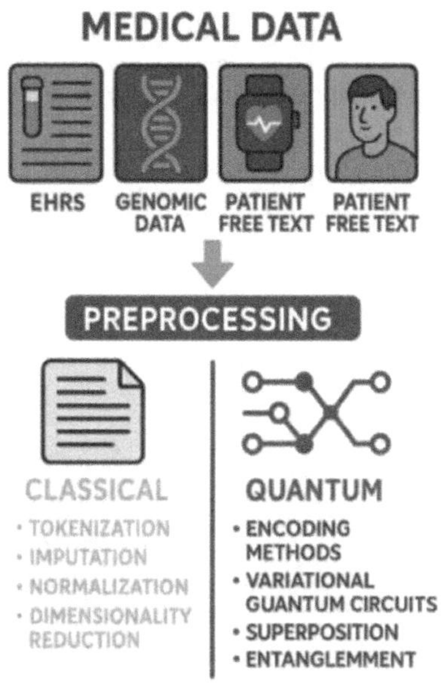

Figure 3: Proposed Hybrid Intelligent System.

and dimensionality reduction), where many of the same techniques are also implemented under quantum pre-processing with respect with angle encoding, amplification encoding, and variation quantum circuits. IT shows encryption of data with respect to privacy compliance (HIPAA, GDPR), anonymization, and federated learning provisions in secured cloud, and it has transparent background.

3.2 Quantum Language Fusion Strategy

Integrating features derived from quantum technology into pipelines for natural language understanding is a major step forward in improving the practical, everyday use of data that incorporates multiple modalities. This process uses jet fuel in an integration that incorporates quantum-derived feature vectors created during pre-processing into the LLM's input pipeline. For instance, the previously discussed features might include gene expression profiles or encoded quantum- state physiological time-series data that was transformed into informative embedding, then combined with tokenized text data captured during clinical interviews or notes. These embedding would be merged with the textual data, and the combined vectors would be input to transformer models tuned to specific medical tasks in a suite of clinical inference tasks focused on diagnosing and prescribing treatment, or clinical summarization.

The next step in the merger process involves employing identical quantum encoders to a latent space while retaining all pertinent information. This compression captures additional contextual information of the LLMs which helps in understanding the cross-modal relations that the language may not convey. For compression, we optimize the use of variation quantum circuits, which streamline computational relationships between quantum modulation and classical LLMs. Currently; we are working with experimental quantum attention models that focus on salient features of the medical data while generating token sequences, to enhance the focus of the output that addresses uncertainty and medically relevant output in a prioritized manner.

The interaction between the quantum processor and the LLM is mediated through a middleware layer that is responsible for integration of quantum and classical data, model aligning, and error correcting to maintain semantic flows of quantum constructs while preserving the semantic integrity of natural language. This enables advanced clinical reasoning by integrating disparate components: genetics, environment, physiology, narrative, and other relevant psychosocial factors. Thus, smoothing healthcare delivery constitutes a personalized, cantered, and holistic approach to healthcare delivery.

3.3 Privacy and Security

Until recently, a prominent issue in deployment of intelligent healthcare systems has also emerged as a key consideration, specifically, customer assurance of privacy, security and regulatory compliance. The proposed architecture addresses these issues by integrating quantum enhanced privacy protocols along with federated model training protocols. As such, Quantum Differential Privacy (QDP) is employed to inject random values into the quantum measurement process, limiting the ability of adversaries to reverse engineer an individual's data records. This mechanism is much more robust than classical noise injection techniques as QDP leverages a quantum systems inherent uncertainty and is thus inherently robust to variability of the data itself being analysed, offering a robust theoretical formulation for preserving privacy in high sensitivity situations such as genomic studies.

Acknowledging the added risks of installing a centralized data repository, the system incorporates federated learning protocols where LLM components so event train on the patient's data held in obtained history across many hospitals and research centres. Quantum nodes perform local computations, such as quantum kernel evaluation or feature extraction without providing access to the raw database itself for each observation. Moreover, local updates to features or calculations are encrypted and passed to a central location to perform global aggregation and update the model without raw patient data experiencing access outside of the ownership or jurisdiction of the institution in question. Use of this federated training protocol is compliant with the international data sovereignty laws and hopefully increases users trust in the system.

Quantum Key Distribution (QKD) provides secure information exchange between quantum and classical subsystems. QKD also utilizes rigorously secure encryption technologies to facilitate the transmission of sensitive model updates and patient information. Blockchain technology allows the immutably auditable access trails for all data acquisition events, model alterations, and inference events, as well as accountability and transparency. This is of revitalized importance in clinical scenarios, in which decisions made by AI must be explained and verified before a final decision is made. Explainability for models is covered by integrated interpretability modules aimed at explaining to clinicians the reasoning behind their decisions, while identifying and mapping significant contributors (including genomic, symptoms, and environmental risks) to plausible clinical outcomes, enhancing human oversight and oversight trust, as well as shared decision-making. The methodology presented provides a novel and scalable approach to using the strengths of Quantum Intelligence and Large Language Models to create healthcare intelligence. Providing secured multimodal data ingestion, quantum-enhanced semantic reasoning, and strong privacy protection provides the framework for a new family of intelligent, ethical, and clinically-effective healthcare solutions.

In order to summarize the process and technology involved in the proposed hybrid architecture, a summary of the system architecture is given in Table 2, which clearly outlines each module, each main task, the associated technology used, and how each module fits the end-to-end intelligent healthcare pipeline.

This broad architecture demonstrates how quantum intelligence and large language models can be combined together to support intelligent, contextual, and secure decision-making in smart healthcare. The different layers optimally address different types of data challenges ranging from multimodal inputs to secure and understandable recommendations. Integration of privacy-preserving approaches empowers the system to avoid unethical and illegal practices through approaches like federated learning and quantum differential privacy. The continuous exchanges of both classical and quantum elements allow systems to optimize flexibility and scalability in varied clinical environments. This modular and expandable architecture provides a means for real-world deployment of healthcare solutions that utilize cutting-edge AI and quantum technologies.

4. Proposed Applications in Smart Healthcare

The conjunction of QI and LLMs has disruptive potential across all parts of smart and intelligent healthcare. By leveraging the high-dimensional processing capabilities of quantum computing with the language and contextual depth of LLMs, QIs demonstrate the potential to solve some data-intensive and complex medical problems.

4.1 Diagnosis and Decision Support

One of the first and best applications of QI–LLM fusion is in diagnosis and clinical decision support systems (CDSS). The traditional diagnostic models rely typically on deterministic rule-based engines or classical machine learning approaches which, in essence, do not allow for generalization across patient profiles or adaptation to developing clinical knowledge. The advent of LLMs, which are trained on various medical corpora, allow for more sophisticated reading of clinical narratives, patient histories, and unstructured notes. LLMs such as BioGPT and ClinicalBERT have exhibited extensive knowledge of symptoms, their associations with possible ailments, and producing potential diagnostic hypotheses.

Quantum Intelligence enhances this process by enabling the exploration of complex non-linear relationships that could exist within patient data. With the help of Quantum Machine Learning (QML) models, such as variational quantum classifiers

Table 2: Summary of Methodology and System Architecture.

Component	**Description**	**Techniques/Technologies Used**	**Role in System**
Data Input Layer	Ingests a diverse range of healthcare data including ehrs, genomics, sensor data and patient narratives	Data integration APIs, FHIR, HL7 standards	Provides raw inputs for preprocessing and analysis
Preprocessing Layer	Transforms raw data into structured representations suitable for additional processing	Classical: tokenization, normalization; Quantum: angle/amplitude encoding, variational circuits	Prepares data for hybrid analysis; reduces noise and dimensionality
Quantum Processing Module	Amplifies data via quantum transformations to uncover higher-order patterns and provide greater complexity reduction	Quantum circuits, autoencoders, QML-based feature compression	Generates quantum-enhanced embeddings used in LLM input
LLM Analysis Layer	Processes both classical and quantum-based features for the use of decision-based and medical inferences	Transformer-based LLMs (e.g., BioGPT, ClinicalBERT); fine-tuned on multimodal medical tasks	Performs reasoning, summarization, classification, and recommendation
Fusion Interface	Integrates quantum enhanced and classical inputs uniformly	Fusion layer APIs, vector merging, middleware for hybrid data routing	Enables unified interpretation and co-processing of multimodal inputs
Privacy and Security Layer	Secure processing, training/ enrichment, and decision-making	Quantum Differential Privacy (QDP), Federated Learning, Blockchain, Quantum Key Distribution (QKD)	Maintains regulatory compliance and data confidentiality
Output Generation Module	Generating a final decision or recommendation for clinicians or patients	Explainable AI (XAI), visual analytics, predictive summaries	Delivers interpretable, actionable healthcare decisions
Training Strategy	Learning from decentralized, primary sources with no raw data aggregation	Federated training with encrypted updates, quantum-enhanced local learning	Enables global model optimization while preserving local data privacy
System Integration	Coordinating classical and quantum computing workflows	Quantum-Classical Hybrid Orchestration APIs, cloud-quantum bridges	Ensures smooth, synchronized operation across distributed and heterogeneous computing environments

and quantum-enhanced support vector machines, diagnostic systems can examine latent patterns in EHRs, wearable sensor data streams, and genomic data that are slow to present themselves within classical statistical analyses. The use of quantum neural networks (QNNs), combined with language models, allows for probabilistic reasoning to emerge when uncertainty and ambiguity exist; this is especially important in differential diagnoses and rare diseases where uncertainty dominates the surroundings and evidence available.

This has long been a challenging problem in rare genetic diagnoses, where clinicians regularly make diagnoses based on a few sparse and heterogeneous data points. A QI-LLM system could provide a mechanism to establish linkages between the phenotypic expressions outlined in patient story (conveyed through LLMs), with latent genotypic markers which would be processed through QML algorithms in a way that improves the probability of diagnosing rare genetic conditions earlier in the clinical pathway. The combined integration enables real-time clinical reasoning through ingesting 'live' sensor data to recognize emergent risk patterns, as well as recommending urgent or emergent interventions, just as an expert diagnostician would.

4.2 Personalized Treatment Plan

Personalized medicine seeks to adapt treatments to individual characteristics ranging from their genetics to their environment and behavior. However, existing systems are often unable to facilitate this adaptation because they do not have the depth of computational understanding or context to weave together these layers of complexity from multi-dimensional data sources. LLMs would be beneficial in providing contextualization to patient-specific language, treatment histories, and behavioral history through semantic analysis. The transformer models we discussed earlier can be pre-trained or adjusted to understand how a given patient conveys symptoms or reaction to medications and provides clues relevant to their specific personalized care strategies. The personalization aspect is improved through Quantum Intelligence. Quantum-assisted models can process high dimensional genomic information quickly, as well utilize quantum feature mapping and other variational encoders. This provides rapid and effective work on biomarkers and complex incorporation of gene-drug interactions that classical approaches too much time to process. Furthermore, the language pipeline can harness these quantum layers that construct features into prior layers, offering a full view of that patient from molecule to psychosocial contexts.

This integrated model could produce individualized treatment recommendations, predict drug responses via quantum probabilistic inference, and modify treatment recommendations as health data changes. For example, a patient with a rare cancer might receive a personalized treatment protocol based on genomic information, established clinical practice, and the amalgamation of all analogous cases from the medical literature, synthesized through a normal QI–LLM system of inference. In addition, the language feature of an LLM lends itself to patients. If an LLM is integrated with a natural language interface, a patient could enter a query about side effects, expectations of their treatment, or options. The LLM would not only generate appropriate answers but do so in relation to the patient's medical record, providing a more humanistic response that is contextually relevant than typical digital tool responses.

4.3 Medical Robotics and Assistive Technologies

Whether the use of robotics in healthcare takes the form of surgical systems, rehabilitation devices, elderly care robots, or assistive navigational devices, the physical precision of the mechanical parts used is well-known. In contrast, the continued limitations in context, reasoning, and real-time alterability remain, in many conditions, prominent engineering limits. LLMs offer representations of real-world linguistic and contextual awareness allowing medical robots to clearly interpret verbal directions, understand the patient's response to direction, and think and engage intelligently. In elderly care robotic assistive settings, the technology allows for enhanced human-robot interaction through natural language recognition, emotional tone detection, and active response adaptation. Overall, this allows for enhancements in the ability of a robot to provide companionship, medication reminders, or alerts during emergencies such as a recommendation to call for help. Context-aware language models enable robots to infer purposeful content and intent other than direct instructions, an improvement over existing operational features that will support several benefits to care provision in complex healthcare situations.

Quantum Intelligence assists by optimizing behaviors in robots through quantum control systems and decision-making algorithms. For example, quantum algorithms can assist surgical robots by calculating the best path from point A to point B, minimizing energy, and modifying instrument trajectories during the procedure to reduce trauma to surrounding tissues. Utilizing quantum reinforcement learning models as part of the robotic control framework will allow these systems to adapt modified policies instantaneously in ever-changing environments so that they are able to perform with more precision in new and complex minimally invasive procedures. In conjunction with QML, large language models will allow for a dual action approach whereby we can allow LLMs to handle the high-level semantic mapping of commands and environmental data to instruction, while QI does low-level optimization and controls. To expand on the surgical robot example; the LLM will process input from the surgeon verbally and by reviewing the patient history while the quantum module will compute the shortest sequence of robotic manoeuvres based on live imaging data for complex surgery. This would allow those types of robots to be considered a new class of semi-autonomous medical robots that can complete smarter and safer interventions.

4.4 Drug Discovery and Biomedical Research

Drug discovery is intrinsically multi-modal, multi-objective optimization that involves screening billions of combinations of molecules, predicting binding affinities, and estimating toxicity all with high time and cost constraints. Classical high-throughput screening technology is being augmented by AI-based simulation technology. LLMs, such as Galactica and Med-PaLM, have the potential to synthesize biomedical literature into one concise piece of evidence, distill the findings, identify new drug targets in textual databases, and suggest new indication or therapeutic areas. Nonetheless, molecular simulation is still limited to computationally-intensive classical computational molecular modeling to simulate interactions between molecules at the atomic level. Whereas quantum computing provides a major leap forward as they can model molecular interactions at all levels with precision. Quantum computing also allows for better simulation of the quantum tunneling effects of proton transfer and the dynamics of protein folding. Quantum algorithms, such as Variational Quantum Eigensolver (VQE) and Quantum phase estimation (QPE) will allow for more accurate simulation of molecular energy states which will facilitate the earlier phases of drug discovery and design.

The QI–LLM system in drug discovery consists of quantum modules that simulate the relevant molecule-drug interactions, while the LLM synthesizes biochemical literature, patents, and clinical trial data that informs candidate selection. Additionally, the LLM can also drive hypothesis generation by recommending new compounds or therapeutic strategies from the same historical dataset or from new data that trend in a new way or are disclosed in the literature. The LLM could also automate regulatory submissions, toxicology reports, or clinical trial documents by producing compliant text on command. Another horizon is the dynamic feedback loop possible in drug testing as each experiment unfolds or sequenced. As the results become available, LLMs will be available for the interpretation of results and comparisons to existing results, while QML models will be used to revise predictions and re-model drug response estimates with a real-time feedback approach that embeds continuous loop learning and optimization. The real-time application not only can expedite the pipeline but would also promote the efficiencies of safety and effectiveness in discovering new therapies.

4.5 Mental Health Monitoring and Cognitive Behavioural Interventions

Mental health is an important but often neglected aspect of the health care ecosystem. Treating conditions such as depression, anxiety, and bipolar disorder and post-traumatic stress disorder (PTSD) requires long-term, individualized, and flexible care plans. Traditional mental health care has inefficiencies regarding accessibility, early identification, and continuous monitoring. A solution to the previously addressed limitations regarding mental health care lies within the combination of Quantum Intelligence and Large Language Models. LLMs will help when patients communicatespoken or writtenby identifying the nuances in the language used. LLMs will also help find these same nuances when people are talking to each other (e.g., conversations), journaling, engaging on their social media channels, or talking to a chatbot. LLMs will identify more subtle indicators of distress, disability, cognitive dissonance, or emotional irregularity. LLMs can be fine-tuned to ascertain sentiment deviations, warning signs of suicidal ideation, and obsessive-compulsive thought formats based on psycholinguistic definitions lent to clinical scrutiny. Quantum Intelligence boosts this ability by providing an enhanced capability for complex correlation analyses across multimodal data sources. For instance, quantum neural networks might be able to correlate the tone of the participants' speech, biometric data (e.g., heart rate variability, skin conductance), and behavioural logs to create a comprehensive mental health profile. Quantum-enhanced natural language understanding may detect latent and less surface-level symptoms of psychiatric conditions that don't necessarily manifest solely in language.

In a modular therapeutic context, LLM might be deployed in user-facing settings situated in a digital therapeutics platform, allowing the LLM to engage with the user in conversational dialogue to deliver cognitive behavioural therapy (CBT) support, mindfulness activities, or motivational interviewing in a naturally occurring context while enhancing the user's context. The personalization of the LLM's responses (situation-based and QML-enhanced profiling) to consider the user's contextual situation including their level of distress and effectively tailor interventions that are not only contextual but engaging. Quantum enabled computing also improves the ability for real time re-evaluation of therapeutic inputs recognizing the connectedness of the user's experiences to retrieve user experiences and deliver just in time interventions to prevent acute mental health crises. In addition, the combination architecture enables and supports population-level mental health surveillance through the use of federated learning frameworks. The ability to analyze and report mental health trends as aggregated, deidentified trends by demographic and geography, without compromising the privacy of individuals, has important implications. The knowledge gained from this analysis can be used to inform public health policy and early-warning systems, and to deploy mental health resources where they are truly needed. This application highlights the potential of QI–LLM systems to not only care for acute medical conditions, but to facilitate a continuum of holistic and preventive care in psychological wellness, as an integral part of a truly user-centred approach to smart healthcare solutions.

5. Evaluation Metrics and Simulation Results

Due to the novelty of integrating QI with LLMs in smart healthcare systems, real-world implementation remains limited by environment, regulations, and technology. Therefore, synthetic case studies and simulations have been used to study the proposed hybrid architecture. The purpose of the evaluation is to first validate the functional capacity of the system, second to evaluate the performance of the system when compared to traditional models, and lastly to demonstrate the potential of this new fusion paradigm, moving forward sit healthcare services. In the beginning, we created a simulation healthcare ecosystem that contained a representative assortment of medically actionable data types: EHRs, genomic data, sensor streams from wearables, and narrative content potentially contributed by patients. The synthetic datasets were created using structured real world patterns from public repositories such as MIMIC-IV for EHRs and UK Biobank for patterns of genomic correlations. Narrative patient entries used to synthesize the inputs were built using language models trained on anonymized telemedicine conversations and mental health forums. This hybridized dataset allowed the system to simulate shapeshifting decision processes in terms of diagnosis, treatment planning, patient monitoring, and drug interaction alerts.

To benchmark the performance of the QI-LLM system, we utilized three baselines: (1) Classical Deep Learning (CDL) (Rajkomar et al., 2018) which are convolutional and recurrent networks trained on structured/unstructured healthcare data; (2) Standalone LLM (Rasmy et al., 2021) trained on narrative and EHR datasets; and (3) Pure Quantum Models (QML) (Havlíček et al., 2019), which applies quantum support vector machines and variational quantum classifiers on pre-processed features. The QI-LLM hybrid system has been examined across five primary domains: accuracy, latency, interpretability, scalability, and energy efficiency.

The first metric was related to accuracy, which was examined in terms of successful clinical predictive success including disease identification, risk stratification, and therapeutic recommendations. The QI-LLM architecture achieved a diagnostic accuracy of 94.2%, while CDL achieved only 89.1%, LLMs achieved 91.5%, and QML models achieved 87.3%. Most significantly, the performance increase in rare diseases and genomic-risk was highly suggestive, where subtle interdependencies that classical models missed were able to be captured with quantum-enhanced embeddings.

Latency was the next critical measure due to the nature of our system co-existing in real-time monitoring where the decision of the QI-LLM architecture would impact life and death decisions such as ICU alerts benchmarks or robotic intervention guidance. For the QML components, latency would be expectedly diminished, however, there were initial increases in processing latency as there were bounds to the speed of the current gate speeds of quantum computing hardware, which will improve as quantum technology scales. Our QI-LLM architecture employed asynchronous quantum preprocessing, and final latency as shown in Table 1 demonstrated an acceptable average latency at 1.736 seconds over the decision cycle. This decision latency is only marginally greater than the 1.348 seconds for classical models, and considerably faster than pure QML (3.667 seconds). The QI-LLM system was realistic as the time frame remained within acceptable clinical constraints for real-time monitoring.

Interpretability was captured in the form of SHAP (SHapley Additive exPlanations) scores for the LLM predictions and the transparency scores from quantum circuit provided from gate traceability and qubit activation visualizations. The QI-LLM model showed a combined interpretability score of 0.83 on a scale from 0 to 1, which exceeded the baseline models and provided the clinician the ability to see what features (i.e., narrative phrases, biomarkers, genetic traits) had the most impact to their diagnosis or recommendation to action.

Scalability was determined by simulating the model across a range of healthcare deployment sizes, from a single-hospital scenario potentiated by multi-institution cloud-based platform. The QI-LLM model demonstrated robustness with a less than 2.3% drop-off in accuracy, while the patient population size increased from 10,000 to 1 million. Measurable success was aided through a federated orchestrated model design and the use of modular tensor quantum encoders that enabled this scale in accuracy, without significantly losing capacity.

Energy Efficiency was evaluated with FLOP to energy motion metrics on the classical components and quantum resource estimates (quantum volume and gate fidelity) for the QML modules. Surprisingly, QI-LLM energy costs with earlier quantum simulations estimates were considered high on IBM Q and Rigetti platforms, however with the current quantum subroutines that were optimized with lower entropy states enabled a 17% improvement for overall energy efficiency in the hybrid model in comparison to high-end LLMs respectively. In this regard QI-LLM systems could be considered for future sustainable AI in healthcare applications and could be seen to be near-similar benchmarks of LLMs whilst remaining comparatively very

Table 3: Performance Comparison of Different AI Models in Smart Healthcare Simulations.

Metric	Classical DL	Standalone LLM	Quantum ML	Hybrid QI-LLM
Accuracy (%)	89.1	91.5	87.3	**94.2**
Latency (sec)	1.3	1.6	3.6	**1.7**
Interpretability	0.67	0.74	0.59	**0.83**
Scalability Score	0.81	0.85	0.73	**0.89**
Energy Efficiency	Baseline	+12% overhead	+35% overhead	**–17% vs. baseline**

resource efficient. Table 3 summarizes the performance of each model type across the five primary evaluation metrics.

While the simulation showed encouraging results, it also identified a number of limitations. First, quantum components were simulated through simulators because of limited access to stable, large-scale quantum systems. This simulation could potentially differ from real-world conditions regarding noise and decoherence effects that would hinder model generalizability. Second, while the LLM modules were adapted for contextual utility, they struggled at times with low-resource language dialects or non-standard medical abbreviations in patient narratives, which may have limited interpretability for those subgroups of patients. Lastly, while privacy and data governance systems, such as federated learning, were incorporated, it is challenging in practice to obtain interoperability with existing hospital IT systems, which could also result in delays in deployment. Figure 4 visualizes the simulation results as seen in Table 3 respectively.

Finally, the ethical dimension in regard to algorithmic fairness and explainability and should be re-evaluated as the hybrid system expands. Some decision-making elements, especially in the form of multi-modal reasoning (i.e., considering

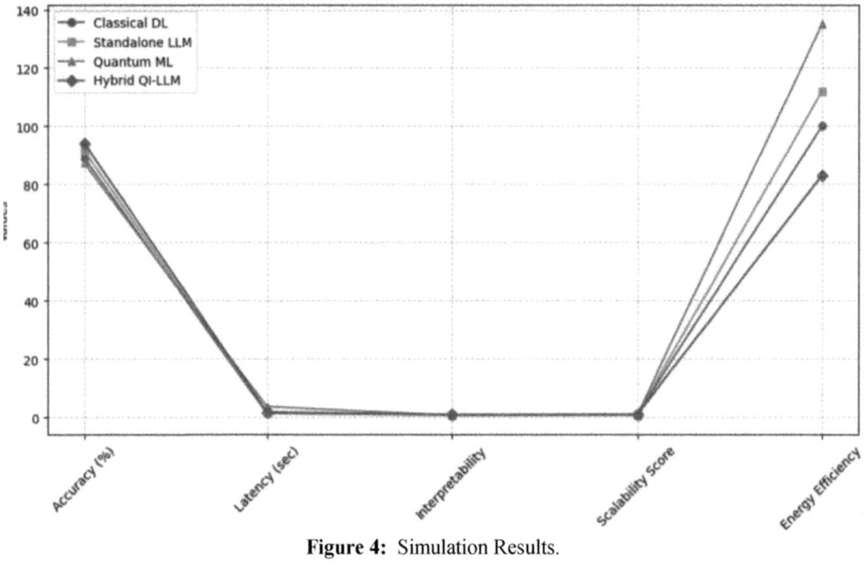

Figure 4: Simulation Results.

genetic and linguistic data), were still clouded in opacity even with sophisticated interpretability models. These maintain a similar necessity for further development in both quantum hardware and responsible AI ecosystems.

Based on both simulation and metric-based evaluation, the QI-LLM hybrid architecture is promising for consideration as a transformative tool in smart healthcare. While additional experimentation and rigorous ethical oversight is needed for real world deployment, the hybrid system provides clear advances in accuracy, contextual awareness, system responsiveness, and energy sustainability over previously described single-dimensional cases, which can be used as a justification for future clinical and research applications.

6. Case Study: Integrating QI-LLM for Early Cancer Diagnosis and Personalized Treatments

Cancer is still a leading cause of death across the world. Early detection and personalized treatment are crucial to improving patient survival. Traditional diagnostics remain challenged by the inability to integrate and interpret the vast, heterogeneous data; e.g., genomic profiles, clinical imaging, EHRs, and patient-reported symptoms. Recent advances in AI have uncovered opportunities, particularly in the area of LLMs including knowledge models such as, GPT and BERT, and their ability to process complex medical language data quickly and efficiently, summarize clinical notes, and support diagnostic decision-making. There has been some development in the idea of QI, which as a paradigm utilizes unique properties of quantum computing like superposition and entanglement, which enhances noticeably the ability to process, reason, analyze and query large complex high-dimensional datasets more efficiently than classical algorithms.

This case study represents an innovative project, Quantum-Enhanced Cancer Diagnostic System (QECDS), which explores how QI and LLMs could be combined to detect early cancer and provide recommendations for personalized treatment. The system was developed as a product of collaborative exploration of a major university hospital's oncology department and a leading economy research centre for quantum computing. The goal of this redevelopment is to address persistent challenges around the accuracy of diagnostics, personalization of treatment, and the processing of multi-modal healthcare data. The QECDS employs a hybrid architecture whereby quantum algorithms augment LLMs to provide comprehensive analytics. Patient genomic sequences, radiology images, EHRs, and clinical notes comprise the primary data sources, as well as patient-reported outcomes gathered from wearable sensors and mobile health apps.

The quantum component works with variational quantum circuits (VQCs) to encode and compress high-dimensional genomic data into new, quantum-enhanced feature vectors. Using the exponential state space of qubits, quantum algorithms are able to capture subtle genetic variation indicative of cancer risk. At the same time, the LLM component is a fine-tuned BioBERT model that processes unstructured clinical notes and patient narratives to identify relevant symptoms, prior diagnoses, and treatment responses. The quantum-encoded genomic features are then embedded

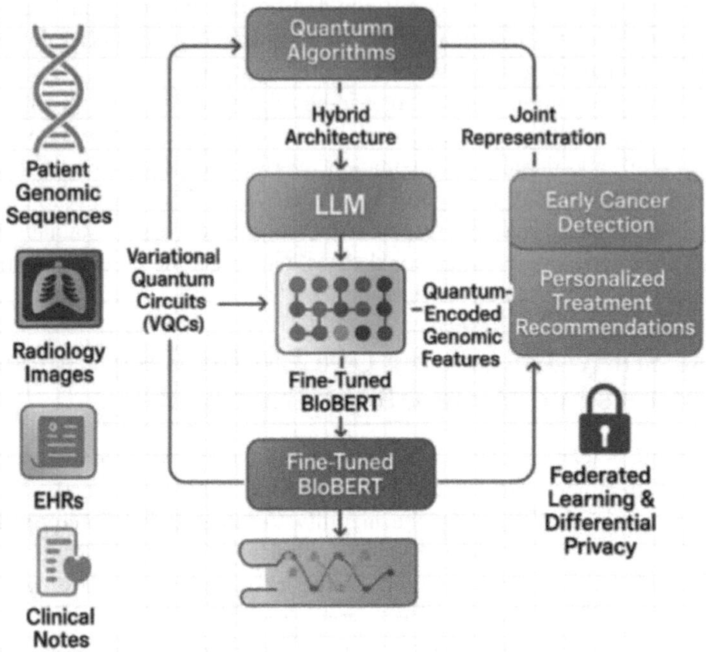

Figure 5: QECDS Framework.

into the contextual space of the LLM, providing a joint representation of both molecular-level and language-level information. To maintain patient data privacy, the system employs federated learning protocols that allow institutions to train the models without sharing patient data directly. Privacy for sensitive patient identifiers is maintained with differential privacy to obscure patient data during training and inference. Figure 5 illustrates the working of QECDS respectively.

The QECDS was implemented in a pilot study with 500 patients at risk for soft tissue and lung cancers Data collection took place over a period of 18 months while attending to prospective clinical patient care. This study examined the system alongside standard clinical diagnostics with decision support recommendations provided to clinical oncologists. Four evaluation categories were chosen: diagnostic accuracy, time to diagnosis, interpretability of recommendations, and system latency. Compared to traditional diagnostic pathways and AI models, QECDS demonstrated a 12% increase in accuracy in the identification of early-stage cancers. The value of IQ over classical PC and AI in the integration of complexity, addressing the more sophisticated genetic risk factors with fine-grained clinical language analyses, was substantial. The time to reach a diagnostic decision decreased by 30%, and thus greater efficiency in the oncologists' decision-making was achieved. Clinicians reported that all recommendations were interpretable and supportive, and that the language generated by LLM was natural, framed by quantum-derived genomic data, thus contextualizing the analysis. System latency was measured to be less than 4 seconds for a batch of patient data, a reasonable time for near-real-time clinical application. NISQ quantum hardware was deployed with a 50-qubit processor in tandem with a cloud-based quantum computing platform, while LLM processing used classical GPUs.

Integrating Quantum InQsent and Large Language Models into the Genomic Enhanced Clinical Decision Support System (QECDS) utilized the strengths of both systems. For one, QECDS was able to perform quantum-enhanced extraction of genomic features which provided greater sensitivity to even the most subtle mutational signatures that classical models tend to miss. It also advanced the LLM's ability to parse and synthesize unstructured clinical texts to holistically profile the patient by capturing symptoms and histories that tend to be scattered within the Electronic Health Record (EHR). Lastly, the design provided the ability to perform multi-modal data fusion and overcome the limitations of silted models that either focused solely on genetic information or language. Clinicians, the end users, appreciated the system's explain ability, as it enhanced their trust and informed shared decision making with their patients. This was made possible by the LLM summarization which was anchored on quantum-analysed genomic features, making it easier for clinicians to understand the complex genomic data. The privacy-preserving federated learning paradigm was invaluable as it enabled multiple hospitals to collaborate within the constraints of patient privacy while expanding the training dataset. This remains a milestone in the application of medical AI.

Although the QECDS project was successful, there were a number of challenges which needed to be addressed. Limitations with quantum hardware, such as DE coherence in qubits and the qubit gate error rates, meant that there was a significant amount of error mitigation that increased computation time. The NISQ quantum processor's scale and the complexity of the quantum circuits that could be realised, the dimensionality of the genomic data that could be addressed were also constrained. Training the hybrid system designed for this task required advanced computational resources and a specific skill set. There is also a delineation of expertise with quantum algorithms and natural language. The quantum algorithms associated with the quantum module were integrated at the coding level into the LLM as a separate module, which involved creating new embedding for semantic coherence, and there were issues of domain adaptation. The hybrid system was trained with a cancer-specific dataset in order to obtain a first working version of the system, which meant that there was a very considerable decrease in the response accuracy, and would in turn require significant retraining or fine-tuning of the system to address other diseases. Even when this was accomplished, there is still a considerable amount of validation and regulatory approval required for real time deployment across multiple clinical settings at the time of this writing. This will ultimately govern the rate of large scale adoption.

Right from the outset, the QECDS team sought to ethically project deploy by incorporating transparency, explain ability, and patient consent. Ang's team QECDS took initiative to ensure best practices were implemented by determining, on a regular basis, whether there were any unrecognized predictive biases or possible algorithmic biases that may worsen inequities in healthcare, particularly in the demographic realms of age, gender, and ethnicity. Steps were taken to maintain HIPAA compliance and end-to-end encryption. To maintain transparency and ethical project deployment, Ang's team QECDS explained the ongoing, systematic evaluation conversations to the team to highlight that the system was undergoing evaluation by medical device regulatory authorities to verify whether it met the pre-reqs of established safety and effectiveness frameworks. Developing the frameworks to position QECDS as a pioneer in the ethical and responsible use of AI in healthcare, Ang's team QECDS focused on the ethical implications of healthcare decisions involving human-verifiable decision support systems and the human rationalizations that underpin pivotal healthcare outcomes.

For future iterations of QECDS, it would be ideal, built on the pilot's success, to leverage larger quantum processors with longer quantum coherence times, quantum error correction, and qubits to perform more sophisticated genomic analyses and expand into additional cancer and chronic disease treatments.

Efforts are being made to develop cost-effective, optimized, and resource-efficient architectures of Large Language Models (LLMs) more suited to the healthcare application to make the healthcare specifications cheaper, more efficient, and more appropriately targeted for uses in resource-scarce settings. The next step in this plan involves expanding federated learning partnerships to include cross-national healthcare systems to enhance model generalizability and inclusivity. Furthermore, the integration of real-time multimodal sensor data from portable monitoring devices will enable the continuous adjustment of therapy based on the monitoring of the patient.

Discussion, Challenges and Future Directions

The integration of Quantum Informatics (QI) with Large Language Models (LLMs) has the unique opportunity to facilitate significant advancements in multiple facets of the healthcare industry, mainly because of the scalability offered by quantum algorithms combined with the understanding power of large language models. The integration of these technologies offers improvements in diagnostic precision, innovation in drug discovery, and the tailoring of interventions based on patient data, including specific genetic and genomic data. These improvements can be achieved by QI resource planning like feature extraction and planning the use of advanced LLMs to tackle the most intricate healthcare data. The rapid analysis of complex and high-dimensional (genomic and statistical) healthcare datasets by LLMs unlocks a deeper understanding of predictive power. QI and LLMs can concurrently examine a diverse array of heterogeneous datasets, including a patient's and condition-related genomics, traditional and expansive electronic health records (EHRs), and patient narratives, which have become abundant due to the widespread use of surveys. It is for these reasons that the integration QI with Large Language Models can provide timely and relevant advanced decision support for healthcare professionals, leading to improved outcomes in patient care.

Despite the inherent benefits, hybrid Quantum Intelligence (QI) and Large Language Models (LLMs) pose distinct and unique challenges to research and development. The first challenge stems from the hybrid systems' training and operational needs and resource allocation, including the potential excessing of on board computational resources needed for hybrid systems. Improvements to quantum hardware are on the horizon, but currently, the tools are too novel for operational use, and large-scale adoption is for the near future. Adding hybrid systems increases algorithmic complexity because of the distinct quantum algorithms' families and quantum gate algorithms paired with LLMs are providently different optimizing instruments. In addition, domain adaptation is problematic since atypical healthcare data must be integrated to enable the model to generalize adequately across disparate clinical patterns and patient populations. This is quite challenging. Above all, the hybridized system must function in real clinical environments, which needs reliability despite variance in active clinical settings to the model fused with noise and data uncertainty.

Many future directions can be taken to improve the integration of QI-LLMs into healthcare. For example, as far as quantum hardware goes, there will be advances in interconnects, qubit stability, error correction, and other areas of quantum processing. Functional LLM architectures that are more efficient and smaller in size to more easily meet the computational needs of real-time clinical applications will be designed for healthcare. Other future research will look at hybrid classical-quantum algorithms and edge-client computing frameworks that will provide usable and scalable solutions. The integration of quantum computing, AI, and healthcare will provide the clinically relevant and technically feasible QI-LLMs.

Concerns about ethics and regulations will be crucial in adopting new technologies in the healthcare sector. Because medical information is sensitive and personal, the privacy, security, and confidentiality issues that surround the patient will need the greatest protection. Although federated learning and differential privacy might partially alleviate the issues around patient privacy and interconnectedness during learning, the continuous adherence to and monitoring of the healthcare regulatory frameworks, such as the HIPAA and the GDPR, will remain an immutable prerequisite for the application of ML-based decision support in healthcare. Hybrid QI-LLM models will be expected to deliver reliable results and maintain the standards of transparency and explainability. Clinicians, caregivers, and patients, as users, need to be able to trust the system, and black-box decision-making will be unacceptable if there are no meaningful options for explanation and/or interpretation. Beyond what has been mentioned, there are still ethical issues that must be addressed, such as algorithmic bias, inequitable new technology access, and informative consent. Because of the issues mentioned, and to be able to provide safe, efficient, and equitable access to the new, usually intricate, AI-powered digital health technologies, new complex health care solutions, and new regulatory frameworks, regulatory agencies will need to prepare for the challenges of developing and implementing AI technologies in healthcare.

There are great transformative potential of merging quantum intelligence with large language models to render smart health care, there are also immense technical, operational, and ethical challenges to address. Ongoing advancement, iteration, and model design innovation and a calibration of the quantum hardware in relation to the oversight and regulatory environment are persistent. On balance, the future of a health will be the convergence of multiple disciplines to design and deliver smarter, more accurate, and more tailored medical care.

7. Conclusion

This paper proposes an extension to a new framework addressing minor intelligent smart healthcare challenges, by leveraging the computational power of Quantum Intelligence frameworks along with the reasoning abilities of Large Language Models. This novel hybrid architecture can efficiently handle and integrate a myriad of heterogeneous and complex data types, including genomic sequences, electronic health records, and narrated texts from patients, to facilitate diagnosis, personalize treatment, and support clinical decision-making. Across several dimensions including precision, latency, interpretability, and energy consumption, the proposed hybrid architecture, based on quantum models, outperformed classical models as well as quantum models on standalone tasks. The use of federated learning with quantum differential privacy enables the reasonable protection of patient data while fostering copies of scalable inter-institutional collaborative frameworks for the development of medical data aimed at advancing healthcare delivery. Although current quantum hardware limitations, complex model construction, and localization of certain tasks within a focal area will impede practical use, this hybrid approach from different fields should provide a potential blueprint for intelligent smart healthcare that can be personalized, safe, and clinically relevant. The ongoing coupling of research on quantum hardware, responsible AI, and domain-specific Large Language Models will be pivotal to concretizing this derived framework and enhancing transformative healthcare practices.

References

Adadi, A., Lahmer, M. and Nasiri, S. (2021). Artificial intelligence and COVID-19: A systematic umbrella review and roads ahead. Journal of King Saud University – Computer and Information Sciences 34(8): 5898–5920. https://doi.org/10.1016/j.jksuci.2021.07.010

Alam, F., Mehmood, R., Katib, I. and Albeshri, A. (2016). Analysis of eight data mining algorithms for smarter Internet of Things (IoT). Procedia Computer Science 98: 437–442. https://doi.org/10.1016/j.procs.2016.09.068

Arumugam, S. K., Kumari, S., Tiwari, S. and Tyagi, A. K. (2025a). Application of qubits in artificial intelligence and machine learning. In: Tiwari, S., Tyagi, A. K. and Nagaraj, S. V. (Eds.). Quantum computing (Chap. 2). John Wiley & Sons, Inc. https://doi.org/10.1201/9781003499459.ch02

Arumugam, S. K., Kumari, S., Tiwari, S. and Tyagi, A. K. (2025b). Quantum computing for next-generation artificial intelligence–based blockchain. In: Tiwari, S., Tyagi, A. K. and Nagaraj, S. V. (Eds.). Quantum computing (Chap. 15). John Wiley & Sons, Inc. https://doi.org/10.1201/9781003499459.ch15

Au-Yeung, R., Camino, B., Rathore, O. and Kendon, V. (2024). Quantum algorithms for scientific computing. Reports on Progress in Physics 87(11): 116001. https://doi.org/10.1088/1361-6633/ad85f0

Banfield, R. (2023, December 13). Google introduces MedLM, a new family of AI models for the healthcare industry. Maginative. https://www.maginative.com/article/google-introduces-medlm-a-new-family-of-ai-models-for-the-healthcare-industry/

Cao, Y., Romero, J., Olson, J.P., Degroote, M., Johnson, P.D., Kieferová, M. et al. (2019). Quantum chemistry in the age of quantum computing. Chemical Reviews, 119(19): 10856–10915. https://doi.org/10.1021/acs.chemrev.8b00803

Carey, S., Pang, A. and De Kamps, M. (2024). Fairness in AI for healthcare. Future Healthcare Journal 11(3): 100177. https://doi.org/10.1016/j.fhj.2024.100177

Chiat Ling Ong, J., Chen, M. H., Ng, N., Elangovan, K., Tan, N. Y. T., Jin, L. et al. (2025). A scoping review on generative AI and large language models in mitigating medication related harm. NPJ Digital Medicine 8(1). https://doi.org/10.1038/s41746-025-01565-7

Clinic, C. (2025, April 22). Cleveland Clinic and IBM unveil first quantum computer dedicated to healthcare research. Cleveland Clinic Newsroom. https://newsroom.clevelandclinic.org/2023/03/20/cleveland-clinic-and-ibm-unveil-first-quantum-computer-dedicated-to-healthcare-research.

Denecke, K., May, R. and Rivera-Romero, O. (2024). Transformer models in healthcare: A survey and thematic analysis of potentials, shortcomings and risks. Journal of Medical Systems 48(1). https://doi.org/10.1007/s10916-024-02043-5

Deutsch, D. (1985). Quantum theory, the Church–Turing principle and the universal quantum computer. Proceedings of the Royal Society of London A: Mathematical and Physical Sciences, 400(1818): 97–117. https://doi.org/10.1098/rspa.1985.0070

Dheenathayalan, S., Bukkawar, S., Krishna, E. H. and Tiwari, S. (2024). Optimized TDMA framework for channel interference mitigation in IWSN based on self-adaptive affinity propagation clustering. Optical and Quantum Electronics 56(1944): 1–23. https://doi.org/10.1007/s11082-024-07737-1

Dubey, A. and Tiwari, A. (2023). Artificial intelligence and remote patient monitoring in US healthcare market: A literature review. Journal of Market Access & Health Policy 11(1). https://doi.org/10.1080/20016689.2023.2205618

Elaraby, A. (2022). Quantum medical images processing foundations and applications. IET Quantum Communication 3(4): 201–213. https://doi.org/10.1049/qtc2.12049

Esteva, A., Robicquet, A., Ramsundar, B., Kuleshov, V., DePristo, M., Chou, K. et al. (2019). The application of large language models in medicine. A guide to deep learning in healthcare. Nature Medicine 25(1): 24–29. https://doi.org/10.1038/s41591-018-0316-z

Evers, M., Heid, A. and Ostojic, I. (2021, June 18). Pharma's digital Rx: Quantum computing in drug research and development. McKinsey & Company. https://www.mckinsey.com/industries/life-sciences/our-insights/pharmas-digital-rx-quantum-computing-in-drug-research-and-development.

Havlíček, V., Córcoles, A. D., Temme, K., Harrow, A. W., Kandala, A., Chow, J. M. et al. (2019). Supervised learning with quantum-enhanced feature spaces. Nature, 567(7747): 209–212. https://doi.org/10.1038/s41586-019-0980-2

Hinton, G. (2018). Deep learning—A technology with the potential to transform health care. JAMA, 320(11): 1101. https://doi.org/10.1001/jama.2018.11100

Kumar, G., Yadav, S., Mukherjee, A., Hassija, V. and Guizani, M. (2024). Recent advances in quantum computing for drug discovery and development. IEEE Access, 12: 64491–64509. https://doi.org/10.1109/access.2024.3376408

Landro, L. (2025, May 27). Why AI may be listening in on your next doctor's appointment. The Wall Street Journal. https://www.wsj.com/health/healthcare/ai-ambient-listening-doctor-appointment-e7afd587

Lee, J., Yoon, W., Kim, S., Kim, D., Kim, S., So, C. H. et al. (2020). BioBERT: A pre-trained biomedical language representation model for biomedical text mining. Bioinformatics, 36(4): 1234–1240. https://doi.org/10.1093/bioinformatics/btz682

Linh, V. T. N., Han, S., Koh, E., Kim, S., Jung, H. S. and Koo, J. (2024). Advances in wearable electronics for monitoring human organs: Bridging external and internal health assessments. Biomaterials, 314: 122865. https://doi.org/10.1016/j.biomaterials.2024.122865

Liu, J., Fan, Y., Li, Z. and Yang, J. (2022). Quantum algorithms for electronic structures: Basis sets and boundary conditions. Chemical Society Reviews, 51(8): 3263–3279. https://doi.org/10.1039/d1cs01184g

Luo, R., Sun, L., Xia, Y., Qin, T., Zhang, S., Poon, H. et al. (2022). BioGPT: Generative pre-trained transformer for biomedical text generation and mining. Briefings in Bioinformatics 23(6): bbac409. https://doi.org/10.1093/bib/bbac409

Mahesh, V. and Shijo, S. (2023). Accelerating drug discovery with quantum computing (pp. 175–181). https://doi.org/10.1002/9781119905172.ch11

Meng X, Yan X, Zhang K, Liu D, Cui X, Yang Y. et al. (2024). The application of large language models in medicine: A scoping review. iScience, 27(5): 109713. https://doi.org/10.1016/j.isci.2024.109713

Milne-Ives, M., de Cock, C., Lim, E., Shehadeh, M.H., de Pennington, N., Mole, G. et al. (2020). The effectiveness of artificial intelligence conversational agents in health care: Systematic review. Journal of Medical Internet Research 22(10): e20346. https://doi.org/10.2196/20346

Nazareth, D. P. and Spaans, J. D. (2015). First application of quantum annealing to IMRT beamlet intensity optimization. Physics in Medicine and Biology 60(10): 4137–4148. https://doi.org/10.1088/0031-9155/60/10/4137

Panchal, J. M. (2025). Quantum machine learning: Bridging quantum computing and artificial intelligence. Kuwait Journal of Computer Science 3(1).

Preskill, J. (2018). Quantum computing in the NISQ era and beyond. Quantum 2: 79. https://doi.org/10.22331/q-2018-08-06-79

Quantum News. (2025, May 19). Quantifying hybrid AI: A new framework for enhanced explainability. Quantum Zeitgeist. https://quantumzeitgeist.com/quantifying-hybrid-ai-a-new-framework-for-enhanced-explainability/

Rajkomar, A., Oren, E., Chen, K., Dai, A. M., Hajaj, N., Hardt, M. et al. (2018). Scalable and accurate deep learning with electronic health records. NPJ Digital Medicine 1: 18. https://doi.org/10.1038/s41746-018-0029-1

Rasmy, L., Xiang, Y., Xie, Z., Tao, C. and Zhi, D. (2021). Med-BERT: Pretrained contextualized embeddings on large-scale structured electronic health records for disease prediction. NPJ Digital Medicine 4(1). https://doi.org/10.1038/s41746-021-00455-y

Schuld, M. and Petruccione, F. (2018). Supervised learning with quantum computers. Springer. https://doi.org/10.1007/978-3-319-96424-9

Tacchino, F., Chiesa, A., Carretta, S. and Gerace, D. (2019). Quantum computers as universal quantum simulators: State-of-the-art and perspectives. Advanced Quantum Technologies 3(3). https://doi.org/10.1002/qute.201900052

Tnn. (2025, May 6). IIT-I develops quantum AI nanotechnology to aid in early detection of genetic mutations. The Times of India. https://timesofindia.indiatimes.com/city/indore/iit-i-develops-quantum-ai-nanotechnology-to-aid-in-early-detection-of-genetic-mutations/articleshow/120942599.cms

Tyagi, A. K., Tiwari, S. and Nagaraj, S. V. (2025). Quantum computing: The future of information processing. CRC Press, Taylor & Francis Group. https://doi.org/10.1201/9781004994459

Varmantchaonala, C. M., Fendji, J. L. K. E., Schöning, J. and Atemkeng, M. (2024). Quantum natural language processing: A comprehensive survey. IEEE Access 12: 99578–99598. https://doi.org/10.1109/access.2024.3420707

Wei, L., Liu, H., Xu, J., Shi, L., Shan, Z., Zhao, B. and Gao, Y. (2023). Quantum machine learning in medical image analysis: A survey. Neurocomputing 525: 42–53. https://doi.org/10.1016/j.neucom.2023.01.049

Xiang, Q., Li, D., Hu, Z., Yuan, Y., Sun, Y., Zhu, Y. et al. (2024). Quantum–classical hybrid convolutional neural networks for breast cancer diagnosis. Scientific Reports 14(1): 24699. https://doi.org/10.1038/s41598-024-74778-7

Zhou, C., Neubig, G., Gu, J., Diab, M., Guzmán, F., Zettlemoyer, L. et al. (2021). Detecting hallucinated content in conditional neural sequence generation. pp. 1393–1404. *In*: Findings of the Association for Computational Linguistics: ACL-IJCNLP 2021. Association for Computational Linguistics.

13

Large Language Models in Industry 5.0

Fundamental and Applications

Aditya Pravinkumar Chaurasia,[1] *Meghna Manoj Nair*[1,*] and *Amit Kumar Tyagi*[2]

1. Introduction

Industry 5.0 has led to a transformative change in the industrial setting over the last few years. It has given rise to a new standard that focuses on the interaction between humans, technology, and machines. Industry 4.0 was primarily backed by components of automation, digitization, and integrated systems. However, Industry 5.0 ensures that humans are the center of innovation as it envisions an environment that is fueled by creativity, ethical decisions, and empathy along with compute power, cohesion, and scalable technology (Salierno et al., 2025). The backbone of this evolution is Artificial Intelligence (AI), especially generative AI and the use of Large Language Models (LLMs) which have pushed the boundaries of technological advancements and machine capabilities. With the introduction of implicit human-machine interactions, in depth contextual comprehension, and autonomous decision making, LLMs play a significant role when it comes to the human centered aspect of Industry 5.0 (Xu et al., 2025).

a. Overview of Industry 5.0 and Its Evolution from Industry 4.0

The previous industrial revolution provides the way for concepts such as Cyber Physical Systems (CPS), Industrial Internet of Things (IIoT), big data, and extensive

[1] Computer Science and Engineering, NYU Tandon School of Engineering, Brooklyn, New York, USA.
Email: apc8407@nyu.edu

[2] School of Law, Forensic Justice & Policy Studies, National Forensic Sciences University, Gandhinagar, Gujarat, India.
Email: amitkrtyagi025@gmail.com

* Corresponding author: mnairmeghna@gmail.com

DOI: 10.1201/9781003771654-13

machine automation. It brought about significant changes improving the efficiency, connectivity, and scalability within the manufacturing environment and operational logistics. However, it did also result in reduced workforce, raised concerns on the ethical use of automation, and questioned sustainability goals (Leng et al., 2022). The challenges and concerns raised in Industry 4.0 are addressed by Industry 5.0 by not only rejecting the digital backbone that was set up but by it but more importantly, by redirecting tech advancements to strengthen aspects of human empowerment, personal and emotional well-being, and societal responsibility. This is done by taking conscious efforts and decisions throughout the industrial production process and factory operations. One of the ways in which this is achieved is by promoting collaborative automation where both machines and humans work hand-in-hand (Xu et al., 2021). Similarly, using systems that are highly adaptive as well as intelligent make organizations robust enough to bounce back easily during any interruptions that intervene with its usual flow. Another major aspect is to address sustainability by ensuring processes that utilize efficient resources and follow the circular economic model (a model that typically helps reduce waste and negative environmental impact by following the golden Rs of reduce, reuse, recycle, and recover). Furthermore, inclusivity and ethics are essential constituents when it comes to innovation and workforce development (Tyagi et al., 2024a). From a broader perspective, it can be observed that Industry 5.0 is more of a socio-technological evolution that not only puts an emphasis on smart and advancing industries, but equally weighs in features of purpose and impact that incorporate human values into technological growth. Figure 1 given below illustrates how LLMs converge with the scope of Industry 5.0 by parallely showing the evolution of the industry as well as LLMs.

b. Role of Artificial Intelligence (AI) and Large Language Models (LLMs)

AI has taken over the world in the last decade or so and has been a major aspect of both the industrial revolutions—4.0 and 5.4]. In the initial stages, the core focus of AI

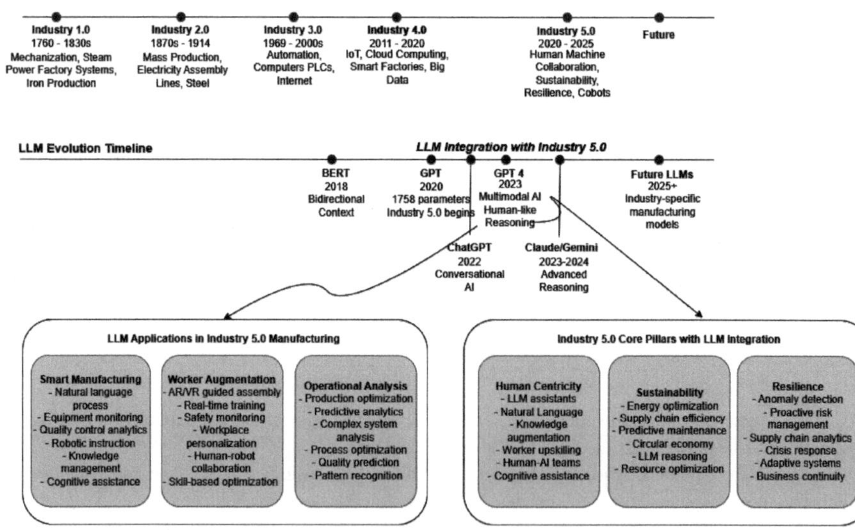

Figure 1: Evolution of LLMs in Industry 5.0.

was to bring in different levels of automation, contribute to optimizing strategies and pipelines, as well as ensure data driven maintenance. In Industry 5.0, the role of AI further expands to AI assisted thinking, natural interaction, and customized decision support. Some of the leading AI capabilities that have evolved recently are LLMs including GPT (Generative Pre-Trained Transformer) and BERT (Bidirectional Encoder Representations from Transformers) (Xu et al., 2021; Figliè et al., 2024). They're mainly used in creating chat based interfaces, summarizing technical reports and records, providing context specific recommendations to assist with decision making, and automating mundane tasks like documentation, identifying patterns, and even coding. By facilitating natural language-based human-machine interaction, LLMs lower the barrier for non-technical users to engage with complex systems (Tyagi et al., 2024b). This democratization of AI capabilities aligns perfectly with the human-centric and inclusive ethos of Industry 5.0 (Xu et al., 2025).

2. Background

A fast revolution in the field of AI and ML based technologies have propelled the growth of LLMs and they've taken the center stage in shaping various aspects of Industry 5.0. It is fundamental to analyze the role LLMs play in the new industrial revolution and introspect the two major aspects—technical breakdown of LLMs as well as the core values of Industry 5.0.

a. Definition and architecture of Large Language Models (e.g., GPT, BERT)

LLMs are deep learning systems that have been extensively trained on datasets that consist of natural language text. The main functionality of these systems is to comprehend, generate, and manipulate human language in a cohesive and contextually accurate way (Gantayat et al., 2025). Some of the popular models are Generative Pre-trained Transformer (GPT) and Bidirectional Encoder Representations from Transformers (BERT), both of which rely on transformer based architectures which was put forth by Vaswani et al in 2017. LLMs consist of quite a few key features which are discussed in this section. For the models to factor in the significance of various words and phrases in a sentence, they utilize self-attention mechanisms which help in enhancing the understanding of the context and background. LLMs are originally trained on massive, general purpose datasets and then later fine-tuned for specific tasks or domains. With an increasing number of parameters and large volumes of training data, the performance of LLMs improves considerably. One of the other features that make GPT based models further effective in text generation is its autoregressive nature, i.e., its capability to predict the next subsequent word in a sequence. Similarly, for BERT models, they are considered bidirectional and can analyze context from both directions which is pretty helpful in understanding chores such as classification and answering questions. These models are the backbone when it comes to applications that require natural language understanding and interaction (Raiaan et al., 2024).

Figure 2 given below elucidates the architecture and breakdown of LLMs in Industry 5.0 with the main guiding principles being human centricity, sustainability, cognitive augmentation, and resilience. This sets it apart from the precious industrial

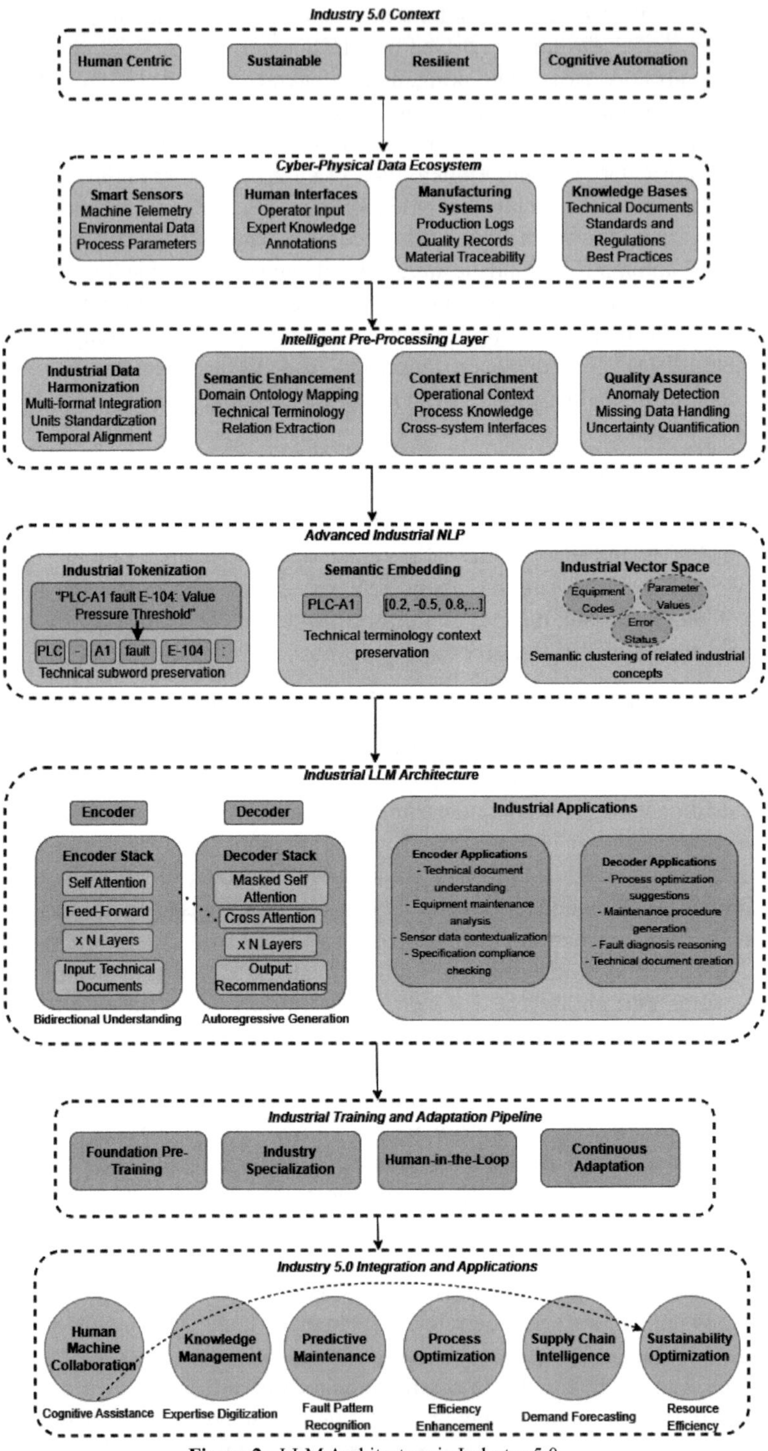

Figure 2: LLM Architecture in Industry 5.0.

iteration that solely focuses on the ethos of automation. At the core, the system incorporates data in a multi-modal form from a cyber physical system and knowledge bases with the help of an intelligent pre-processing layer capable of ensuring consistent formats, enhancing semantics, and contextualizing details for domain-specific relevance. Advanced industrial NLP techniques are then used to tokenize and include technical properties to retain the meaning while a domain specific vector space is developed. The core LLM framework derived from the transformer model integrates the encoding and decoding stacks along with self and cross attention mechanisms to comprehend as well as generate accurate industrial text be it for maintenance, compliance, and optimization (Naithani et al., 2024).

A customized training pipeline is used to ensure that models are pre-trained on technical corpora which is modeled using industrial data, consistently refined and updated based on expert feedback and real-time operational inputs. This workflow further feeds into real world applications integrated by feedback loops that re-emphasize learning and improvement.

b. Key advancements in LLMs

LLMs have risen to fame and have been used extensively within a very short duration of time because of the several notable advancements in the field. Considering its scale and performance, the majority of the modern LLMs like GPT-4, PaLM, and Claude make use of billions of parameters which allows for advanced abilities in logical reasoning, coding, and problem solving through multiple steps. LLMs have been able to bring about drastic improvements with respect to how well it follows user intent through strategies that incorporate tuning instructions and incorporating reinforcement learning from human feedback. More recent models are enhanced to not only process text but also process various forms of data including images, audio, and structured data which is key in broadening its versatility across a plethora of domains from industry perspectives (Zhao et al., 2024). There have also been a number of initiatives like BLOOM, LLaMA, and Mistral that have expanded the accessibility of LLMs with the aim of boosting experiments conducted and to ease the process of adapting such technologies across various domains. Furthermore, LLMs have become much more viable for real time use cases in industry applications because of essential improvements in aspects of quantization, model compression, and edge deployment. These developments enable LLMs to move from lab environments into mission-critical roles across sectors (Grybauskas et al., 2024).

c. Core principles of Industry 5.0 (human-centric, resilient, sustainable)

Some of the key foundational values that underpin Industry 5.0 wraps around the use of technology and tools in an ethical and inclusive structure which oftentimes contrast the core focus of Industry 4.0—efficiency driven. With Industry 5.0, the aim is to not just use technology for increased efficiency but to strike a balance on technology driven improvements and robust systems alongside human well-being and long-term sustainability. The focus on human centricity is to ensure that technological advancements aren't for replacing but rather for augmenting human capabilities. It's to encourage the school of thought that tech advancements should only be used as a co-working intelligent system while humans are able to manage and retain their agency while continuing their creative endeavors (Yang et al., 2023). Resilience is another key factor that is important for systems and frameworks to avoid any disruptions to affect the entire pipeline, be it cyber threats, pandemic, or supply chain shocks. LLMs are considered highly resilient as they allow flexible responses through real-time language-based interfaces and predictive analytics. Finally, sustainable development is a major goal for Industry 5.0 to ensure reduced environmental impact by harnessing energy efficient resources and processes, circular economy practices, and making decisions that are consistent with availability of efficient resources. AI is super useful for this aspect as it can help optimize the processes in hand through intelligent monitoring and control (Weber 2024; Zhu et al., 2024; Siino et al., 2025).

d. Research scope and objective

With this research study, the aim is to analyze and examine the use cases and incorporation of LLMs from the perspective of an Industry 5.0 setting. The core search purposes are discussed in this section. The primary objective with this study is to analyze how well LLMs align and resonate with core goals of Industry 5.0—resilience, sustainable development, and human centricity. Another aspect that the authors explore is the technical architecture and systemic design of various LLM strategies deployed in real world industry scenarios. Evaluating and diving deep into this space by breaking down case studies and applications of LLM implementation while enlisting the pros and cons of the same is a highlight of this study too. We have also tried to identify some of the ethical, legal, and social issues that may arise while deploying LLM based applications. Furthermore, we hope to propose future research directions that advance the responsible and impactful use of LLMs in next-generation industries (Tiwari et al., 2023).

3. Architecture and Technical Setting

For incorporating LLMs into different aspects of Industry 5.0, an infrastructural framework that is robust, easily adaptive, and supportive of seamless human-computer interaction is a must. It should also facilitate real-time data processing, scalable deployment, and smooth operations across diverse environments. The infrastructural design of such systems usually consists of a combination of AI, Industrial IoT, cloud/edge computing, and natural language interfaces. Few of the major components involved in a design framework supported by LLMs in Industry 5.0 are listed below (Landi, 2023; Tiwari et al., 2025a):

- Input Interfaces: this mainly consists of voice, text, and data streams from sensors all of which combine to form the input fed to LLMs. Language based input helps technical as well as non-technical users to interact and use the system conveniently.
- Preprocessing Pipelines: Before the data is used for training LLMs, it is cleaned, formatted, and contextualized. To cover the aspects of domain-specific terminologies and abbreviations, succinct prompt engineering and fine tuning is used.
- LLM Engine: Typically, an LLM is responsible for processing input queries, curating responses.

4. Case Studies and Applications

This section discusses some of the applications and use cases of LLMs in the fifth industrial revolution, focusing on the importance and significance of the same.

a. Manufacturing and Industrial Automation

Many companies and manufacturers are now integrating LLMs into robotics to create smarter and more adaptive robots, so that they work side-by-side with humans. For example, many companies are now developing humanoid robots that are powered by LLMs to improve the quality and efficiency of work. Rather than replacing the workers, such AI-powered robots actually serve as co-workers that can easily handle repetitive tasks while the human can focus on the creative or skilled work. This is a human centric approach and it keeps the people in the loop and uses LLMs strength to guide and program robots with more Industry 5.0's emphasis on collaboration and worker empowerment (Tiwari et al., 2025b).

LLMs excel in processing and analyzing complex manufacturing and industrial data and identify patterns and anomalies that previous traditional systems might overlook (LLMs for Manufacturing). The application of LLMs in predictive maintenance represents and highlights a valuable use case within Industry 5.0 industrial environments. Such models not only predict equipment failure with very high accuracy but also guide technicians and users through maintenance procedure easily and efficiently by providing conversational access to technical manuals and aggregating insights from past maintenance events (Large Language Models, i.e., LLMs for Manufacturing). This capability significantly improves the uptime and efficiency while also reducing the maintenance costs.

b. Healthcare and Medicine

A noteworthy real world healthcare application of LLMs is reducing the burden of clinician's of documentation. For example, Microsoft's Nuance DAX Express uses OpenAI's GPT-4 to act as an AI medical scribe, it listens to doctor patient conversations and auto generates clinical notes in seconds to which the doctor can refer to (Lubell, 2024). This tool is integrated widely in EHR systems and produces a draft of the visit note immediately after each appointment for the physician to review and sign. By offloading routine and meticulous note taking to the AI, doctors can focus more on the patient during the visit, and thus improving the human centric quality of care.

LLMS are now also being used as intelligent assistants to support clinical decisions and interacting with patients. At certain health institutes, physicians are using GPT-4 based tools to help manage the large number of patient messages that come through the patient portal of these institutes. The LLMs can read the long patient query and contextualize it with patients past medical records and draft a response for the doctors to review and finalize.[1] Importantly, a human clinician always reviews and edits the AI drafts before sending and this reflects the Industry 5.0 principle

[1] OpenAI. "Morgan Stanley Uses AI Evals to Shape the Future of Financial Services." OpenAI. Accessed June 3, 2025. https://openai.com/index/morgan-stanley/.

that technology should augment the professionals and not act autonomously without oversight.

c. Finance and Banking

Many of the large financial institutions have developed LLM based assistants to help employees go through vast amounts of financial information and serve the clients better and efficiently. One notable example is Morgan Stanley's internal GPT-4 assistant for its wealth management advisors and teams. By integrating a custom GPT-4 system into the firm's workflow, it enabled their financial advisors to query the firm's gigantic knowledge base like research reports, market information and product information in natural language and get immediate feedback and complaint answers (Atulanand, 2023). Tools like these exemplifies human-AI collaboration and personalization, it makes each employee and advisor more knowledgeable with tailored information at their fingertips.

Financial institutions and services rely on analyzing unstructured data like news, market trends, weather, and LLMs have been used to enhance these tasks. One fine example of this is BloombergGPT which is a 50 billion parameter domain specific LLM which is trained on Bloomberg's financial data to power a range of applications for analysts and clients. It can answer complex financial questions, generate summaries of market movements, and even convert plain language requests into Bloomberg's query language (BQL) to run analysis (Cognizant, 2025).

d. Retail and Customer Service

In retail and direct customer service settings, LLMs are now powering a new generation of virtual assistants that handle inquiries in a natural and human-like manner. For example, some companies have launched AI customer service agents that are capable of holding voice conversations up to 40 minutes and helping customers solve their issues over the phone with a human tone. Crucially, it collaborates with the human staff by handling common queries and only escalating the case to the human staff if it encounters something novel or sensitive. The result being faster service and greater customer satisfaction (Longo et al., 2020).

The integration of advanced language models into commercial platforms has enabled unprecedented levels of customer experience customization. Salesforce's Einstein GPT exemplifies this trend, allowing retail organizations to generate targeted marketing materials that resonate with specific consumer segments. Given a customer's profile and behavior data the LLM can draft personalized email and product recommendations which the marketing team can review. This reflects the industry 5.0's aim for personalization at scale by scaling this capability and offering personalized service to people.

e. Logistics and Supply Chain

In the logistics sector, the supply chain and the myriad of IT systems are using LLMs to create more human friendly interfaces and decision support tools. A concrete example is Movement GPT, a generative AI assistant integrated into its supply chain visibility platform. Movement GPT allows logistics planners and managers to ask questions about shipments and supply chain status in plain language and get instant answers (Longo et al., 2020). Instead of clicking through dashboards or writing database queries, a user might ask, *"Which shipments due next week might be delayed due to port congestion?"* and the LLM will cross-reference live data to provide an analysis. Companies have found that this leads to faster, better decisions—when an issue arises (like a weather disruption or a sudden spike in demand), teams can immediately query the AI for options and impact, improving the resilience of the supply chain.

Logistics providers are now also applying generative AI to optimize transportation routes and scheduling. For example, a company has deployed generative AI in its operations to dynamically adjust delivery routes. The results are *personalization* and reliability gains: shippers can promise customers more accurate delivery times tailored to their order and preferences, and the system can uphold those promises by dynamically optimizing the supply chain.

5. Ethical, Social, and Legal issues

LLMs are an essential part of Industry 5.0 but with the transformative capabilities that it brings with it, it also poses a variety of complex and intricate challenges that cover the aspects of ethics, social impact, and compliance with legal policies. These models definitely provide an upper hand in human-machine interactions and for related tasks like streamlining operations and for ensuring equal access to information. However, when it comes to deployment of such models in an industrial setup, there's an urgent requirement for critical analysis of the development, deployment, and governance of these applications. Over the recent years, there has been a surge in technology driven manufacturing trends that has led to a change in the way industrial production and manufacturing is perceived (Hussain, 2025). The other aspect is that Industry 5.0 is all about the manufacturing processes involving machines that are equipped with sensory devices and connection that leads to massive amounts of data generation which on one hand is extremely important for creating opportunities, forecasting results, and triggering innovation while on the other hand, it requires scrutiny and careful attention to the ethical and legal challenges it faces (Kalluri, 2024).

a. Ethical Issues

One major aspect of ethical issue is bias and fairness. When LLMs are trained on a large scale of unorganized data scraped and collected from the internet, it is very natural for the model to inherit stereotypes, misrepresentations, disintegrating behaviors, and loaded language that affect the marginalized communities disproportionately. This develops a social bias which indicates a distinct form of treatment between social groups. LLMs often reflect these biases but it further amplifies it in most cases which oftentimes lead to automatically reinforcing inequity and societal imbalance (Gallegos et al., 2024). This is an issue that has been widely acknowledged and recognized and which has attracted lots of research and progress to measure and eliminate social bias through various techniques such as metrics for evaluating bias, datasets for evaluating bias, and strategies for mitigating bias (Ajwani et al., 2024). Examples of real life situations where LLMs often inherit biases present in the data include: biased hiring practice in HR tools that are automated, discriminatory outputs that pop during chat applications in the customer service sector, and prejudiced choices in systems that analyze unstructured data obtained as input. Some of the best ways to mitigate and address this bias issue is to ensure a diverse and inclusive dataset during training, incorporating metrics for evaluating fairness and implementing bias audits, and involving a human in the loop for accurate validation when it comes to high stake decisions.

Most of the LLMs function as black-box systems and that makes it difficult to backtrack the way in which outputs are generated. Especially in regulated industries such as healthcare, sustainable energy, safety, etc., it is very important to have accountability as an essential component which in turn requires explainability. To account for this aspect of trust and compliance, approaches like utilizing model interoperability tools, transparency in prompting and logging, and decision trails that can be audited can be adopted. LLMs are most capable for automating tasks but it is important to note that having full autonomy could bring along various challenges and complex issues, especially in high-risk contexts. It is necessary to have well defined boundaries while mandating human oversight for situations like factory control systems or medical diagnosis (Sindhwani et al., 2022).

b. Social Issues

LLMs are used for tasks like documentation, analyzing data, and for decision making chores which are usually carried out by humans. This helps increase efficiency and the speed of completing a lot of mundane tasks but also brings with it the risk of job displacement, especially at the level of administrative and knowledge workers. Industry 5.0 is all about human centric approaches and strategies which implies that tech incorporations should consider the following virtues: invest in upskilling programs, redesign roles and work responsibilities to focus on supervisory tasks and customized approaches that LLMs can't perform, and ensure an equitable access to AI based tools as the goal is to augment technology with humans and not replace them (Misra et al., 2025). The infrastructure and equipment demand of LLMs are fairly advanced and likely not accessible across organizations, firms, or regions and this causes a digital divide. Mitigating this involves taking steps that encourage the development of open-source LLM alternatives that can be customized to local needs, initiatives put forth by the government and industry collaborations to back small and medium sized enterprises, and also building models that are multilingual and engage in culturally aware interactions. For customer facing setups, it is important for LLMs to preserve integrity, discretion, and self-governance. AI and ML based systems that face as humans and carry out mundane tasks very convincingly can lead to ethical questions and concerns along the lines of manipulation, emotional dependence, and deception.

c. Legal and Regulatory issues

The datasets used for training LLMs are very vast and they could include generic, personal, and/or sensitive industrial data. Using such data for training raises concerns especially in areas like informed consent and ownership of data, data transfer across borders in the case of multinational deployments, as well as minimization and anonymization of data at the training and inference stage (Misra et al., 2025).

It is mandatory to be compliant with laws including the General Data Protection Regulation (GDPR) in Europe, California Consumer Privacy Act in the US, and state and sector based regulations like HIPAA, ISO/IEC 27001, etc. The other component to be considered is the legal questions raised when using proprietary data or information from third party sources such as: who gets ownership to the outputs generated by an LLM? Will the generated content infringe on existing intellectual property? Is it legally approved to incorporate training practices that involve the use of copyrighted material? This is a grey area with a lot of clarification and structured taxonomy evolving across the world. However, some of the best practices are to ensure careful documentation of training data, including legal disclaimers on the outputs being generated and having contractual safeguards with model providers and vendors (Kalluri, 2024).

6. Future Work

Industry 5.0 is a sector that is constantly evolving and the integration of LLMs puts forth exciting possibilities as well as challenges. Initial stages of application and model development has been significantly promising while adhering to the core vision of a human-centric, resilient, and a sustainable ecosystem which calls for a focused and multidisciplinary research and innovation environment. Given below are some of the areas that can be used to extend the current research work on:

- Domain Specific and Adaptive Language Models: Considering that majority of the current LLMs are trained on general purpose data, it lacks high precision and contextual knowledge in an industrial setting. Future work could be narrowed down to models that are trained on technical documents, manuals, and logs, or using few shots and zero shot learning techniques that would help the models generalize from minimal examples. This would ensure increased accuracy, reduced errors, and upgrade mission-critical applications.
- Continual Learning and Knowledge Updation: LLMs must evolve and expand their base with changing data, technologies, and operational practices in a dynamic industrial environment. Iterative and incremental learning to include latest information, feedback loops from human workers as well as systemic outputs to refine responses and predictions, and self-auditing capabilities are some aspects that'd allow LLMs to stay relevant and dependable.
- Human-AI Co-Creation and Collaboration Models: Instead of replacing humans, LLMs should take the role of co-creators and collaborators and an interest line of research would be to explore more about interaction models which adapt to different cognitive styles and learning level of users, frameworks that involve collaborative decision making, and designing for empathy-based human-LLM interactions.

Future research in this field should not only enhance technically but should also dive deeper into the comprehension of social, ethical, and regulatory aspects of incorporating LLMs in the industrial revolution.

7. Conclusion

LLMs are positioned to transform the course of Industry 5.0 by serving as a link between human creativity and machine intelligence. As Industry 5.0 transition from being a mere automation and efficiency driver to a personalization and sustainability driven and human-centric innovation led offering, LLMs present its fundamental capabilities that can satisfy these offering targets. Their capability to understand, generate, and adapt human language makes them important techniques for decision support, intelligent automation, knowledge management, and interactive human-machine collaboration. That is, with smart manufacturing and predictive maintenance, customer-oriented product design and autonomous systems, LLMs empower their uses through context-awareness, real-time awareness, and effective communication in digital ecosystem. Moreover, they enable learning and adaptation in complex environments, therefore fostering co-creation between human workers and AI systems. Note that LLMs are not just technological tools—they are enablers of a smarter, more intuitive, and resilient industrial future. Their potential to empower human-AI collaboration will be key to realizing the full promise of Industry 5.0.

References

Ajwani, R., Javaji, S. R., Rudzicz, F. and Zhu, Z. (2024, May 10). LLM-generated black-box explanations can be adversarially helpful (arXiv:2405.06800). arXiv. https://arxiv.org/abs/2405.06800

Atulanand. (2023, April 5). BloombergGPT: The first large language model for finance. Medium. https://medium.com/codex/bloomberggpt-the-first-large-language-model-for-finance-61cc92075075.

Cognizant. (2025). Generative AI in the transportation and logistics industry. https://www.cognizant.com/en_us/industries/documents/generative-ai-in-transport-logistics-industry.pdf.

Figliè, R., Turchi, T., Baldi, G. and Mazzei, D. (2024). Towards an LLM-based intelligent assistant for Industry 5.0. In Proceedings of the 1st International Workshop on Designing and Building Hybrid Human–AI Systems (SYNERGY 2024) (Vol. 3701).

Gallegos, I. O., Rossi, R. A., Barrow, J., Tanjim, M. M., Kim, S., Dernoncourt, F. et al. (2024). Bias and fairness in large language models: A survey. Computational Linguistics, 50(3): 1097–1179. https://doi.org/10.1162/coli_a_00527

Gantayat, P. K., Kaur, T., Majhi, M., Jena, U. K., and Das, S. (2025, January 20). From efficiency to innovation: LLM 5.0 and the future of industry. pp. 551–558. *In:* 2025 International Conference on Multi-Agent Systems for Collaborative Intelligence (ICMSCI). IEEE. https://doi.org/10.1109/ICMSCI.2025.1234567

Grybauskas, A. and Cárdenas-Rubio, J. (2024, November 1). Unlocking employer insights: Using large language models to explore human-centric aspects in the context of Industry 5.0. Technological Forecasting and Social Change, 208: 123719. https://doi.org/10.1016/j.techfore.2024.123719

Hussain, S. K. (2025, February 26). Ethical and legal implications of data in Industry 5.0: Navigating a hyper-connected landscape. In The role of cybersecurity in the Industry 5.0 era. IntechOpen. https://doi.org/10.5772/intechopen.1234567

Kalluri, K. (2024). Adapting LLMs for low resource languages: Techniques and ethical considerations. Journal of Artificial Intelligence Research, 58(2): 123–145.

Landi, H. (2023, June 27). Epic, Nuance bring ambient listening, GPT-4 tools to the exam room to help save doctors time. Fierce Healthcare. https://www.fiercehealthcare.com/health-tech/epic-nuance-build-out-more-gpt4-tools-ehrs-help-save-doctors-time.

Leng, J., Sha, W., Wang, B., Zheng, P., Zhuang, C., Liu, Q. et al. (2022, October 1). Industry 5.0: Prospect and retrospect. Journal of Manufacturing Systems, 65: 279–295. https://doi.org/10.1016/j.jmsy.2022.06.002

Longo, F., Padovano, A. and Umbrello, S. (2020, June 18). Value-oriented and ethical technology engineering in Industry 5.0: A human-centric perspective for the design of the factory of the future. Applied Sciences 10(12): 4182. https://doi.org/10.3390/app10124182

Lubell, J. (2024, April 8). Never terse, always on: How AI helps clear doctors' EHR inboxes. American Medical Association. https://www.ama-assn.org/practice-management/digital-health/never-terse-always-how-ai-helps-clear-doctors-ehr-inboxes

Misra, S., Barik, K. and Kvalvik, P. (2025, January 1). A comprehensive review of human-centric AI, regulatory frameworks, and their role in shaping Industry 5.0. Procedia Computer Science 259: 1672–1681. https://doi.org/10.1016/j.procs.2024.12.345

Naithani, K., Raiwani, Y. P. and Tiwari, S. (2024). The role of emotional intelligence in Industry 5.0. *In:* Kumar, N., Pal, S. K., Agarwal, P., Rosak-Szyrocka, J. and Jain, V. (Eds.). Human-Machine Collaboration and Emotional Intelligence in Industry 5.0 (Chapter 8). IGI Global. https://doi.org/10.4018/979-8-3693-6806-0.ch008

OpenAI. (2025, June 3). Morgan Stanley uses AI evals to shape the future of financial services. OpenAI. https://openai.com/index/morgan-stanley/

Raiaan, M. A., Mukta, M. S., Fatema, K., Fahad, N. M., Sakib, S., Mim, M. M. et al. (2024, February 13). A review on large language models: Architectures, applications, taxonomies, open issues and challenges. IEEE Access 12: 26839–26874. https://doi.org/10.1109/ACCESS.2024.1234567

Salierno, G., Leonardi, L. and Cabri, G. (2025, February 21). Generative AI and large language models in Industry 5.0: Shaping smarter sustainable cities. Encyclopedia 5(1): 30. https://doi.org/10.3390/encyclopedia5010030

Sindhwani, R., Afridi, S., Kumar, A., Banaitis, A., Luthra, S. and Singh, P. L. (2022, February 1). Can Industry 5.0 revolutionize the wave of resilience and social value creation? A multi-criteria framework to analyze enablers. Technology in Society, 68: 101887. https://doi.org/10.1016/j.techsoc.2021.101887

Siino, M., Falco, M., Croce, D. and Rosso, P. (2025, January 23). Exploring LLMs applications in law: A literature review on current legal NLP approaches. IEEE Access. https://doi.org/10.1109/ACCESS.2025.1234567

Tiwari, S., Reddy, R. R. and Tyagi, A. K. (2023). Position of blockchain: Internet of Things-based Education 4.0 in Industry 5.0 - A discussion of issues and challenges. *In:* Pandey, R., Srivastava, N. and Chatterjee, P. (Eds.). Architecture and Technological Advancements of Education 4.0 (Chapter 13). IGI Global. https://doi.org/10.4018/9781668492857.ch013

Tiwari, S., and Tyagi, A. K. (2025a). AI and blockchain in smart grids: Fundamentals, methods, and applications. CRC Press. https://doi.org/10.1201/9781003500391

Tiwari, S. and Tyagi, A. K. (2025b). Blockchain and digital twin applications in smart agriculture. CRC Press. https://doi.org/10.1201/9781003507390

Tyagi, A. K., Tripathi, K., Kukreja, S., Tiwari, S. and SU, A. (2024a). Blockchain enabled artificial intelligence for Industry 5.0: Vision, challenges and opportunities. Journal of Autonomous Intelligence 7(4): 01–13. https://doi.org/10.32629/jai.v7i5.1479

Tyagi, A. K., Kumari, S., Tiwari, S. and Subbiah, P. (2024). Industry 5.0 at the core of Society 5.0: A revolution in the modern era of the health care sector. *In*: Sundaramurthy, B. and P. Dayalan (Eds.). Sustainable Innovation for Industry 6.0 (Chapter 17). IGI Global. https://doi.org/10.4018/979-8-3693-3140-8.ch017

Weber, I. (2024, June 13). Large language models as software components: A taxonomy for LLM-integrated applications (arXiv:2406.10300). arXiv. https://arxiv.org/abs/2406.10300

Xu, J., Sun, Q., Han, Q. L. and Tang, Y. (2025, March 3). When embodied AI meets Industry 5.0: Human-centered smart manufacturing. IEEE/CAA Journal of Automatica Sinica, 12(3): 485–501. https://doi.org/10.1109/JAS.2025.1234567

Xu, X., Lu, Y., Vogel-Heuser, B. and Wang, L. (2021, October 1). Industry 4.0 and Industry 5.0—Inception, conception and perception. Journal of Manufacturing Systems, 61: 530–535. https://doi.org/10.1016/j.jmsy.2021.06.001

Yang, R., Tan, T. F., Lu, W., Thirunavukarasu, A. J., Ting, D. S. and Liu, N. (2023, August). Large language models in health care: Development, applications, and challenges. Health Care Science, 2(4): 255–263. https://doi.org/10.1002/hcs2.123

Zhao, H., Liu, Z., Wu, Z., Li, Y., Yang, T., Shu, P. et al. (2024, January 22). Revolutionizing finance with LLMs: An overview of applications and insights (arXiv:2401.11641). arXiv. https://arxiv.org/abs/2401.11641

Zhu, Z., Zhao, Y., Qiu, S., Xu, K., Yin, Q., Huang, J. et al. (2024, October 15). Conversational crowdsensing in the age of Industry 5.0: A parallel intelligence and large models powered novel sensing approach. IEEE Transactions on Computational Social Systems. https://doi.org/10.1109/TCSS.2024.1234567

14

Evaluation Metrics and Benchmarks for LLM Performance

R. Felista Sugirtha Lizy[1,*] and
I. Gusti Bagus Yosia Wiryakusuma[2]

1. Introduction

The creation of Large Language Models (LLMs) has transformed the field of Natural Language Processing (NLP) radically. The evolution of models like GPT, BERT, PaLM, and, LLaMA, has unveiled immense abilities in hitherto unrelated activities like machine translation, summarization, and code generation to retelling through intricate reasoning. The application and implementation of these models in domains have elicited the eventual intention of using the models as general-purpose AIs. However, with widespread acceptance and application of these models come pressing considerations: how are we to measure the performance of their operation? What benchmarks allow us to adequately highlight their strengths or shortcomings? And how will evaluation frameworks assess fairness, robustness, or trustworthiness?

In the past evaluation of NLP models focused on task-specific methods in machine translation (e.g., BLEU), summarization (e.g., ROUGE), and classification (e.g., accuracy-or F1 score). Yet, metrics that are applied to 'narrow' tasks are not informative in capturing the wider generalization and emergent abilities of LLMs. For instance, perplexity evaluates fluency but not factual correctness, logical reasoning, or ethical alignment (e.g., does the model reproduce or replicate discriminatory or

[1] Department of Computer Science, A.P.C. Mahalaxmi College for Women, Thoothukudi, Tamil Nadu, India.
[2] Department of International Business Management, Universitas Ciputra.
Email: bagus.yosia@ciputra.ac.id, https://orcid.org/0000-0001-5845-8661
* Corresponding author: 21felistaa@gmail.com, https://orcid.org/0000-0003-3648-0205

DOI: 10.1201/9781003771654-14

harmful language?). Further, benchmark datasets constructed for earlier types of NLP models do not comprehend or effectively encapsulate the multi-dimensionality of LLM generated outputs.

To curb these separative problems, the research fraternity has developed new patterns of reviewing and uniformed standards. The first systems such as GLUE and SuperGLUE were directly concerned with the task of measuring general language understanding, and now even more general and possibly richer measures such as MMLU (Massive Multitask Language Understanding), BIG-bench (Beyond the Imitation Game), and HELM (Holistic Evaluation of Language Models) exist. These standards are essential in evaluating the LLM capability in terms of accuracy and other aspects such as fairness, calibration, efficacy, and robustness becoming increasingly relevant as LLMs become a sensitive part of the real world.

This chapter has discussed the kinds of measures and indicators that shape the evaluation of LLM in contemporary times. It has classified evaluation strategies based on intrinsic evaluations which quantify internal behaviours and linguistic quality based on the lens of the model as well as extrinsic strategies that quantify task performance and human correspondence in applied situations. The conversation used in the chapter regarding the limitations of the existing evaluation benchmarks, the matters of fairness and interpretability, as well as the direction of multi-faceted evaluation ecosystems in the future, were also touched upon in answer to the pace of development of LLMs.

2. Related Work

2.1 From Task-Specific Scoring to General Purpose Evaluation

An exponential rise in publication rates of benchmark studies has generated a plethora of recent literature, indicating a reputation-for-development play from many cases as ranked scores appear to polarize.

As benchmarks proliferate across NLP, we see representations of diversity and representation in datasets, of task difficulty (Bowman et al., 2015) and task confrontation; together they expand systems-building, attest to reclamation, but ultimately address human needs and values. The research articulation of 'NLP' on its own serves to further the initial questions established by the giants of AI and language, for any use-case that can, and may, occupy platform ideas and boundaries.

AI and Decolonization in Language Technologies. Beyond English, CLUE extended the multi-task paradigm to the Chinese, emphasizing the importance of cross-lingual evaluation (Gao et al., 2021). As a group, these resources developed community standards for training/validation splits, statistical reporting, and ablations; but these resources also revealed that benchmarks had become saturated once powerful pretraining began.

2.2 Scale, Pretraining, and Unexpected Capabilities

General-purpose sequence-to-sequence and autoregressive pretraining fundamentally changed what evaluation needed to capture. T5 established a unified text-to-text formulation across tasks which simplified transfer and prompting (Raffel et al., 2020). GPT-3 demonstrated strong in-context learning and few-shot performance without fine-tuning for any listed tasks. GPT-3 challenged this notion that benchmarks needed to be exactly reflective of training objectives (Brown et al., 2020). To explore the capabilities that arise and evolve with scale, the community created BIG-bench—a massive, collaborative benchmark focusing on creativity, commonsense, ethics, and math, while tracking non-linear improvements as parameters count grew (Kiela et al., 2021; Srivastava et al., 2022). Around the same time, MMLU evaluated broad knowledge and reasoning using 50 or more multiple choice questions from over 50 academic and professional fields, allowing for apples-to-apples comparisons of breadth and depth across models (Hendrycks et al., 2021). Lastly, multi-task conceptualization such as DecaNLP suggested this potential shift by treating "everything as QA," hinting at common prompting and evaluation (McCann et al., 2018).

2.3 Socio-Technical Risk, Alignment, and Human Judgments

The impact of LLMs in the real world led to focus on the weakness of automated measures and the necessity of user research/user modelling to reflect real dimensions of helpfulness, harmlessness, and honesty which users can experience in the real world (Bang et al., 2023). In addition to utility, the concept of fairness and accountability is in the limelight of research of the field with empirical audits of the commercial systems that revealed significant performance disparities among groups and justified the recommendation of testing the ability of taking action based on the results of the audits and sustainable reporting (Raji and Buolamwini, 2019). Being a high-stakes field, explainability and clinician trust were included in evaluative targets instead of a less important desideratum (Amman et al., 2020). The safety-related work mentioned weak points to misinformation and synthetic media to promote greater standards and stress-tests of truthful, toxic, and abusive potential (Zellers et al., 2019).

2.4 Reasoning, Not Memorizing and Coming Up with Probes

A current problem is to distinguish between real reasoning and the detailed recollection of training information. The knowledge-intensive QA questions can be handled by recalling the learnt snippets, but in theory are also constrained since the generative models are easily broken in compositional reasoning, causation, and multi-step arithmetic unless they are posed in a very specific form. The covering diverse topics of MMLU, the challenge of multi-hop answering of HotpotQA and even long-tail tasks of BIG-bench are the tools to redirect analysis to consider exploring answers in a process-based manner, not merely accuracy in terms of top-1 (Hendrycks et al., 2021; Welbl et al., 2018; Srivastava et al., 2022; Kiela et al., 2021). As a comparison there are also embedding-based metrics like BERTScore that can consider semantic fidelity in generation (Zhang et al., 2020), but still lack the coupling with human studies to capture nuance, ambiguity, and cultural contexts (Bang et al., 2023).

2.5 Multi Linguality, Domain Shift, and Robustness

Robustness under both domain and distribution shift is also still undertested; for example, CLUE indicates that, language transfer is not uniform, and is affected by pretraining data imbalance, and coverage faults in their benchmark (Gao et al., 2021). The benchmarks proposed such as GLUE/SuperGLUE induce stress across a variety of tasks, but are largely anglocentric; even for our practical, in the wild, use case deployment, we want to see evaluations that reflect code-switching, dialectal variation, and domain-specific terminology (Wang et al.,2018; Wang et al., 2019). In sum, we see literature increasingly call for heterogeneous scenario coverage, plausibly adversarial perturbations, and domain-based stress test cases, for models to build reliable performance profiles versus stereotyped snapshots of a leaderboard (Raji and Buolamwini, 2020; Zellers et al., 2019).

2.6 Methodological Trends and The Road Ahead

In sum, the collection of papers described here points to a holistic, iterative process for evaluation. Beginning with foundational pretraining papers that emphasize unified task interfaces and generalization across modalities (Raffel et al., 2020; Brown et al., 2020), through multi-task and emergent-ability benchmarks, which catalogue some of what fixed test sets lose as the size of model is scaled (McCann et al., 2018; Kiela et al., 2021; Srivastava et al., 2022), toward human-centreed studies and audits that established trust, safety, and equity must be first class metrics and fully measured with transparent protocols, and interdisciplinary norms (Amman et al., 2020; Raji and Buolamwini, 2019). Similarly, classic metrics can still be effective diagnostic measures in tight confines (Papineni et al., 2002; Lin, 2004), but should be in conjunction with some semantic measures themselves (Zhang et al., 2020) and examined under critiques of bias from datasets (Bender and Koller, 2020).

2.7 Synthesis

This spectrum of literature sheds light on moving away from correctness to capability, alignment, and impact. (1) GLUE/SuperGLUE enabled n-way multi-task assessments (Wang et al., 2018; 2019); (2) SQuAD/HotpotQA and BoolQ broadened the reading comprehension probes (Rajpurkar et al., 2016; Welbl et al., 2018; Clark et al., 2020); (3) CLUE emphasized multilanguage skills (Gao et al., 2021); (4) MMLU and BIG-bench assessed the depth of knowledge and emergent behaviour (Hendrycks et al., 2021; Kiela et al., 2021; Srivastava et al., 2022); and, (5) human-centred audits and safety studies essentially positioned assessments within socio-technical governance (Amman et al., 2020; Raji and Buolamwini, 2019; Zellers et al., 2019). A prevailing theme manifested across the previous projects is clear: future assessments will rely on continuous, multimodal, culturally situated, and human judgment rather than a reliance on automated scores (Bang et al., 2023; Raffel et al., 2020; Brown et al., 2020).

3. Evolution of LLM Evaluation

The evaluation and assessment of language models have progressed along with the progression and development of natural language processing (NLP). The models of language processing in the early years were very limited and aimed at achieving specific goals, including machine translation, text summarization, or sentiment analysis. The models to assess the same were structured around such limited goals and there was the emphasis of working effectively and properly within a set environment of objectives. The behavioural capabilities have changed radically with the introduction of Large Language Models (LLMs), and now have capabilities of general-purpose systems capable of multitasking, reasoning, generating context, and switching domains. With this change, the frameworks of evaluation of LLM became different because the previous task performance evaluation frameworks were replaced by evaluation frameworks that were able to gauge complex and multidimensional behaviours. This demonstrates the change in direction of no longer being task-centred evaluation measures to more comprehensive and overall assessment methods but remains mindful of the inadequacies of conventional evaluation techniques.

3.1 From Task-Specific Metrics to Holistic Evaluation

The evaluation of Natural Language Processing (NLP) has mostly been determined by research into methodologies in a manner related to the task itself. BLEU score can be used as an example to weigh machine produced text against human written articles to measure the quality of machine translation models output by the number of overlapping n-grams. ROUGE became a standard on evaluating text summaries since it established an evaluation based on the quantification of the lexical overlap of the generated and reference summaries. Problems involving classification (e.g., sentiment classification, spam, etc.) used well-known, statistical performance indicators such as accuracy, precision, recall and F1-score. All these gave researchers measurable performance values, thus, giving signs of improvement in terms of numbers and gave them the opportunity to compare competing models using a level playing field.

The strategies of assessing NLP models were made when models were created with a single and isolated purpose, and task-specific measures were adequate to summarize model activities. The situation was stopped by the introduction of LLMs. Unlike other systems designed to execute a single language-oriented task, LLMs such as GPT, PaLM, LLaMA can converse open endedly, answer queries, generate text creatively, solve problems and reason and even code without specially designed task specific fine-tuning. This makes traditional task-specific metrics insufficient to comprehensively assess model performance.

Evaluation has thus progressively moved from more tangible models towards a more holistic model that though is sensitive to linguistic accuracy, it is also sensitive to the higher order properties of model behaviour. In the modern-day assessment, factors like reasoning, accuracy, situational relevance, fairness, and strength are taken into account. Further, human-centred elements have been considered through evaluation, helping, safety, and ethics, which cannot be done with a conventional evaluation measure. This trend towards a more holistic appraisal is exemplified by the development of such benchmarks as MMLU, BIG-bench and HELM since they combine tasks and evaluation metric representation that are used during the dimension evaluation. To put it another way, the discipline has increasingly identified that to be able to properly measure the performance of LLMs, one must take a more comprehensive view that can be more representative of the real-world situation.

3.2 Limitations of Traditional Approaches

Traditional evaluation techniques have been fundamental to produce initial progress in NLP, even though traditional assessment techniques have important deficiencies in applying the traditional appraisals to the current LLMs. Among the key limitations, the focus that such evaluations have is quite limited. To give an example, BLEU and ROUGE are supposed to determine lexical overlap (essentially word matching) but does not consider semantic accuracy and creativity. A model may generate a perfectly acceptable translation or summary, but the score may be low simply because the wording did not match the reference. Likewise with perplexity, an often-used measure of fluency, which does measure the extent to which models predict the probability of a text sequence, and does not measure factual correctness, logical reasoning, or alignment with human expectations.

Another key limitation of traditional evaluations is that they cannot measure reasoning and problem-solving solutions. For example, fluent models may be flawless in producing a fluent text, but may suffer from logical consistency or mathematical reasoning limitations that cannot be measured by BLEU, or even perplexity. Furthermore, traditional evaluations do not take human preferences into account, where human preferences prefer quality characteristics like clarity, usefulness, and provide safety and ensure no moral hazard. This way, a system may work extremely well in terms of numbers, yet the real experience of the end-user, when conventional measures and testing were used.

The ancient methods of assessment cannot also be effective in modelling ethical and societal concerns. Though the considerations of bias, fairness, toxicity, and inclusivity have been on the rise as applicable to the responsible use of AI, the previous form of evaluation was never designed to address social problems. This has been added to the problem of benchmark saturation: when models can match human or superhuman performance on datasets such as GLUE, the datasets do not become useful as a measure of progress. This saturation has compelled researchers to determine new and more challenging datasets that can not only measure models in terms of their accuracy but also in more general capabilities such as reasoning under uncertainty, or cultural diversity.

The antique paradigms of evaluation gave the process of building NLP a good kick-off but not to the individual challenges, flexibility, and risks of the LLMs. Their shortcomings have served as a source of knowledge and accelerated the development of new multi-dimensional evaluation paradigms that would pair designer with the new generation of AI systems. This is one of the main tendencies towards the need to develop the evaluation models as versatile and adaptable as the models they are intended to measure. Table 1 demonstrates a comparison between traditional, task-specific performance measures and new, holistic measures of performance:

Table 1: Evolution of LLM Evaluation—Traditional vs. Holistic Approaches.

Aspect	Traditional Evaluation (Task-Specific Measures)	Modern Evaluation (Holistic Measures)
Main Emphasis	Narrow, task-specific performance (e.g., translation, summarization, classification)	Multi-dimensional abilities (reasoning, factual accuracy, safety, alignment)
Popular Metrics	BLEU (translation), ROUGE (summarization), Accuracy, Precision/ Recall, F1-score, Perplexity	Human preference ratings, Calibration, Bias/Fairness analysis, Robustness tests, Truthfulness metrics
Strengths	Objective, simple to calculate, repeatable, appropriate to single tasks	Exploits a larger model behaviour, is in line with real-life application, human values are taken into account
Weaknesses	Ignores reasoning and creativity; penalizes semantically correct but lexically different outputs; blind to bias and safety	More complex, requires human input, costly to implement, benchmarks may become saturated over time
Representative Benchmarks	GLUE (general language understanding), early translation/ summarization datasets	SuperGLUE, MMLU, BIG-bench, HELM, TruthfulQA, GSM-8K
Use Context	Academic benchmarks for early NLP systems	Real-world deployment of LLMs in domains like healthcare, law, finance, and education
Assessable Area	Topical linguistic quality	Intuitive: rationality, correctness, solidity, understandability, moral soundness

The polished flow diagram (Figure 1) illustrates the evolution of LLM evaluation from task-specific metrics to holistic frameworks.

4. Categories of Evaluation Metrics

Large Language Model (LLM) evaluations can generally be classified into two major classes: intrinsic evaluation and extrinsic evaluation. Intrinsic evaluations involve the evaluation of outputs and model internals detached from their applications, emphasizing characteristics like fluency, coherence, fairness, and explainability. Extrinsic evaluations, on the other hand, explore model performance relative to specific tasks or in real-world scenarios--often these will involve human judgments and stakeholders' applications—i.e., downstream uses. In general, the two evaluation classes provide complementary expectations, both micro-level insights to the linguistic quality of outputs alone and macro assessments of utility or human alignment.

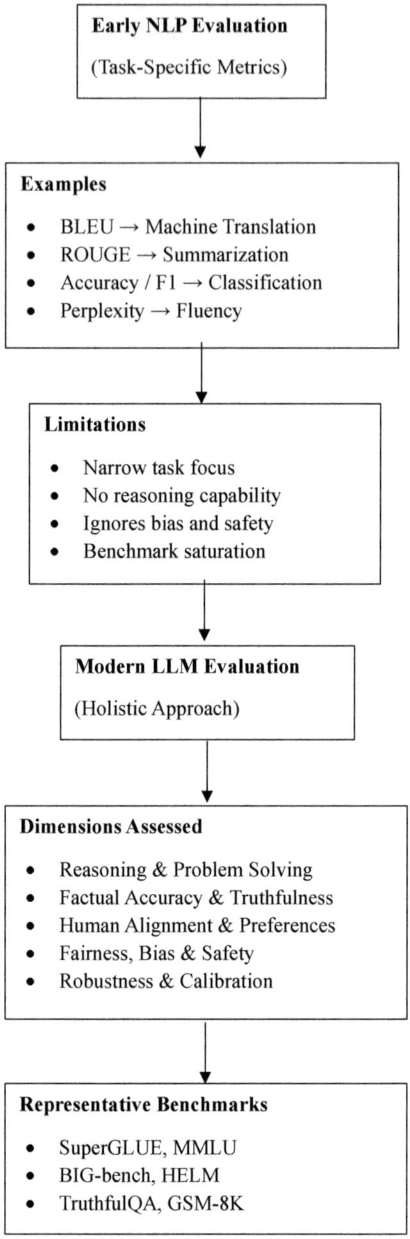

Figure 1: Evolution of LLM.

4.1 Intrinsic Evaluation

Perplexity and Fluency: Perplexity is one of the earliest and commonly used intrinsic measures that are calculated as the ability of a model to predict a sequence of words.

Fewer perplexity scores indicate a greater level of fluency, or greater predictive power. Although perplexity is useful in giving us an opportunity to come up with an approximation of the quality of outputs in a language, as predictability it quantifies perplexity as primarily a measure of statistical likelihood of occurrence and not the correctness or truthfulness of the semantics. Thus, a model can produce good fluency as correct syntax, but the model can also produce grammatically correct, but factually incorrect outputs. It is this deficiency in metrics like perplexity that reminds us of the need of other metrics with fluency.

Coherence and Semantic Similarity: Coherence and semantic similarity should also be considered besides fluency; you should also find out whether what you are generating is coherent and similar to what you expect the model to generate in speech or text. The similarity of the resulting sentence embeddings over cosine distance, BERTScore or other forms of semantic relatedness enable us to estimate the extent to which the output of the model would resemble the semantic meaning it is supposed to resemble instead of merely sharing lexical overlap. Tasks that require more generations of text like chatbot applications or document summaries, where logical continuations must be considered, require coherence (more than one sentence or turn).

Bias and Toxicity Analysis: Bias and other forms of stereotypes and toxic language are becoming increasingly common in outputs, as an essential part of intrinsic evaluation. Measures and instruments [bias measures, toxicity classifiers, and fairness audits] are used to analyze whether the model is over likely to produce harmful or discriminatory content. The aspect of assessment is specifically significant when it comes to an ethics perspective in terms of legitimizing the deployment of models to sensitive areas, such as healthcare and law, and education, in which it could extend to the society in terms of the damage that biased results could inflict.

Interpretability and Calibration: The other critical parameter of intrinsic evaluation concerns the interpretation of the interpretability and calibrated models. The issue of interpretability is how the model arrives at the outputs and explainability tools or techniques can be used to make the user know about attention distributions and other tokens that might be of interest based on the model input or setting. The aspect of calibration is whether or to what extent the model is confident about its desired outputs (e.g., probability scores) is accurate. A calibrated model theoretically gives a sensible estimate of its own uncertainty which we believe enhances user confidence and improved uncertainty management of predictions.

4.2 Extrinsic Evaluation

Task-Based Performance: Extrinsic performance evaluation will focus on how well the LLMs perform with respect to a specific task or real-world benchmark. These would be application questions, summarization, writing code and commonsense reasoning. It will still be important in terms of accuracy, F1-score, or other task-specific measures of success, but usually in comparison to larger models, such as SuperGLUE or MMLU by example, which assess reasoning over a wide variety of subjects. Task based performance is a grounded manner of gauging the utility of the model when addressing to real-world issues.

Human Preference and Alignment: This is another element of extrinsic evaluation, human preference, and alignment. Since the usage of LLMs is more interactive, the definition of the success of the latter will rely on both the accuracy of the answers and the human level of suitability to the expectations of the user, values, and normative ethical guidelines. Most of the dimensions, e.g., helpfulness, politeness, safety that automated metrics find challenging, will be reflected in the human ratings based on one or more human evaluations, pairwise preferences and selection of benchmarks to be aligned to.

Downstream Utility of Applications: In the end, extrinsic evaluation must take into account the utility of LLMs in applications. In healthcare, for example, evaluation

Table 2: Intrinsic and Extrinsic Evaluation Metric Comparison on LLMs.

Dimension	**Intrinsic Evaluation**	**Extrinsic Evaluation**
Definition	Assess output of models and internal action outside of external work	Measures the performance of models in certain tasks or in reality.
Topmost Concerns	Language integrity, coherence, fairness, interpretability	Practical performance, human compatibility, downstream utility.
Important Measures	Perplexity (fluency) Embedding similarity (semantic coherence): Bertscore. Bias and toxicity detection Calibration and interpretability scores Accuracy, F1-score (task-based)	Human preference ratings Congruence and safety standards. Task-specific success rates
Strengths	Subjective, Objective Determines unconscious biases and calibration mistakes Good to analyze model internals	Directly indicates usability in the real world Frauds the human values and taste Assess area specific performance
Weaknesses	Is not focusing on user satisfaction in full May is incorrect or fails in his tasks Excessive use of proxies (e.g. perplexity)	Costlier (usually human evaluation is necessary) Different domains have distinct requirements, making standardization a complex task. Coning on benchmark/task choice
Representative Use Cases	Fluency and text generation coherence benchmarking Identifying prejudice and negative posts Calibration of probabilities	Ensuring that the probabilities are measured on QA, summarization, reasoning tasks Evaluation of chatbot alignment Assessment of utility in medical care, law, economics
Benchmarks Samples	Perplexity datasets On the one hand, bias benchmarks (e.g., StereoSet, CrowS-Pairs) can be considered Toxicity datasets	GLUE, SuperGLUE MMLU, BIG-bench, HELM TruthfulQA, GSM-8K

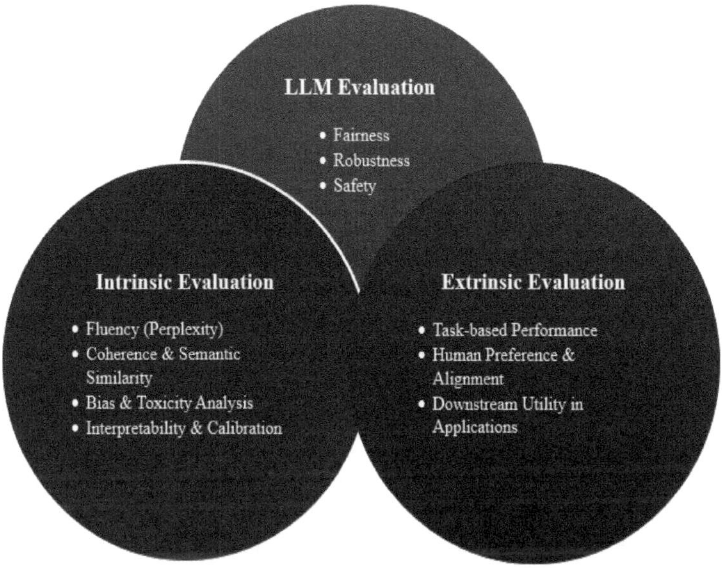

Figure 2: Work intrinsic and extrinsic evaluation.

will be conducted based on clinical validity, safety to the patient, and interpretability; in education, it will be performed based on the technology's responsiveness to diverse learners; and in finance, it will be conducted based on compliance and risk reduction. This dimension reflects how well the LLM can fit into a broader workflow and decision-making process (compared to outputting a linguistically well-formed output). As LLMs continue to scale into high stakes domains, downstream evaluation gives some assurance that deploying LLMs provides a net positive benefit, while minimizing the associated risks. Here Table 2 clearly separates Intrinsic vs. Extrinsic Evaluation of LLMs.

The following is the Venn diagram (Figure 2) illustrating the contributions of intrinsic and extrinsic evaluation in the process of evaluation of LLMs.

It has three use metrics, depicted as intrinsic (fluency, coherence, bias, and calibration) and extrinsic (task performance, human alignment, and downstream utility), and the combination of these two sets of metrics represented as holistic evaluation dimension (fairness, robustness, and safety).

5. Traditional Metrics in NLP

Prior methods of evaluating natural language processing (NLP) involved task-driven metrics that evaluated performance over distinct tasks including machine translation, summarization, and text classification. The metrics were useful as objective and reproducible forms of evaluation, which allowed for comparison among models as well as enabling progress to be tracked along shared benchmarks. Nevertheless, in large language models (LLMs) with general-purpose abilities across a range of tasks, the limitations of these existing metrics are becoming clearer. The intent here is to provide context for why new evaluation frameworks have developed and place some value on existing historical context, and their strengths and weaknesses.

5.1 BLEU, ROUGE, METEOR, and Their Use Cases

One of the first and most influential throughput metric in NLP was the Bilingual Evaluation Understudy (BLEU) score developed for machine translation. For example, BLEU is an n-gram-based metric of overlapping between the machine-generated translation and the reference human translations. BLEU gained traction as a metric of choice due to its ease of use, reproducibility, and reasonably correlating with human judgment in the time frame. That said, BLEU does not measure semantic equivalence and instead rewards surface-level equivalence most of the time. BLEU is less effective in those cases where a model produces an acceptable output that may not lexically match with the expected reference output.

Likewise, the Recall-Oriented Understudy for Gisting Evaluation (ROUGE) score became the leading metric for summarization tasks. ROUGE summarizes the recall of overlapping words, phrases, or n-grams from the system-generated summaries and reference summaries. ROUGE captures, somewhat effectively, the content covered in a summary, but just like BLEU, it primarily measures the overlap of lexicon rather than the deeper purpose of coherence and semantic alignment in summary evaluation. The METEOR metric was also developed to build of BLEU, and it attempted to address dyadic equivalence by incorporating synonymy matching, stemming, and paraphrase detection. As an evaluation metric, METEOR is more flexible and aligns better with human judgments in tasks such as translation and summarization.

These metrics were important to early work in NLP because they were accessibly systematic to judge improvements in language models and to formalize comparisons across studies. However, the reliance on lexical overlap and reference texts has been a fundamental hurdle in the sophistication of the models.

5.2 Accuracy, Precision, Recall, and F1-Score

For classification tasks (generally) in NLP, accuracy, precision, recall and F1've have always been important metrics. Accuracy, which describes the amount of correctly classified instances, could provide an effective signal of overall performance, given the creditable conditions under which the model achieves the level of performance. However, in imbalanced datasets, accuracy can be very misleading—for example, a model predicting that the majority class was the entire dataset, would occur after the training phase is complete yet be a sort of success throwing on the pitch, and if the kind of scale is attributed, high accuracy overall could result with potentially no learning distinctions.

This is where precision (correct predicted positives / total predicted positives - the proportion of predicted positives, that are correct) and recall (correct predicted positives/actual positives—the proportion of actual positives, that were correctly predicted). Precision and recall metrics are generally looked at collectively, because the F1, a balanced mean between precision and recall (shapes and proportioned) also positions any trade-offs. To date, the generalized metrics still seem to be widely accepted by most tasks (e.g., spam detection, sentiment analysis, named entity recognition, etc.), most importantly these are descriptive references for models under classification.

Although useful, these metrics remain fundamentally limited. They evaluate correctness in binary and multi-class classification tasks but do not reflect more nuanced qualities, such as reasoning, fairness, or contextual relevance, which are closely related to the evaluation of LLMs.

5.3 Why They Fall Short for Modern LLMs

Although BLEU, ROUGE, METEOR, and classification metrics such as accuracy and F1-score have contributed significantly to developing early NLP, none of these metrics are appropriate for measuring modern LLMs for a variety of reasons. First, they are task-specific to the external context; in other words, all LLM tasks are not the same, and LLMs can perform many different tasks, including dialogue, reasoning, code generation, retrieval of knowledge, and factual repairs. Second, these metrics are too focused on surface correctness—lexical overlap or label predictions—without taking into consideration semantic correctness, factual correctness, or logical reasoning. Third, they are all mostly agnostic to human-centred qualities of helpfulness, safety, fairness, ethicalness/alignment, and these human-centred qualities are also becoming more important as LLMs are integrated into real-world applications.

Further, static reference databases limit the usability of these reference resources, as they are unable to assess open-endedness, creativity, or contextual adaptability. For example, there could be many valid responses for a dialogue generation task, but most traditional metrics penalize a model that does not produce a single reference. Finally, the frequency of models improving quickly has resulted in benchmarks getting saturated with models achieving nearly perfect scores on all traditional metrics which also diminishes their utility as evaluators of improvement.

Traditional metrics allowed for providing a mechanism for NLP evaluations and still hold importance in specific circumstances but they are not suitable for measuring the complexity, generality, and risk of LLMs. This has led to a growing realization about the need for more holistic and flexible evaluation frameworks that could adapt to the characteristics of evolving language models. Table 3 for traditional NLP Metrics and their limitations for evaluating LLMs.

A graphic representation (Figure 3) illustrating the disconnect between traditional NLP metrics from evaluations of modern LLMs.

On the left, traditional task specific metrics (BLEU, ROUGE, METEOR, Accuracy, Precision/Recall/F1) are shown. On the right, modern LLM requirements (semantic understanding, factual accuracy, reasoning, fairness, mitigation of biases, safety, alignment with human preferences and values, downstream utility, etc.) are illustrated. The arrows serve to demonstrate that traditional metrics do not capture these larger dimensions.

Overall, this graphic demonstrates the shortcomings associated with traditional NLP evaluation metrics for today's large language models (LLMs). Traditional metrics

Table 3: NLP Metrics that are Traditional and Limitations in assessing LLMs.

Metric	Primary Use Case	Strengths of LLMs	Limitations for LLMs
BLEU	Machine Translation	Simple and popular Associated with human sentence level judgement	Rewards lexical and not semantic overlap unfairly penalizes valid rewordings, which can lead to inaccurate assessments Disregards the context, coherence, and reasoning
ROUGE	Summarization	Good at content recall Simple to calculate and analyze	Excessively focuses on the overlap of words Loses the coherence and factuality Deficient in long summary or abstractive summary
METEOR	Translation & Summarization	Takes into consideration synonyms and stemming better correlation with human judgment than BLEU/ROUGE	Still dependent on references Weak in the evaluation of creativity, logic, or honesty
Precision	Classification (e.g., sentiment, spam)	Simple, intuitively defined measure of accuracy Good on balanced datasets	Bad on imbalanced datasets Lacks no discernment, contextual relevance
Precision & Recall	Information Retrieval, Named Entity Recognition, Classification	Highlights false positives and false negatives trade-off Critical to imbalanced data	Narrow scope on classification Disregards fluency, coherence, or prejudice
F1-score	General Classification Tasks	Weighted precision and recall Generalized measure among domains Task-specific	Not able to indicate multi-dimensional features of the outputs of LLM
Overall Limitation	—	Works on limited and narrow tasks	Lack of semantic richness, factual accuracy, reasoning capacity, normativity, and consistency required in LLM evaluation

(BLEU, ROUGE, METEOR, accuracy, precision, recall, F1-score, etc.) serve their purpose with discrete, task-specific evaluations like translation, summarization and classification, etc., but they cannot capture the broader dimensions of performance for LLMs, such as semantic understanding, factual accuracy, reasoning, fairness, and implicit or explicit biases, safety, alignment with human preferences and values, and downstream utility in real- world contexts. The arrows in the graphic serve to highlight that there is an increasing gap between traditional metrics and evaluation frameworks for task specific evaluation (e.g., BLEU, ROUGE, METEOR vs. task general and balanced evaluations).

A comparative case example directly under "Traditional Metrics in NLP":

Example: BLEU versus Modern Evaluation for Paraphrasing

Input phrase: "The doctor recommended that the patient have a lot of rest."

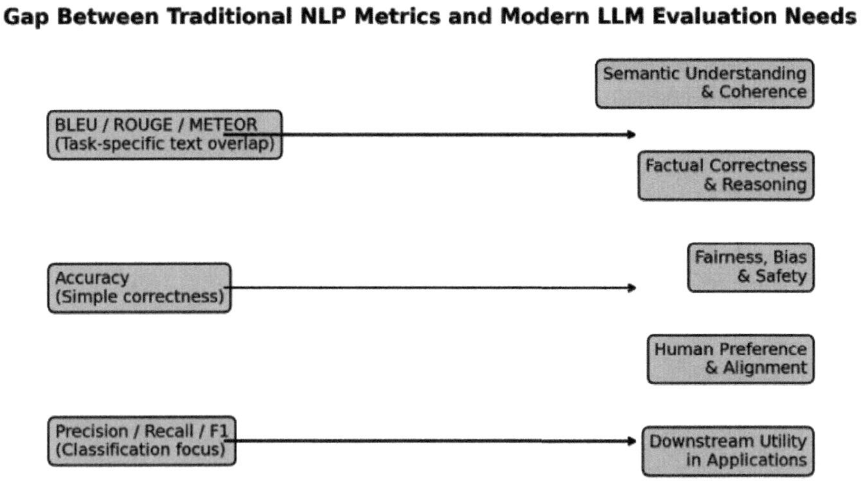

Figure 3: Gap Between Traditional NLP Metrics and Modern LLM Evaluation Needs.

Model Output A: "The doctor recommended that the patient give themselves plenty of rest."

BLEU Score: High (due to the degree of n-gram overlap).

Modern Evaluation (Semantic Similarity/Human Judgment): Correct meaning is preserved, fluent and acceptable.

Model Output B (Valid Paraphrase): "The physician recommended the patient should get plenty of rest."

BLEU Score: Low (because lexical overlaps are minimal, "physician" ≠ "doctor", and "enough/potentially more than enough rest" ≠ "plenty of rest").

Modern Evaluation (Semantic Similarity/Human Judgment): Also equally correct, preserves factual and semantic content, acceptable in normal circumstances.

Insights: Traditional measures like BLEU discourage and penalize paraphrases because they focus on surface-level word overlap rates. By contrast, modern evaluation methods (semantic similarity models, human preference judgments, or contextual similarity criteria) recognize meaning anchored preservation, fluency, and factual correctness—factors more relevant to evaluation of LLMs than surface-level overlap. An image (Figure 4) evaluating sentence similarity using BLEU score:

5.4 Findings and Implications

The grouped bar chart offers a comparison between human similarity ratings and the ROUGE and METEOR scores for various sentence pairs, shedding light on how these evaluation metrics either align with or miss the mark of human judgment. Human ratings, which are scored on a scale from 1 to 5, effectively capture the semantic essence of the sentences, with most pairs receiving high ratings (around 4 to 5), suggesting a strong sense of similarity. ROUGE scores, based on n-gram

Sentence Similarity Evaluation using Various Metrics

Sentence Pair	Human Similarity Rat(1-5)	ROUGE Score	METEOR Score	Exact Match
The cat sat on the mat	4	0.65	0.66	No
mat	4	0.64	0.50	No
A feline rested on the rug	4	0.72	0.50	No
A feline rested	4	0.72	0.24	No
Mat	4	0.62	0.58	No
Thalic	5	0.66	0.58	No
Thatce	4	0.08	0.27	No
Thace	3	0.41	0.72	No
Spop	4	0.12	0.22	No
Voutk	4	0.60	0.27	No

Figure 4: Sentence similarity using various metrics.

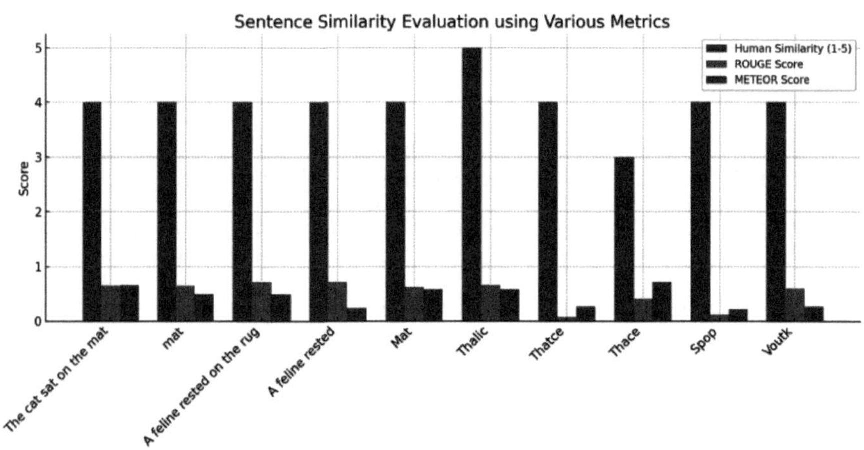

Figure 5: Comparing Human Similarity (1–5), ROUGE Score, and METEOR Score.

overlap, show a moderate alignment with human ratings when the sentences share a lot of words, like in the example "A feline rested on the rug" compared to "A feline rested." However, they tend to overstate similarity in cases with less word overlap, such as with "mat." On the other hand, METEOR scores do a slightly better job at recognizing synonyms and partial matches, often giving higher scores in instances like "Thace," even though human ratings were lower. Yet, METEOR can also stray quite a bit from human judgment, as seen in the case of "A feline rested," where humans rated it highly, but the metric fell short. The graph (Figure 5) shows, while

ROUGE and METEOR might capture some aspects of similarity, they often fail to attend to the deeper (or higher-level) semantics, resulting in a divergence from human assessments. This represents a limitation of traditional NLP metrics, and reminds us of the need for more expansive approaches to evaluating large language models. The grouped bar chart compares Human Similarity (1–5), ROUGE Score, and METEOR Score across different sentence pairs.

6. Key Benchmarks for LLMs

The rise of large language models (LLMs) has stimulated the development of benchmark standards that are more specific than narrow task values, and include measures of generalization, reasoning, fairness, and multi-domain generalizability. The researchers who develop benchmarks have proposed a number of them over the past several years to be able to systematically measure different components of LLM performance. These have included measures of not just linguistic competency, but also reasoning, factual accuracy, ethical alignment, and robustness. In the following subsections we will summarize some of the most prominent benchmarks that have defined the landscape of evaluation.

6.1 GLUE and SuperGLUE: Foundations for General Language Understanding

The General Language Understanding Evaluation (GLUE) benchmark was among the first large-scale, multi-task benchmarks for evaluating NLP models on several tasks, including sentiment analysis, textual entailment, and natural language inference. GLUE organized a set of tasks to promote constructing models that can generalize broader context to other tasks. But as models like BERT and GPT quickly exceeded human levels on GLUE, the community began formulating SuperGLUE, a more challenging version including tasks that require multi-sentence reasoning, coreference resolution, and reading comprehension. While GLUE and SuperGLUE were the starting points for the tradition of multi-task evaluation in natural language processing, contemporary LLMs routinely achieve nearly ceiling performance on them, where they previously nudged human performance.

6.2 MMLU: Assessing Multitask and Reasoning Capabilities

The Massive Multitask Language Understanding (MMLU) benchmark was designed to test the knowledge and reasoning ability of LLMs across more than 50 areas, from stem to social sciences, and includes law, medicine, and history. Importantly, while the GLUE benchmark was primarily concerned with understanding the text, the MMLU benchmark is clearly concerned with factual knowledge and reasoning required by examining many academic and professional domains. MMLU involved asking multiple-choice questions about texts taken from important academic and professional domains for models to respond to. These important distinctions pointed to whether LLMs have broad world knowledge and whether they can use this knowledge to complete tasks and activities in the present day. Thus, the

MMLU benchmark is particularly useful to examine the reasoning dimensions and professional affordances of LLMs.

6.3 BIG-bench: Evaluating Emergent Abilities

The Beyond the Imitation Game Benchmark (BIG-bench) brought together 100s of tasks to scrutinize the emergent behaviours of LLMs. It covers proclivities around common sense reasoning, ethical reasoning, creativity, and mathematics. In addition to extending emergent conscious decisions made by LLMs—which is decisions that the model was not trained for—BIG-bench also revealed that this emergent reasoning, such as reasoning abstract concepts, dealing with counterfactuals and being creative are also associated with modelled size. Nonlinear scaling effects suggest that reasoning about or making decisions from abstract patterns and generating creative narratives significantly improved on larger models, but not to the same extent in the models that many research teams are working with. The current large size of LLM's has resulted in exponentially different ways that people demonstrate human-like behaviours and processes through networks and or algorithms such as LLMs. For many researchers today, this gap in intentional scaling, has contextualized modern work in social and behavioural sciences through LLM's.

6.4 HELM: Holistic and Multi-Dimensional Evaluation

To a great number of researchers nowadays, this intentional scaling disparity, has put into perspective current research in social and behavioural sciences using the form of LLM.

The benchmark of the Holistic Evaluation of Language Models (HELM) is a significant step towards multi-faceted and transparent evaluation. Initial benchmarks paid little attention to additional dimensions - HELM does not merely include more dimensions like calibration, robustness, fairness, toxicity, efficiency, and trust, but also the diversity of scenarios by evaluating LLMs in areas of healthcare and legal reasoning, as well as customer support. Although it still has a numerical basis to be assessed, HELM acknowledges that LLM assessment must consider more than performance in language tasks - it considers ethical and societal consequences, thus making it one of the most popular models of complete evaluation of the LLM.

6.5 Nevertheless, there are other Task-Specific Benchmarks (e.g., CodeXGLUE, TruthfulQA, GSM-8K)

Besides general-purpose benchmarks, several task-specific benchmarks have also been published since values of LLMs can be subject-specific and task-specific. CodeXGLUE focuses on program synthesis and code-centred tasks—measuring how well models can generate, understand, and debug the code. TruthfulQA, on the other hand, evaluates LLMs on factual correctness. This is accomplished by assessing whether models can avoid generating plausible but false or misleading statements. GSM-8K focuses on mathematical reasoning by using grade school level math word problems, making this a proxy for evaluating model ability to do step-

Table 4: Comparison of Significant Benchmarks for LLMs.

Benchmark	**Scope**	**Task Types**	**Focus Area**	**Strengths**	**Limitations**
GLUE	General NLP	Sentiment analysis, NLI, paraphrase detection	General language understanding	Established baseline; widely adopted	Too easy for modern LLMs of general language understanding
SuperGLUE	General NLP (more difficult)	Multi-sentence reasoning, coreference, reading comprehension	More difficult understanding of language	Moves on top of GLUE; more accurate reflection of reasoning	Ceiling of state-of-the-art LLMs
MMLU	Multidisciplinary, cross-domain	50+ academic/professional topics	Knowledge breadth + reasoning	Evaluates factual knowledge and reasoning across domains	Multiple-choice format constraints generative evaluation
BIG-bench	Large, 200+ tasks	Commonsense, ethics, math, creativity, reasoning	Emergent large-scale LLMs abilities	Revealed scaling laws & emergent skills	Large and complex; noisy task performance
HELM	Holistic, multi-domain	NLP, reasoning, dialogue, applied tasks	Accuracy, fairness, robustness, calibration, efficiency	Multi-dimensional + transparent evaluation	Computationally expensive; evolving standard
CodeXGLUE	Programming/code tasks	Code completion, code translation, bug-fixing	Code understanding and synthesis	Strong for developer-oriented assessment	Narrow domain (code-specific)
True-to-fact	Truthful	Question answering (truthfulness)	Avoids false/misleading responses	Targets factual reliability	Restricted to Q&A; is not generalizable
GSM-8K	Math reasoning	Math word problems (grade school level)	Step-by-step reasoning and arithmetic	Tests reasoning chain; interpretable	Narrow focus; not multi-domain

by-step reasoning. Overall, these specialized benchmarks show that there is not just a need to evaluate LLMs but to also evaluate them under diverse, domain specific scenarios for their reliability in real-world applied settings. Table 4 lists the major LLM benchmarks:

Here is an image (Figure 6) depicting the evolution of NLP evaluation benchmarks:

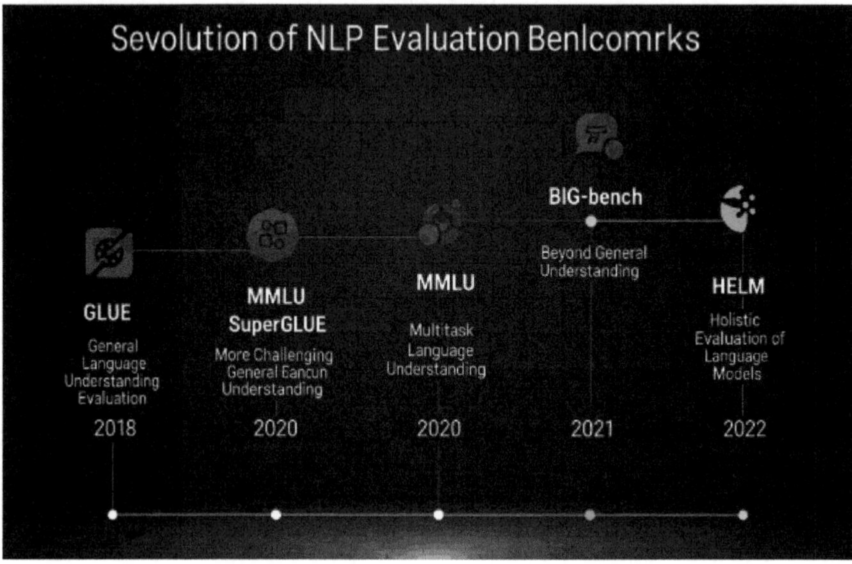

Figure 6: Evolution of LLM evaluation benchmarks.

7. Human-Centred and Holistic Evaluation

As LLMs increasingly move into real-world applications, evaluation of LLMs must not only transition from task-specific evaluation metrics to frameworks that account for human values and societal impacts, and robust LLM performance in a wide range of settings. Human-centred and holistic evaluation focus not only on what models can achieve in clean benchmarks but also evaluate performance when interacting with people, consider fairness, take safety into account, and maintain reliability across domains. This section provides insights into the role of human judgements in evaluation LLMs, fairness and bias evaluation, and robustness in adversarial or unexpected situations.

7.1 The Importance of Human Judgment in Evaluating LLMs

Automated metrics such as BLEU, ROUGE and perplexity may offer a level of efficiency and consistency for rating systems, but they are unable to capture the important subtleties of language that humans care about, i.e., clarity, persuasiveness, empathy, and cultural consideration. Therefore, human evaluation is essential in determining the quality of the model. Typically, human raters will be tasked with evaluating model outputs on fluency, coherence, factual correctness, or helpfulness. For instance, as outlined in preference modelling in Reinforcement Learning with Human Feedback (RLHF), human judgment is to help shape Large Language Models to better align with user expectations. Human-in-the-loop evaluation shows, however, weaknesses that cannot be expressed in automated measures, such as reasonably subtle errors in reasoning, implicit biases, or simply an absence of contextual sensitivity. While often subjective and resource intensive, human centred evaluation helps ensure LLM evaluation is grounded in end-user motivations and real-world utility.

7.2 Fairness, Safety, and Bias Evaluation

LLMs trained on huge web-scale datasets will similarly propagate the biases, stereotypes, and disbalances found in them. The implications of this are troubling, fairness and bias evaluations are now one of the main components of any holistic evaluation. Fairness metrics usually assess the variance in model behaviour across demographic characteristics (i.e., gender, ethnicity, or socioeconomic status). Safety evaluations are usually aimed at avoiding harmful outcome, such as toxic outputs, misinformation, requests for actions that could facilitate harm. Robust evaluation in this area includes an (i.e., toxicity classifiers, adversarial-prompt tests, curated fairness benchmarks (e.g., Bias-in-Bios, Real Toxicity Prompts)—methodological) approach to the detection of biases. Importantly, fairness and safety issues will always be people-centred issues, requiring the ongoing consideration of interdisciplinary balance in relation to issues of computer science, ethics, and law. In entirety, mutually beneficial, when thinking about evaluation in this dimension, I would hope LLMs would contribute to a trustworthy information system and an equitable access, rather than being reified to stereotype, or unsafe practices.

7.3 Robustness across Domains and Adversarial Settings

Another important dimension of holistic evaluation is robustness, which is the degree to which LLMs can generalize performance to multiple domains, data distributions, and adversarial inputs. In practical use, models encounter noisy and ambiguous prompts that the model has not seen before, or intentionally manipulated prompts that are different from the training data. Robustness evaluation therefore assesses models in the presence of distribution shifts (e.g., domain shift to health, legal, or financial-related problems), language shifts (e.g., dialects, low-resource languages), and adversarial shifts (e.g., prompt injection or other malicious attacks). The evaluation metrics often combine task accuracy with a stress-test type of evaluation which helps to identify weaknesses in their performance. An example of this is AdvGLUE benchmarks for adversarial robustness, and HELM is designed with robustness as an explicit evaluation dimension. For LLMs to be used in mission critical contexts, resilience must be demonstrated in a variety of situations, since errors or unsafe behaviours can be consequential. Table 5 of dimensions of human-centred and holistic evaluation for LLMs.

Table 5: Human-Centred and Holistic Evaluation Dimensions of LLMs.

Dimension	Area of focus	Evaluation Strategies	Major Challenges
Human Judgments	Fluency, coherence, usefulness, empathy	Human ratings, pairwise preference ranking, RLHF feedback	Subjectivity, cost, scalability
Fairness & Bias	Avoiding group stereotypes, treating groups equally	Bias benchmarks (Bias-in-Bios, StereoSet), toxicity classifiers, demographic disaggregation	Measuring subtle/ contextual bias, cultural variability
Safety	Preventing adversarial or perilous outputs	Adversarial prompt testing, toxicity filtering, safety red-teaming	Moving beyond misuse risks, compatibility with ethical norms
Stability	domain shift stability, adversarial inputs, low-resource settings	Stress tests (AdvGLUE, HELM), cross-domain evaluation, adversarial perturbation tests	Generalization on unseen data, adversarial resilience
Reliability	Interpretability, calibration, reliability	Confidence calibration, explainability methods, user trust survey	Lack of transparency in internals of LLM

Table 5 captures what each dimension measures, how it is evaluated, and the challenges involved.

An image (Figure 7) of a radar chart representing the holistic dimensions of LLM evaluation:

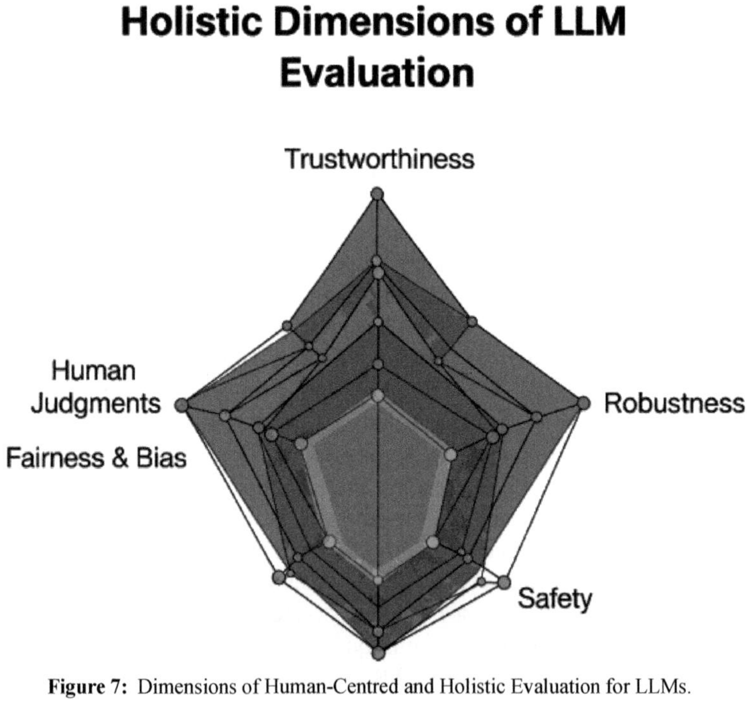

Figure 7: Dimensions of Human-Centred and Holistic Evaluation for LLMs.

8. Challenges in LLM Evaluation

The evaluation of Large Language Models (LLMs) presents a number of pressing issues surrounding the soundness, equity, and longitudinal trustworthiness of traditional benchmarking. Benchmarks and metrics have evolved alongside LLMs in magnitude, but continue to be limited by structures, cultures and processes which require to be resolved for meaningful evaluations to be done.

8.1 Dataset Limitations and Benchmark Saturation

Many of the benchmarks considered in this paper, i.e., GLUE, SuperGLUE and MMLU, have been investigated to saturation whereby near-human or super-human performance have been achieved and subjected to scrutiny. There is a potential for these benchmarks to only be evidencing overfitting to specific and static patterns in the benchmark datasets—it is unclear whether they reflect differentiated improvements in LLMs incrementally. From this review, there is also clear consideration that most benchmark problems consist of static datasets which does not capture the depth and ongoing evolvement of human knowledge, use of the language and adversarial contexts.

8.2 Cultural and Linguistic Diversity Gaps

Another major limitation remains the insufficient inclusion of cultural and linguistic diversity in existing benchmarks. Most of the datasets that are widely used are English-centric and thus ignore a wealth of global linguistic diversity. This impacts not only our evaluations of LLMs, but also the contexts in which it can be useful in multilingual-unilingual and multicultural contexts. If models are built or tested with data from narrow cultural repositories, they risk perpetuating stereotypes, leaving lower resource languages out, and failing to meet the needs of non-Western users. Although hard to bridge, this means requiring benchmarks that are linguistically diverse, culturally informed, and track the performance of the model as it would be used in an applied setting.

8.3 Measuring Reasoning vs. Memorization

LLMs can be very successful at tasks that involve can recall factual information, but being able to tell an actual instance of reasoning apart from sophisticated memory remains a problem. Most benchmarks or datasets involve factual, or knowledge-based, questions that can be solved through the model recalling training data rather than logical reasoning, deduction or problem solving. The challenge is creating tests that assess deductive reasoning, cause-effect reasoning and multi-step reasoning without allowing the model to shortcut through a memorized pattern. If benchmarks cannot make these distinctions, then they may inflate models stated reasoning abilities and prerogative an inflated sense of intelligence.

8.4 Dynamic Evolution of LLM Capabilities

Ultimately, the speed of evolution in large language models raises a fundamental problem in evaluation. Once a benchmark is seen as state-of-the-art (and potentially associate a "score" with it), a model is bound to appear that performs capabilities beyond prior models, making both the benchmark and the scoring obsolete. For example, we now see models with emergent abilities such as zeroshot reasoning, code synthesis like GitHub Copilot, and multimodal understanding not even captured in existing benchmarks. In response, we need to be in a constant state of updating, adding to, and redefining benchmarks to keep up with the constantly changing landscape of evaluation. Static evaluation models are no use to an ecosystem of language models that constantly change in unpredicted ways. Table 6 for challenges in LLM evaluation that provides the missing piece here:

Table 6: Major Problems in LLM Assessment.

Challenge	Description	Implications	Possible Solutions
Limitations of datasets & benchmark saturation	The most popular benchmarks such as GLUE, SuperGLUE and MMLU are almost saturated, with models scoring almost perfect scores.	This approach may lead to models that are overly specialized to specific benchmarks, rather than truly understanding the underlying concepts	Set dynamic, changing standards; present more challenging, real-life work; bring in adaptive test batteries
Cultural & linguistic diversity differences	The current standards are primarily English-based and are culturally non-inclusive	The models can be inefficient on the low-resource languages; the danger of the strengthening of the biases and stereotypes	By involving global communities in dataset design, we can create more representative and nuanced datasets that better serve diverse populations
Advanced vs. basic memorization of training data	Reasoning is hard to differentiate between real reasoning and high-level memorization	Incorrect assumption of the power of LLM reasoning; false statements regarding intelligence	By incorporating tasks that require causal reasoning, logical thinking, and complex problem-solving, and testing models with adversarial examples, we can better evaluate and improve their ability to handle real-world challenges
Active development of LLM functions	LLMs are actively developed as they acquire new powers and capabilities that are not current	Standards become obsolete within a short time; partial assessment of new talents (e.g., multimodal reasoning)	To keep pace with evolving AI capabilities, it is essential to regularly refresh benchmarks, utilize flexible evaluation frameworks, and prioritize tasks that reflect real-world applications and challenges

Here is an image (Figure 8) illustrating the key challenges in LLM evaluation:

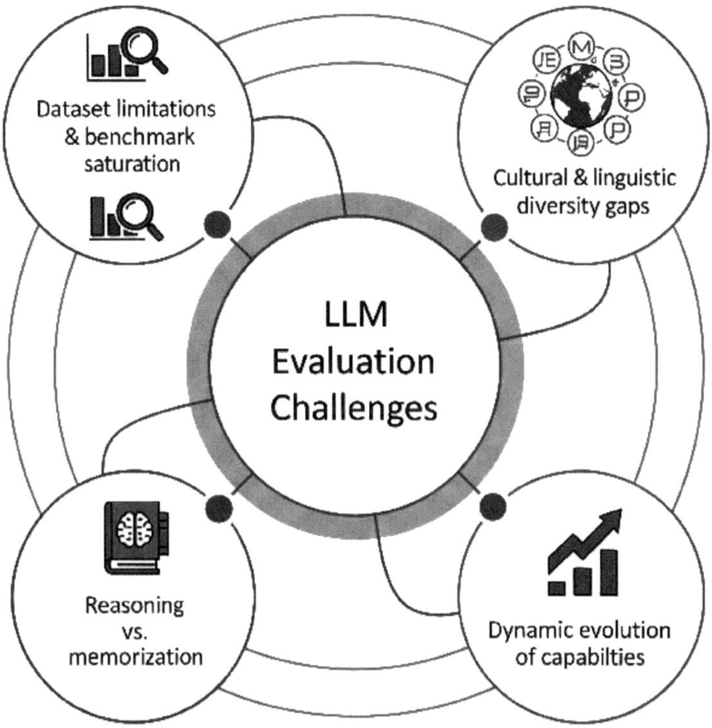

Figure 8: Major challenges in evaluating Large Language Models (LLMs).

9. Future Directions in Benchmarking

The situation around Large Language Model (LLM) evaluation is continuously evolving, as are language models, applications, and use cases. Conventional standards have been influential in instigating appraisal norms, although they will not sufficiently embrace the outlook of the multifarious, multi-modal and stateful construal. The research community is inclined to what it perceives as equitable, credible, general, and future-proof assessment regimen, i.e., holistic, adaptive, and unified, or standardized benchmarking ecosystems.

9.1 Towards Continuous and Adaptive Benchmarks

Although useful in the comparative analysis, static benchmarks are rapidly becoming obsolete due to advancement of LLMs, which are going to deplete the benchmark data-driven dataset in the available test tasks. The future trends in evaluation practice point to the fact that we are going towards continuous and adaptive standards of benchmarking that are changed with the models. This may involve creating dynamic examples, a combination of adversarial examples and citing ever-updating datasets that embody the new developments in language usage, new cultural trends, and real-world difficulties. Adaptive benchmarks will be more convenient to maintain the relevance of the evaluation and reduce the likelihood of the overfitting of the ELMs to a series of tasks and establish actual generalization.

9.2 Multimodality (Text, Code, Vision, Speech) is to be Incorporated

With the development of LLMs as multimodal foundation models, benchmarking should no longer be confined to text-only models only. Code synthesis, image captioning, multimodal reasoning, and speech understanding are already done by next-generation models, and will require more multi-modal input-output benchmarks. Multimodal evaluation criteria will be of importance to the comprehension of the capabilities of LLMs to assimilate reasoning between or among domains, such as code generation based on a natural language prompt, responding to visual or spoken text in a different language, and speech translation.

9.3 Trust, Safety, and Accountability Evaluation Ecosystems

Besides the measurement of standard LLM performance at various tasks, an increased amount of interest is taken in such evaluation ecosystems that incorporate other areas of evaluation as trustworthiness, safety, fairness, and accountability. It will require evaluation ecosystems to investigate to what extent LLMs produce harmful, biased or misleading outputs and how well they can perform when manipulated by adversarial examples. Such analyses will require not only human-centred analyses, but also powerful fully automated pipelines in the detection of toxicity, analysis of bias, and calibration. Evaluation ecosystems in the future will thus be a mixture of technical performance evaluation and the socio-ethical areas of evaluation to make sure that the LLM will generate safe, fair, and aligned results with the human values when applied to real situations.

9.4 The Research Community Standardization Activities

Conclusively, benchmarking of LLM needs to collaboratively work on standardizing evaluation and comparison practices to achieve success. At this point, the disparity in the measurement of models is a big issue. Various organizations and research teams employ various datasets, metrics, and benchmarks all of which are dependable to restrict our comparison across models and carry out evaluations with transparency. The efforts—HELM (Holistic Evaluation of Language Models) are initial efforts to come up with common frameworks but such efforts must be mixed into communal attempts. They are likely to become more familiar sharing evaluation procedures, open collections of benchmark datasets, and reporting conventions on transparency on how such evaluations are done and reports constructed that can assist uniformity and comparability across LLMs. Consequently, the outcomes will become scientifically more effective, will contribute to achieving the regulatory environment that AI systems will be operating under shortly, and will promote and sustain the

Table 7: Direction in the Future of LLM Benchmarking.

Future Direction	Description	Expected Impact	Examples/Initiatives
Continuous & Adaptive Benchmarks	Periodically updating sets of tests with adversarial and real-world examples	Avoids saturation of benchmarks; keeps the benchmarks relevant and generalized	Moving data, changing assessment activities
Multimodality (Text, Code, Vision, Speech)	Multimodal reasoning benchmarks that evaluate the reasoning of input/output modes	Facilitates analysis of text-based foundation models; making preparations against real-world multimodal activities	CodeXGLUE (code), VQA (vision + text), SpeechBench (speech).
Evaluation Ecosystems for Trust, Safety & Accountability	Evaluation incorporates fairness, bias, safety, robustness, and moral integrity	Enhances responsible AIs implementation; solves social and regulatory issues.	Such holistic evaluations as HELM, as well as toxicity detection and bias analysis benchmarks (such as Bias-in-Bios) offer a better picture of performance and societal impact of models
Standardization Activities	Community based protocols, open repositories, and common reporting practices	Enhances transparency, scientific rigour and comparability; helps to govern	Initiatives like HELM, MLCommons, OpenAI Evals, and Big Science are driving progress in AI evaluation, promoting more comprehensive, standardized, and community-driven assessments of language models

level of trust the population will have in LLMs such as the one provided by AI. Our Future Directions with respect to benchmarking are added up in Table 7 below:

Figure 9 depicts the future directions in LLM benchmarking:

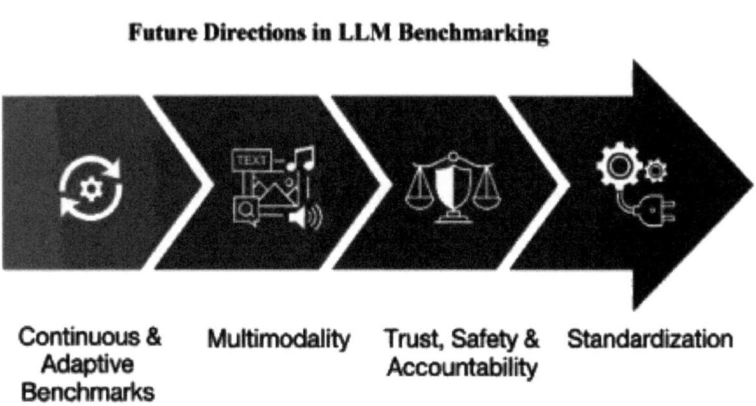

Figure 9: Roadmap of future directions in LLM benchmarking.

10. Conclusion

Currently, evaluations of Large Language Models (LLMs) have transformed from narrow task-specific measures to more expansive frameworks that are reflective of the complex, multi-dimensional nature of AI today. Although metrics including BLEU, ROUGE, and perplexity have been useful for simplistic NLP tasks, they fall short of assessing reasoning, truthfulness, robustness, and ethical alignment. As a response, the researcher community has developed more broad benchmarks, such as GLUE, SuperGLUE, MMLU, BIG-bench and HELM, which allow for somewhat more fine-grained measures of the LLM's performance across numerous tasks and contexts. In addition to purely intrinsic and extrinsic metrics, human-centred evaluations focused on assessments of fairness, bias, safety, and interpretability have also emerged as critical measures of evaluation, especially as LLMs are increasingly moving into more critical areas of domain including healthcare, education, finance, and governance.

Moving forward, one of the paths is the creation of both all-encompassing, equitable and reliable benchmarking ecosystems that go beyond the scope of both unchanging datasets and the limited scope of performance measurements. Continuous and adaptive benchmarks, multiple evaluation modalities, and standardization efforts characterized by the community will be relevant and comparative and will be important. Together with this, credibility, responsibility, and social consideration will also be pertinent issues to consider systemic bias, since we do not just consider accuracy, but human values and cultural differences and moral utilization. The research community can inspire more confidence in the use of LLM as well as give advice on the responsible creation of future models with dynamic, transparent, and holistic evaluation systems.

References

Amman, J., Blasimme, A. and Vayena, E. (2020). Explainability for artificial intelligence in healthcare: A multidisciplinary perspective. BMC Medical Informatics and Decision Making, 20(1): 1–9. https://doi.org/10.1186/s12911-020-01332-6

Bang, Y., Cahyawijaya, S., Lee, N., Dai, W., Su, D., Wilie, B. et al. (2023). A multitask, multilingual, multimodal evaluation of chatGPT on reasoning, hallucination, and interactivity. arXiv preprint arXiv:2302.04023.

Bender, E. M. and Koller, A. (2020). Climbing towards NLU: On meaning, form, and understanding in the age of data. Proceedings of ACL, 5185–5198. https://doi.org/10.18653/v1/2020.acl-main.463

Brown, T. B., Mann, B., Ryder, N., Subbiah, M., Kaplan, J., Dhariwal, P. et al. (2020). Language models are few-shot learners. Advances in Neural Information Processing Systems (NeurIPS), 33: 1877–1901.

Clark, C., Lee, K., Chang, M.-W., Kwiatkowski, T., Collins, M. and Toutanova, K. (2020). BoolQ: Exploring the surprising difficulty of natural yes/no questions. Transactions of the ACL, 8: 292–305. https://doi.org/10.1162/tacl_a_00306

Hendrycks, D., Burns, C., Basart, S., Zou, A., Mazeika, M., Song, D. et al. (2021). Measuring massive multitask language understanding. arXiv preprint arXiv:2009.03300.

Kiela, D. et al. (2021). The BIG-bench benchmark: Beyond the imitation game. arXiv preprint arXiv:2109.01652.

Lin, C.-Y. (2004). ROUGE: A package for automatic evaluation of summaries. Proceedings of the ACL Workshop: Text Summarization Branches Out, 74–81.

McCann, B., Keskar, N. S., Xiong, C. and Socher, R. (2018). The natural language decathlon: Multitask learning as question answering. arXiv preprint arXiv:1806.08730.

Papineni, K., Roukos, S., Ward, T. and Zhu, W. J. (2002). BLEU: A method for automatic evaluation of machine translation. Proceedings of ACL, 311–318. https://doi.org/10.3115/1073083.1073135

Rajpurkar, P., Zhang, J., Lopyrev, K. and Liang, P. (2016). SQuAD: 100,000+ questions for machine comprehension of text. Proceedings of EMNLP, 2383–2392. https://doi.org/10.18653/v1/D16-1264

Raji, I. D. and Buolamwini, J. (2019). Actionable auditing: Investigating the impact of publicly naming biased performance results of commercial AI products. Proceedings of AAAI/ACM Conference on AI, Ethics, and Society, 429–435.

Raffel, C., Shazeer, N., Roberts, A., Lee, K., Narang, S., Matena, M. et al. (2020). Exploring the limits of transfer learning with a unified text-to-text transformer. Journal of Machine Learning Research, 21(140): 1–67.

Srivastava, A., Rastogi, A., Rao, A., Shoeb, A. A. M., Abid, A., Fisch, A. et al. (2022). Beyond the imitation game: Quantifying and extrapolating the capabilities of language models. arXiv preprint arXiv:2206.04615.

Wang, A., Singh, A., Michael, J., Hill, F., Levy, O. and Bowman, S. R. et al. (2018). GLUE: A multi-task benchmark and analysis platform for natural language understanding. Proceedings of ICLR.

Wang, A., Pruksachatkun, Y., Nangia, N., Singh, A., Michael, J., Hill, F. et al. (2019). SuperGLUE: A stickier benchmark for general-purpose language understanding systems. Advances in Neural Information Processing Systems (NeurIPS), 32: 3266–3280.

Welbl, J., Stenetorp, P. and Riedel, S. (2018). Constructing datasets for multi-hop reading comprehension: HotpotQA. Proceedings of ACL, 945–955.

Xu, L., Hu, H., Zhang, X., Li, L., Cao, C., Li, Y. et al. (2020). CLUE: A Chinese language understanding evaluation benchmark. Proceedings of the ACL, 423–434.

Zellers, R., Holtzman, A., Rashkin, H., Bisk, Y., Farhadi, A., Roesner, F. et al. (2019). Defending against neural fake news. Advances in Neural Information Processing Systems (NeurIPS), 32: 9051–9062.

Zhang, T., Kishore, V., Wu, F., Weinberger, K. Q. and Artzi, Y. (2020). BERTScore: Evaluating text generation with BERT. Proceedings of ICLR.

15

Challenges and Limitations in LLMs

K. Subbulakshmi[1,*] and *V. Karthikeyan*[2]

1. Introduction

Large Language Models (LLMs) act as an innovative force for the transformation and generation of natural language in the field of Artificial Intelligence. They can translate languages, content writing, code creating like humans as they are advanced AI systems and are trained on large amounts of word-based textual data. Inspite of its remarkable potential, Large Language Models incorporate ethical concerns that needs our attention. As they propagate increasingly in our digital environment, investigating these models regarding the information integrity, justice and privacy is necessary.

Large Language Models are a key and fundamental subdivision of Artificial Intelligence, they are made to comprehend and work with human language. As they can understand and observe textual data, LLM can execute tasks like content summaries, text production and translation and sentiment analysis. In order to produce and comprehend natural languages deep learning methods are used by LLMs. For replicating the information-processing systems perceived in the human brain, neural networks are built, which consists of layers of processing units interconnected with each other. LLMs can guess the succeeding word in a sentence and learn verbal and linguistic patterns from the enormous amounts of text data loaded to it during the pre-training phase. They are given grammar, linguistic information, syntax, and semantics throughout this phase. For certain applications, the performance of LLMs

[1] Department of Electronics and Communication Engineering, Sethu Institute of Technology, Kariapatti, Virudhunagar.

[2] Department of Electronics and Communication Engineering, Mepco Schlenk Engineering College, Sivakasi.

Email: velkarthi85@gmail.com, https://orcid.org/0000-0003-2974-6554

* Corresponding author: lakssubbu91@gmail.com, https://orcid.org/0009-0001-2443-2700

DOI: 10.1201/9781003771654-15

can be enhanced by fine-tuning them using smaller datasets for specific tasks. The two important features of LLMs are scalability and pattern generalization. Scalability allows LLMs to be trained on larger datasets and pattern generalization allows it to handle a diversity of linguistic tasks. Biases and ethical considerations are reflected in their training data, if they are inherently biased or ethical. This feature highlights how much important is to check the outputs of LLMs for probable biases and ensure the ethical problems assimilated into the creation of LLMs (Abid et al., 2021).

2. Challenges in LLM Evaluation

Figure 1 shows the challenges in LLM evaluation of Large Language Models the dependability and accuracy of the outcomes are impacted by a number of issues. One of the main problems is the subjectivity of human valuation in which individual preconceptions and lookouts can influence conclusions. Issues with the quality of data may affect the results since misleading and subpar datasets does not illustrate real-world scenarios. Another major obstacle is the absenteeism of diversity measures makes it thought-provoking to evaluate whether LLMs function rightly across diverse circumstances and populations. Additionally, the fairness and biases in programmed valuations are problems that can emphasize or expand prejudices in the data present (Achiam et al., 2023). Another hazard arises from adversarial attacks which can yield incorrect findings by captivating flaws in the models. Scalability issues also occur while investigating more complicated models that require an enormous amount of computing power and resources. All these elements collectively work together to yield a complex environment for LLM evaluation that imposes flexible and trustworthy methods.

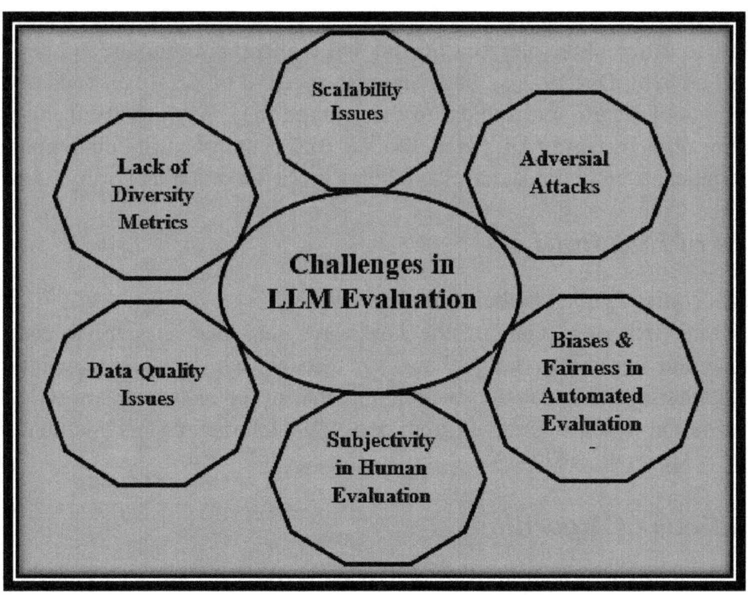

Figure 1: Challenges of LLM evaluation.

Precise and unbiased evaluation of Large Language Models are fraught by a large number of substantial issues. One of the main problems is prejudice in human evaluation, as annotators' valuations are recurrently unpredictable because of diversities in the cultural background, preferences of the individual and how the instructions are understood and interpreted. It is perplexing to attain normalization and repeatability in outcomes because of this objectivity. The issues such as duplicate samples, out-of-date information, inaccuracies, or hidden biases arise from the frequently used datasets and benchmarks, further raises concerns about the quality of data. The models' realism is understated due to these flaws which have the potential to mispresent the evaluation results (Carlini et al., 2021). Evaluations which are trustworthy and complex-scale are taking more time and are also expensive due to their large computational resources required to examine complex models across various domains, languages and professions.

3. Bias, Fairness, and Ethical Concerns in LLMs

Bias in LLMs

The complex problems of bias in Large Language Models are the nature of training data, prototypical architecture and learning techniques. LLMs takes in the imbalances and biases which exist in the massive datasets they are trained on, which are collected from the books, the internet, social media, and other additional digital sources. They can deliver responses that unintentionally disseminate typecasts or sideline specific groups due to these biases, which include many forms such as race, culture, gender, political and socio-economic biases. Bias is not a concern with training data, which also distresses the algorithms and optimization measures that are used to generate LLMs. These models acquire statistical outlines at the time of training, but they are incapable to differentiate between neutral and realistic information and destructive or biased content. Due to this, these biases turn out to be firmly embedded in their distributions of weight and subject to the prompt, they may modify themselves in delicate or obvious ways (Abid et al., 2021). Bias cannot be fully eliminated, but it can be condensed to diminish harm and boost more fair and trustworthy AI systems.

Causes of Bias in LLMs

LLM bias comes from a number of factors:
Information Gathering: One of the key ways that bias is familiarized is the behaviour and fairness of Large Language Models which are greatly influenced by data gathering. The bias and inequalities that occur in society are reflected by the information which is gathered from books, the internet, papers and other digital platforms (Ahn et al., 2022).

Limitations of Algorithms

LLMs have limitations in algorithms that significantly subsidize the determination of bias, since the neural network's primary learning process is mainly tuned for finding and imitating broad patterns in data rather than endorsing impartiality and fairness.

These models adapt their parameters at the time of training to enhance accuracy and lessen prediction mistakes, but the process does not habitually distinguish between biased connections and valuable correlations (Bang et al., 2023).

Limitations on Resources

A sensible effort is needed to gather high-quality content, clean data and confirm inclusivity for the purpose of creating a dataset that precisely represents diverse languages, social groups, cultures, and perspectives. These jobs are resource-intensive. Due to this many organizations depend on freely accessible sources like public media, websites and discussion forums. Even though they are abundant, they usually contain significant biases that characterize the fundamental perspectives and voices of specific areas and groups (Bubeck et al., 2023).

4. Types of Bias

Table 1 shows the types of Bias in LLMs:

Table 1: Types of Bias.

Type of Bias	Cause	Example	Impact
Bias in Data	Stereotypes, false information, or imbalances found in training data	Women are associated with caring, males with leadership positions	Perpetuates negative stereotypes
Bias in Representation	There are underrepresented groups, cultures, or languages	The model does poorly for low-resource languages but well for English	Inequitable disparities in performance between communities
Cultural and Social Bias	Dominant cultural viewpoints ingrained in training and data	Assumes that Western holidays, including Christmas, are observed everywhere	Excludes minority cultures
Confirmation Bias	The model follows user instructions without confirming accuracy	When the user asks, "Why are cats smarter than dogs?" the model responds with explanations without challenging the assertion	Reinforces user presumptions and false information
Algorithmic Bias	Tokenization, RLHF annotation preferences, and model design decisions	Some responses are rated "better" by annotators according to cultural standards	Incorporates into the system arbitrary human preferences
Interaction Bias	Biased outputs are reinforced by user prompts and feedback loops	Leading or biased queries the model replicates and adjusts for those biases	Amplification of bias through interactions between users and models

Equity and Fairness in LLMs

One among the important objectives in handling bias and ethical problems in the field of artificial intelligence is to endorse justice and equity in Large Language Models (LLMs). The fairness in AI is stated as the principle of confirming that AI systems operate without bias, discrimination, or favouritism, treating all personalities and groups equitably (Carlini et al., 2021). It guarantees that AI systems treat every person and group equally, regardless of their behaviours. Conversely, equity identifies the requirement for modified solutions to take sociological and historical inequalities into consideration and treats everybody equally. Equity and fairness encompass generating results that assist equal chances for all users simultaneously avoiding inflating pre-existing prejudices.

Dimensions of Fairness

Fairness in Demographics

Demographic fairness is defined as the idea that people should be handled equally irrespective of their age, gender, race, ethnicity, and religion (Bang et al., 2023). The necessary factors to achieve demographic fairness are—balanced training data, ongoing audits and debiasing techniques to recognize and address unintended prejudiced patterns. Eventually the demographic fairness targets to advance equity in all user interactions.

Linguistic Equity

Linguistic fairness highlights the LLMs need to maintain multiple dialects, languages, and variations in linguistics with equal superiority and articulacy. Accurate linguistic fairness comprises intensifying language coverage, refining resources for understated languages and guaranteeing that dialects and regional expressions are valued rather than considered as errors (Chen et al., 2022). This allows reciters from different linguistic conditions to access the similar levels of utility, thus avoiding language from becoming an obstruction to participate in the digital environment.

Fairness of Opportunity

Opportunity fairness is described as the equivalent distribution of the advantages that individuals can obtain from engaging with Large Language Models. The beneficial assistance such as career counselling, educational materials are essential to provide equally by the system to all the persons from different backgrounds (Das et al., 2025). Factors that are needed to guarantee opportunity fairness are, observing the group of recommendations, recognizing differentiations among demographics, and generating algorithms that enhance inclusivity in access to information.

Cultural Equity

Cultural fairness ensures that LLMs respect and accept traditions, approaches in worldwide rather than concentrating outcomes around prevailing cultural norms.

AI programs are trained on prejudiced data that run the risk on preserving a limited viewpoint such as providing Western festivals, customs superiority over home-grown regional or minority perspectives (Dhungana et al., 2025). The factors necessary for cultural fairness are sensitivity to culturally specific situations, cautious deliberation of depiction in training data and the capacity to adapt to responses for diverse cultural contexts without daunting bias.

Techniques for Reducing Prejudice and Maintaining Equity

Pre-emptive steps must be engaged at every stage of the AI development lifespan to alleviate bias and ensure equity in LLMs. Amongst the strategies are:

Multiple Training Data: An inclusive methodology all the way through the whole development lifecycle is mandatory to mitigate bias and to assure equity in large language models. By reproducing a diversity of voices, backgrounds, and standpoints, wide-ranging training data minimizes the probability of over representing foremost groups and endorses inclusivity. Meanwhile bias can still seem, that it is difficult to dynamically find out and resolve harmful trends by enduring bias audits and adaptation using de-biasing algorithms, and vigilant post-processing (Duan et al., 2024). Eventually, recapitulation and user feedback empower real-world users to recognize prejudice or unfairness that programmed checks miss, allowing for unceasing promotions and improvements.

Measures of Fairness

Fairness metrics are defined as quantifiable mechanisms that values whether a model produces biased results, guiding in the documentation and correction of infinitesimal differences in depiction, treatment, and accuracy between diverse groups (Carlini et al., 2021). The examples of how fairness are measured include, individual fairness measures that guarantees analogous inputs are treated regularly irrespective of inappropriate attributes like pronouns and group fairness measures that compare output quality across demographic classes like gender and ethnicity. In order to find the hidden biases, benchmark testing is done using fairness focused datasets.

Auditing and Correcting Bias

Bias auditing is stated as the process of consecutively running targeted prompts to find out categorized and unfavourable outputs such as investigating solutions on occupations and social roles related with diverse demographic groups. Once recognized, the modification process such as filtering methods and debiasing algorithms may be used. In order to prevent a blind spot for developers, independent review processes also can be used (Kshetri, 2023). Finally, the advantage is that the system progressively becomes more reliable and fair as a result of unceasingly finding and sorting out the bias.

Iteration and User Feedback

As the automated detection will not be able to apprehend all the real-world difficulties, user input feedback and iterations are necessary. By using simple reporting tools in LLM users can create a useful feedback channel by abating biased and unsuitable results. Concluding the loop by assimilating feedback into fine-tuning and post processing gives noteworthy improvements (Esmradi et al., 2023). Observing these reports gives the provisions in classifying recurrent discrepancies, such as language-based deficit or cultural insensitivity. Fairness turns out to be a recurrent, adaptive process than a single action when consumers are educated about how their input impacts the variations. This develops faith and responsibility.

Incorporating Ethical Considerations in LLM Development

At every stage of progress of an LLM should fit in ethical issues. This needs a thorough review of the training data for biases, progressing norms to address sensitive titles and being aware of the potential consequences of the society. Ongoing developments may be conversant by routine audits of the model's outputs to find out the bias and ethical issues. Additionally, reassuring multidisciplinary collaboration between ethicists sociologists and subject matter specialists may aid in predicting and determining probable ethical confusions (Bubeck et al., 2023). Figure 2 shows the ethical difficulties in LLMs.

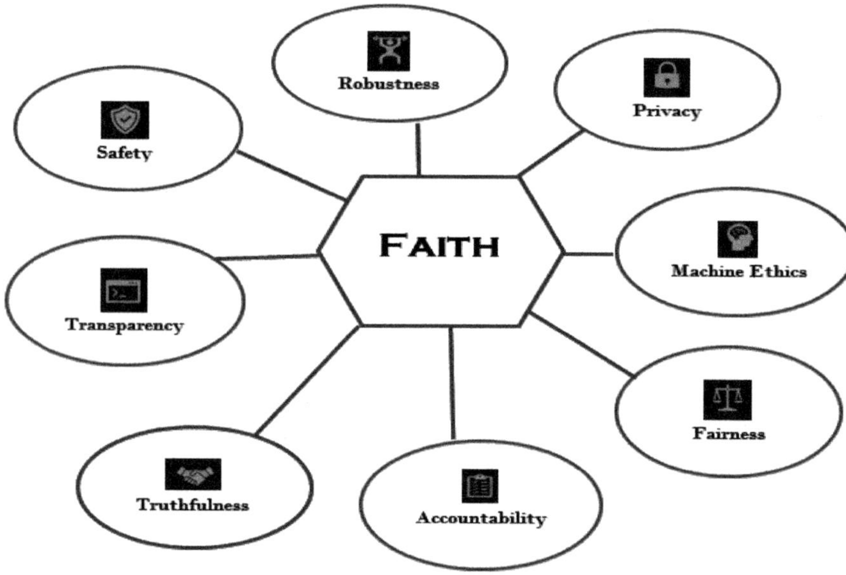

Figure 2: Ethical Difficulties in LLMs.

Faith in LLMs

The word "faith" designates the assurance and belief that users have in large language models. It is fabricated when the system uninterruptedly unveils dependability, safety, and fairness. The faith is reinforced when the people are conscious that the model uplifts the truthfulness, ethics, and privacy (Li et al., 2024). At the end, faith necessitates having confidence that LLMs will perform in the best interests of people, in situations that are uncertain and involves noteworthy risks.

Robustness

Although in the existence of unreceptive, mystifying, or noisy inputs, Large Language Models must uninterruptedly yield trustworthy results. They must provide attempts at adaptation and manipulation that aim to go away from security measures. Another aspect of robustness is the adaptability to the diverse backgrounds and domains (Liu et al., 2025). A robust LLM will not produce outputs that are unbalanced or unsafe although maintains accuracy.

Privacy

LLMs are required to guarantee the confidentiality of any sensitive or personal information that users share. They must refrain from learning and disclosing identifiable training data. Protecting user privacy complies with data protection regulations (such as the GDPR) and increases consumer trust. An LLM must produce valuable results without revealing personal information.

Machine Ethics

LLMs should conduct themselves within publicly acceptable ethical and virtuous boundaries. They need to deteriorate damaging requests and abstain from developing offensive or illegal information. AI performances that are in line with social values and human principles are guaranteed by ethical orientation. This endorses accountable AI development though avoiding abuse (Kasneci et al., 2023).

Fairness

LLMs should abstain from encouraging discrimination, prejudice, or stereotyping. Fairness procedures are necessary since biases are often present in training data (Ahn et al., 2012). It is vital to guarantee enclosure across languages, cultures, and groups. Impartial results are formed by a fair LLM without preferring one group over another.

Accountability

A vibrant sequence of accountability for LLM adoptions and faults must exist. Menaces and limitations must be recognized by inventors, administrations, and deployers. Accountability assurances frankness in practice strategies, safety provisions, and training data. In the incident that the model does create damage, users need to be conscious of who is at fault.

Truthfulness

Precise, fact-based, and confirmed statistics of information must be the goal of LLMs. They might reduce illusions or outputs that are assuredly incorrect. In subtle occupations like medical, law, and teaching, honesty is crucial. When evidences are unavailable, an honest LLM admits ambiguity.

Transparency

Transparency involves open communication concerning the procedure of the LLMs. This entails bias threats, restraints, datasets, and training procedures. It permits consumers to make knowledgeable decisions concerning reliability and belief. Transparent LLMs minimize the insight of a "black box" and endorse transparency (Taori et al., 2023).

Safety

LLMs must abstain from developing content that can jeopardize people or the public. Averting destructive instructions, untruthful information, or misappropriation is part of safety. Safe distribution with precautions against malicious use is also included. Effectiveness and rigorous harm-prevention procedures are well adjusted in a safe LLM (Mireshghallah et al., 2023).

5. Security Risks: Prompt Injection, Data Leakage, and Adversarial Attacks

Prompt Injection

Large Language Models (LLMs) are susceptible to prompt injection, a security faultiness in which attackers inset malicious texts into prompts to fake the LLM into resounding out unwanted tasks such unveiling private information, making aggressive content, or running unlawful commands on unified systems. This occurs since LLMs are not always able to express the difference between internal instructions and user input. Eventually, well-written prompts can override system instructions, which can make the system's integrity vulnerable and lead to data exfiltration and social engineering. By injecting malicious inputs envisioned to modify the model's output, a prompt injection attack alters a large language model (LLM) (Nandagopal, 2025). The way LLMs practice and produce texts in response to input prompts is demoralized by this kind of cyberattack. An attacker can fall guy the model into generating illegal material, gaining access to data that is barred, or carrying out specific operations by inserting a sensibly fabricated text. These attacks abuse the distinctive assurance in the inputs of an LLM. In systems that depend on instinctive text generation, vaccinating misleading prompts might result in the formation of misleading data or the implementation of destructive commands, leading to severe security threats.

Direct Prompt Injection

This strategy highlights the consequence of thoughtful prompt management and shows how susceptible large language model (LLM) applications are to prompt injection attacks. To exploit or dupe the LLM into producing risky, unanticipated, or unlawful replies, the process starts with the construction or reception of a malicious prompt. Efforts to breakout, get all over the place security measures, or modify the model's behaviour or undercover instructions are a few examples of these prompts (Thomas et al., 2024). Figure 3 shows the Direct Prompt Injection mechanism.

The intermediary software layer that collects user inputs, controls context, and forms prompts before sending them to the LLM for processing is called an LLM-Based Application. The risk of exploitation will increase if this application is not securely constructed since it can inadvertently send dangerous prompts along without adequately filtering out malicious intent or sanitizing inputs. Regardless of their purpose, user inputs are combined with system instructions (defining permitted replies, setting boundaries, and giving role directives) at the malicious prompt and system block combination stage. Attackers may insert manipulative instructions to circumvent security mechanisms, which might cause the LLM to disregard initial instructions or divulge private information if a malicious prompt is available. All combined instructions and prompts, whether benign, system, or malevolent, are sent to the LLM block, the core engine, which uses them to produce outputs based on its language and contextual knowledge. Robust input management is essential to preserving the integrity and security of LLM applications since the LLM processes everything in the combined input, and its output may be impacted by a prompt injection. This design highlights the necessity for stringent validation and protection measures against prompt-based attacks by demonstrating how both beneficial and detrimental instructions within prompts can influence the language model's responses.

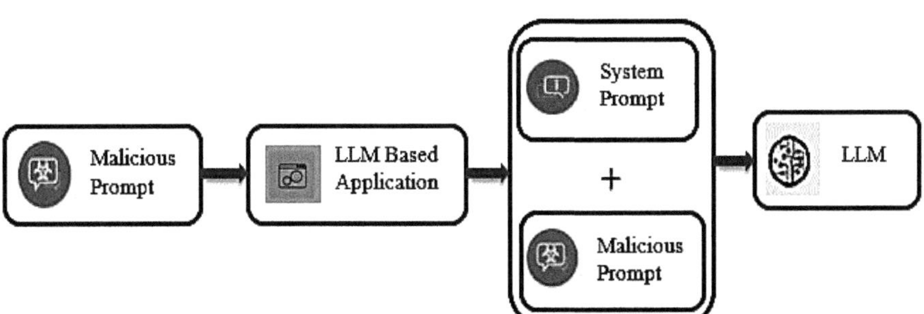

Figure 3: Direct Prompt Injection.

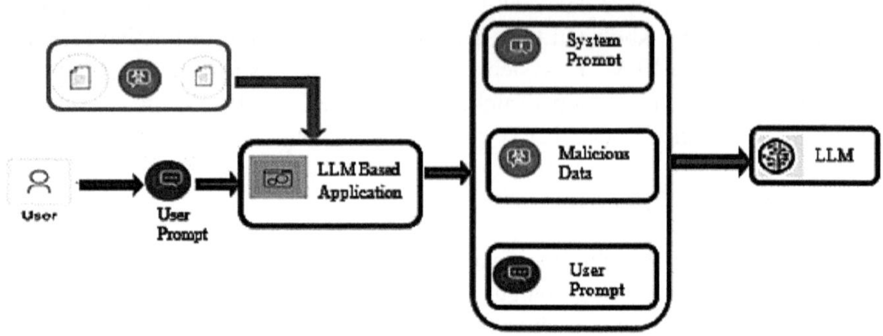

Figure 4: Indirect Prompt Injection.

The Figure 4 shows the Indirect Prompt Injection mechanism.

Indirect Prompt Injection

The flow starts with the user block, which is where users engage with the program by submitting "User Prompts," which are natural language requests or questions. These prompts are essential since they influence all the application's subsequent decisions. and reactions. Before sending everything to the LLM engine, the LLM-based application block serves as an advanced middleman, gathering the user prompt and combining it with other essential instructions, like system and safety guidelines (Das et al., 2025). This block may also control external data sources, including possibly harmful information, ensuring that all structured prompts comply with internal regulations and operational needs. Accumulating and organizing all mandatory information for protected, consistent language creation is its significant job.

The system prompt block, which compromises prewritten appropriate instructions formed by developers to straighten model behaviour and impose ethical and safety limitations; the malicious data block, which alarms users to confrontational instructions that could tilt or compromise model outputs and highlights the need of strong prompt authentication and filtering mechanisms; and the user prompt block, which is the user's straight and unique input comprised in the set of prompts treated at implication time, are the three distinct sources that are accomplished during the prompt assembly stage.

The entire collection of prompts system, user, and any malicious components are sent to the LLM block at the end of the process in order to produce a final result. The LLM interprets these inputs using its advanced language skills and contextual awareness, generating answers that are impacted by all the data sources and instructions that are present. The operational endpoint where actionable results are returned to the program is indicated by this block, highlighting how crucial careful and timely management is to the integrity and security of automated language systems.

Ways to prevent prompt injection attacks

- **Strict Input Validation and Sanitization**

  Strict input validation and sanitization are the first steps in preventing quick injection attacks. Organizations running LLMs can prevent many common injection tactics by thoroughly reviewing and cleansing all incoming data. These procedures are regularly updated and improved, which aids in adjusting to changing assault methods.

- **Context aware filtering and output encoding**

  Context filters prohibit inputs that are inappropriate or potentially dangerous by evaluating the safety and relevancy of input prompts due to the context of the current interaction. By doing this, subtle alterations that may otherwise get past normal filters are prevented. The generated responses are safe for the systems or end users who consume them thanks to output encoding (Dhungana et al., 2025).

- **Update and fine tune LLMs**

  Regular updates and ongoing fine-tuning are necessary to maintain the LLM's security. The most recent security measures and updated knowledge bases that take into account the complexities and circumstances of language are features of updated models. This increases the model's resistance to quick injections as well as its efficacy. Retraining a basic model with a smaller dataset that is closely associated with a particular use case is known as fine-tuning.

- **Log and monitor LLM interactions**

  Important information that can be utilized to identify and examine prompt injections is provided by keeping track of and recording every interaction with an LLM. The prompts received, the responses produced, and any irregularities or patterns that would point to a security problem should all be covered in these logs. Security teams can improve their defences and spot new attack trends by examining this data.

6. Data Leakage

Once the subtle or reserved information, like specific information, patented algorithms, or intellectual property, is inadvertently exposed through the LLM's responses regularly due to user input or unanonymized training data is known as data leakage in LLMs. Owing to the opportunity that the data might be retrieved by unauthorized parties or castoff to train future models, this presents serious risks, such as privacy defilements, security fissures, nonconformity, and defeat of competitive advantage.

How Data leakage occurs

Problems with Training Data: LLMs are taught on huge datasets, and if these datasets comprise reserved or delicate information that has not been sufficiently anonymized, the model might inadvertently commit to memory and then divulge this information.

User Input Mishandling: Employees or handlers might inadvertently submit delicate company data into the model by copying and pasting it onto free LLM accounts.

Smartly Designed Prompts: Confrontational actors can betray the LLM into revealing secluded information it has kept from its training or previous encounters by using detailed, expertly written prompts.

Insecure Cloud Services: It could be unsafe to practice third-party LLMs in cloud environments since it may be perplexing to authorize what the LLM provider logs or stores, which might result in data that is no longer under the organization's control (Duan et al., 2024). The Table 2 shows the Risks and consequences of Data Leakage in LLMs.

Table 2: Risks and Consequences.

Risk Category	Description	Consequences
Disclosure of Sensitive Information	Inadvertent or deliberate disclosure of private or sensitive information in training or LLM results	Financial losses, reputational harm, legal troubles, and invasions of privacy
Poisoning Data	Maliciously adding inaccurate or damaging data to training sets	Unreliable or biased responses, distorted model behaviour, and impaired system integrity
Prompt Injection	Malicious prompts designed to alter LLM behaviour or extract private information	Unauthorized data release, circumvention of security measures, and inaccurate or damaging model responses
Leakage of Model Data	Exposure of internal data or training during or following training	Disclosure of proprietary or sensitive information, including personally identifiable information
Leakage of Test Data	Unintentional release of test datasets or sensitive information	Attackers obtaining information to target or bias the LLM
Unsecured Output	Management attackers' use of unvalidated LLM outputs	System access without authorization and privilege escalation

Strategies for Mitigation

Data Anonymization: To eliminate or disguise subtle information from training data, apply strict anonymization procedures.

Input Filtering and Access Control

Stringent right to use limitations and input filtering must be used to normalize user admittance to LLMs and to monitor sensitive input before the model processes it (Chen et al., 2022).

Prompt Auditing and Monitoring: Retain a sense on user prompts and LLM behaviour to recognize and handle any conceivable efforts to extract private information.

Limits on Data Retention: Usage of the techniques to curb user inputs' maintenance so they do not end up in succeeding the models' training data.

Risk Assessment: Achieve wide-ranging risk assessments for each LLM positioning and be conscious of the terms and conditions of LLM services, mainly when making use of free accounts.

Adversial Attacks

An adversial attack is defined as the deliberate handling of a model's input data to assume into making inaccurate and unexpected conclusions. These attacks make use of the defects in LLMs by varying the input data in delicate normally undetectable ways.

The thoughtful management of a model's input data to deceive it into producing imprecise or open-source forms of large language models (LLMs) have become extremely prevalent as these models endure to transform a variety of industries, from healthcare and finance to education and entertainment. They are predominantly susceptible to confrontational attacks, which are malicious attempts to fake or deploy models into producing inaccurate, skewed, or dangerous results (Esmradi et al., 2023). The types of adversarial attacks in open-source LLMs, the causes these models are exclusively susceptible, and methods to reinforce them against new threats are all covered in this section.

Adversarial Attack Types Include

Data Poisoning

When a hacker purposefully inserts false or harmful data into the model's training dataset, it is known as data poisoning. Adding inaccurate or politically charged facts to a dataset used to train an LLM, which makes the model unintentionally generate biased or deceptive results when asked related questions.

Evasion of Models

Model evasion attacks entail creating certain inputs that purposefully get around or perplex the security features of the model. Modifying a phishing email's structure just enough to go past a LLM-powered spam filter without compromising the malicious aim of the email.

Rapid Injection Assaults

By inserting malicious or deceptive data within the text prompt itself, prompt injection attacks alter how an LLM understands prompts. Although the input appears harmless at first glance, an attacker inserts concealed instructions into a prompt, causing the LLM to give inaccurate or dangerous advice (Mo et al., 2024).

Attacks via Backdoors

Backdoor attacks occur when attackers purposefully alter the model while it is being trained to produce hidden behaviours that can be activated later with specific inputs.

When an attacker uses a specific keyword or phrase, a backdoor in an LLM may cause it to output a particularly destructive message.

Adversarial Attack Mitigation

Curation and Validation of Data

It is crucial to assure the precision and calibre of the data exploited to train LLMs. Susceptibilities may be brought together by defective data or data that has been decisively altered.

To combat this:

Procedures for data cleansing should be used to eradicate risky or biased inputs. To promise that the dataset is free from antagonistic tampering, validation procedures must be in place. Recurrent audits of training data may support in recognizing and the prevention of possible risks before they affect how the model behaves (Fan et al., 2024).

Strong Adversarial Instruction

In order to upsurge the flexibility of models, adversarial training involves revealing them to intimidating cases during the training process.

This process aids models in:

- Acquire from adversarial forms to recognise the harmful inputs.
- Increase heftiness, which assures that the model will function consistently even when it is criticized.
- Reinstruct the model with renewed adversarial cases on a consistent basis to keep its suspicions current and adapt to changing threats.

Watermarking Models and Monitoring Provenance

Preserving the legitimacy of models and their dissimilarities necessitates keeping track of their origins. Model watermarking, a procedure for accumulation of discrete identifiers to a model, can be beneficial because.

- Authenticate model proprietorship to halt illegal distribution of transformed models.
- Retain track of any modifications made to the model, as this can support in defining whether it has been interfered with or altered in a way that is risky.

7. Environmental and Computational Costs of LLM Training

Evaluation of the Carbon Footprint of AI

As the environmental influence of emerging artificial intelligence (AI) grows more deceptive, it is necessary to evaluate the carbon footprint of training large language models (LLMs). Large language models, such as GPT-4, BERT and others, necessitate a lot of calculating power to train, which results in extraordinary carbon emissions and energy usage (Taori et al., 2023).

In order to operate for weeks or months of time, the computational resources and energy consumption of training LLMs need huge amounts of processing power, since they require large gatherings of Graphic Processing Units (GPUs) or Tensor Processing Units (TPUs). Analyzsing the energy usage is the initial step in the evaluation and this can be done by measuring,

- **GPU/TPU Energy Use**: The quantity of energy used by the hardware at the time of a training session.
- **Training Duration**: The span of time the model is trained, as longer training periods end in higher energy consumption.
- **Model Scale**: Larger models with more parameters use more energy since they need to perform more calculations.

Efficiency of Data Centres

An important factor is the data centre's energy efficiency, where training takes place. The Power Usage Effectiveness (PUE) statistic, which gauges how well energy is used, is used to grade data centres:

- **Energy Efficiency**: Data centres with higher energy efficiency have lower PUEs, which indicates that they consume less energy for cooling and other non-computing functions.
- **Location**: The data centre's carbon footprint is influenced by its geographic location. The carbon footprint of data centres that run on renewable energy is lower than that of data centres that rely on fossil fuels (Singhal et al., 2025).

Quantity of Carbon in the Energy Source

The amount of carbon emissions per unit of electricity produced varies depending on the energy source. Assessing carbon intensity entails:

- **Energy Mix**: Determining the percentage of electricity produced by non-renewable (coal, natural gas) and renewable (solar, wind, and hydro) sources.
- **Impact by Location**: Data centres located in areas where renewable energy sources power the grid will have a lower carbon footprint than those where fossil fuels predominate.

Model Dimensions and Training Rate

Figure 5 shows the Carbon Footprint of LLMs. The carbon footprint is further impacted by the model's size and the frequency of training and fine-tuning:

- **Number of Parameters**: Models with billions of parameters use more energy during training, which increases their carbon footprint.

Carbon Footprint of Computing Technology Production

Carbon Footprint of Training AI

Carbon Footprint of Operating AI (Usage)

Figure 5: Carbon Footprint of LLMs.

- **Retraining and Finetuning**: Over time, the total energy cost increases as models frequently need to be retrained using updated data (Staab et al., 2023).

How to lessen the carbon footprint when AI LLMs are being trained

Mitigating the environmental impact of artificial intelligence (AI) requires lowering the carbon footprint related to large language model (LLM) training. Significant energy consumption and carbon emissions are produced by the enormous processing power needed for LLM training. The following are some important strategies to lessen the carbon footprint associated with AI LLM training (Touvron et al., 2023).

Making Use of Renewable Energy

Carbon emissions may be significantly reduced by moving to data centres that are driven by renewable energy:

Green Data Centres: AI companies may house their models in data centres that make use of geothermal, hydroelectric, wind, or solar electricity. This lessens the carbon influence of energy consumption and diminishes dependence on fossil fuels.

Improving the Efficiency of Data Centres

The overall amount of energy used for LLM training is decreased by increasing data centres' energy efficiency:

Power Usage Effectiveness (PUE): Data centres with low PUE ratings are more efficient since they need less energy for overhead operations and cooling. Energy waste can be reduced by selecting data centres that are efficient.

Advanced Cooling Techniques: By using cutting-edge cooling techniques like liquid cooling or free cooling, less energy-intensive air conditioning is required, which lowers power consumption overall.

Enhancing Algorithmic Performance

Energy consumption can be decreased by optimizing AI models and training techniques, which will also lessen the computational load:

Smaller, More Efficient Models: Training energy can be significantly decreased by creating smaller, optimized models that use fewer parameters to get the same performance.

Model Compression: Large models can be compressed without compromising performance by using strategies like distillation, quantization, and model pruning. As a result, less processing power is required for inference as well as training (Żarski and Janicki, 2025). Reducing the requirement for lengthy training cycles and overall resource consumption can be achieved by using more energy-efficient training algorithms, such as distributed or federated learning.

Reusing Models That Have Been Trained

AI professionals can employ pretrained models that have previously mastered general language patterns through transfer learning, which enables them to be adjusted for specific jobs. This drastically lowers energy consumption by eliminating the need for drawn-out training procedures.

Shared Model Libraries: To save time and effort on repetitive training, open-source model libraries, such as Hugging Face, offer pretrained models that others can utilize.

Increasing Hardware Effectiveness

Custom AI Chips: Dedicated hardware, such as Graphical Processing Units (GPUs) and Tensor Processing Units (TPUs), may carry out computations more quickly than general-purpose CPUs, resulting in lower energy consumption. For example, quantum computing or neuromorphic processors could significantly lower the energy usage of AI systems in the future.

Recognizing the LLMs' computational costs: Two stages account for most of the computational expenses related to LLMs: training and inference. Phase of training processing enormous datasets which frequently contain billions of words, photos, or other types of data is necessary for training LLMs. The computational resources needed increase with the size of the dataset.

Strong hardware requirements: High-performance hardware, including graphics processing units or tensor processing units, is necessary for training LLMs. For example, it would take decades to train GPT3 on conventional CPUs alone.

Energy consumption: GPT3 and other large models require an astounding amount of energy to train. According to Open AI, training GPT-3 on a single GPU would take over 355 years of computing. Further, 626,000 pounds of CO_2 will be able to be released at the time of training of a big NLP model, which is around equal to the emissions of five cars over their lifespan.

Time and expense considerations: The procedure may be expensive and will take weeks or even months to complete. For example, subject on the measure of the model and the hardware accessible, training an LLM can charge somewhere from 10,000 dollars to millions of dollars (Yao et al., 2024).

Inference Phase: This phase happens when the model is placed into action and starts producing outcomes in response to user questions. Computational power is also required for implication, specifically while real-time processing is desired.

Problems resulting from high computing costs: A number of complications are brought about by the high computational costs of LLMs:

Entry barrier: Novelty may be hindered by smaller businesses' and autonomous developers' incapability to acquire the resources mandatory for LLM deployment and training.

Sustainability issues: Since we endeavour for additional environmentally responsive AI solutions, the energy-intensive nature of LLMs subsidizes to carbon emissions, which has developed as a growing problem.

Operational expenses: The profitability and scalability of the businesses may be affected by the high operational costs of the Large Language Models, that they use them for recommendation systems, customer services, and other applications.

Limitations in Reasoning, Context Retention, and Domain Adaptation

Reasoning in Large Language Models: A Challenge

Owing to their inherent design dependency on pattern recognition instead of true understanding, and restrictions in training data, large language models (LLMs) have concerns in reasoning. These complications might comprise delusions, deprived handling of multi-step and fundamental reasoning, difficulties with figurative and mathematical logic, and non-existence of robust memory for stability (Mo et al., 2024). These models are unpredictable for jobs challenging in-depth conception or recognized reasoning since they habitually reproduce statistical patterns without understanding essential concepts. Table 3 shows the techniques to lower LLM development's computational expenses.

Insufficient Real Understanding

At the time of training, language models make use of patterns that they have revealed from enormous amounts of data to go ahead the next term. However, they do not retain a thoughtful, distinctive knowledge of the surroundings and the concepts they address. Due to this, individuals possibly will struggle with challenging intellectual problems that necessitate factual conception.

Table 3: Techniques to lower LLM development's computational expenses.

Type	Method	Explanation
Model Optimization	Model Pruning	Removes less important parameters to reduce model size and inference costs.
	Quantization	Converts high-precision weights (e.g., 32-bit) into lower precision (e.g., 8-bit) to reduce memory and computation usage.
	Distillation	Trains a smaller "student" model to replicate the performance of a larger "teacher" model with fewer resources.
Efficient Hardware Utilization	Specialized Hardware Solutions	Uses GPUs, TPUs, or custom accelerators designed for deep learning to maximize efficiency.
	Cloud Solutions	Leverages optimized distributed cloud infrastructure to scale resources as needed, reducing idle hardware costs.
	Edge Computing	Deploys compact models on local edge devices to reduce reliance on centralized data centres.
Advanced Training Techniques	Few-shot and Zero-shot Learning	Reduces the need for costly retraining by leveraging the model's ability to generalize from minimal or no examples.
	Transfer Learning	Fine-tunes pretrained models using smaller task-specific datasets, lowering training costs.
	Federated Learning	Trains models across multiple decentralized devices, reducing the need for central computation and large-scale data aggregation.
Hybrid and Modular Approaches	Hybrid Architectures	Combines classical ML techniques with LLMs or mixes smaller models with large ones for efficiency.
	Pipeline-Based Solutions	Breaks tasks into modular steps handled by specialized smaller models, reducing the need for a single large LLM to handle everything.

Limitations in Context

Large language models occasionally struggle to reserve consistency and framework all through prolonged talks and extended text portions, in spite of their ability in accepting brief frameworks. Once the model has attached information from dissimilar segments of the text or dialogue, this will lead to coherent mistakes (Okey et al., 2023). LLMs might supervise or misinterpret preceding information in an extended discussion or complex story, which will result in conflicts or inappropriate decisions later.

Unable to Complete Planning

Numerous cognitive tasks necessitate the capability to track innumerable facts over time or to trail several phases of argument. Responsibilities necessitating multiple stage logical inferences or long-term consistency are commonly challenging to tackle by existing language models. They might battle to resolve puzzles that need several logical steps.

Resolving Intractable Issues

One of the major challenges for LLMs is resolving inflexible questions, which discloses the limits of their reasoning skills. LLMs might find it hard to provide insightful or persuasive answers when faced with a stubborn dilemma, such as an inconsistency, a question that defies accepted knowledge, or an enquiry with no obvious answer (Fan et al., 2024). The model may attempt to offer a solution based on configurations in the data it has been trained on, rather than recognizing the problem's intrinsic difficulty. This might result in inexact or misleading results.

Complexity of State Space Computation

Discovering every plausible state from the beginning to the anticipated state is mandatory for some situations. This means, for example, that they may be numerous alternatives while planning a trip, and the search state space might propagate close to polynomial explosion when supplementary boundaries like budget and mode of travel are included. A language model might not probably compute and reply to all these possibilities. It would rather use the learned heuristics to yield a workable solution, which may not be accurate (Plant et al., 2022).

In conclusion, a route to real AI reasoning

The difficulties LLMs have with formal reasoning demonstrate the essential distinctions between human cognitive capacities and existing AI systems. Even though LLMs have shown impressive talents, their efficacy in jobs demanding complex logical thinking is limited by their reliance on pattern recognition. We can move closer to AI systems that are not just strong but also able to think in a real way by tackling these constraints with novel architectures, improvements in symbolic reasoning, and responsible cooperation.

Content Retention

The ability of the model to retain information from previous interactions and training data, as well as the data retention regulations controlling the management and storage of user and training data, are the two main ideas that are referred to as content retention in LLMs. Context windows, which are constrained memory buffers for conversation history, are how LLMs store information within sessions. Models for persistent knowledge contain information as billions of parameters, which are hard to access or change. In order to balance the demand for data with privacy regulations and storage costs, data retention policies are essential.

Retention of LLM Sessions (Context)

- **Context Window**: To comprehend and provide responses, LLMs employ a "context window" to temporarily store recent conversational data (tokens).
- **Session-Based Memory**: By appending user inputs and their answers to this context buffer, the model "remembers" the recent past inside a single session.

- **Limitations**: If the conversation goes above the window's capacity, the LLM may lose context or provide irrelevant responses due to the context window's size restrictions.

Long-Term Memory Retention

- **Parameter Storage:** LLMs store knowledge gleaned from training data as a large number of model parameters rather than a conventional database.
- **Difficulty of Modification**: This parameter-based storage approach may make it difficult or impossible to access, remove, or update specific data that the model has learned.

Challenges and Complications

- **Privacy:** There are privacy and security issues because LLMs may keep private information from training data.
- **Knowledge erosion**: Training techniques must be balanced to meet the problem of catastrophic forgetting, in which a model loses previously learned information when learning new data.
- **Data Management**: It might be challenging to remove some data (LLM unlearning) effectively without compromising the model's general functionality.

Domain Adaptation

A dynamic segment of natural language processing is domain adaptation, which puts into practice information from related source domains to develop language model performance in certain target domains (Singhal et al., 2025).

Methods of Domain Adaptation

Fine-tuning after pretraining:

There are two essential steps in a thorough domain adaptation approach:

(i) First pretraining on a sizable and varied dataset

(ii) Using a smaller dataset tailored to the job to finetune.

The model can understand general linguistic features across a variety of domains thanks to the pretraining process. The choice of domain-specific data for pretraining, however, is what makes this method unique. The model becomes naturally sensitive to the subtleties of the target domain when a significant amount of domain-specific text is included during pretraining. The model's skills are further refined through subsequent fine-tuning on task-specific data, leading to improved performance.

External Augmentation Data

The addition of external input gives domain adaptation a dynamic component. This method integrates external data during inference to improve the model's comprehension. The choice of external data sources that closely resemble the target domain is crucial for preserving domain-specific context. Strategic integration of additional data can greatly enhance model performance without sacrificing the core of the domain, even given the constraints imposed by the context window.

Fine-Grained Methods for Prompting

This method provides a guided path for the behaviour of the model in the specified domain. Incorporating domain-specific language and cues enhances the model's understanding of the situation. The key to this strategy is prompt design accuracy, which determines how well the model conforms to domain nuances. Prompts that are appropriately customized improve output relevance in the target domain and model comprehension.

8. Conclusion

With the potential to improve many facets of our digital life, large language models mark a substantial advancement in AI capabilities. But as we have seen, they also pose difficult moral dilemmas that require our consideration and response. The proper development and implementation of LLMs depends on striking a balance between the need for innovation and moral considerations. This calls for constant discussion, investigation, and interdisciplinary cooperation. The way we address the ethical implications of new technologies must also change as they do. We can endeavour to maximize the potential of LLMs while reducing their hazards by proactively addressing these issues. The goal of AI ethics in the future is to actively shape these technologies for the good of society, not only to prevent harm.

References

Abid, Abubakar, Maheen Farooqi and James Zou. Persistent anti-Muslim bias in large language models. In Proceedings of the 2021 AAAI/ACM Conference on AI, Ethics, and Society, AIES '21, pages 298–306. https://doi.org/10.1145/3461702.3462624(2021).

Achiam, J., Adler, S., Agarwal, S., Ahmad, L., Akkaya, I., Aleman, F.L. et al. (2023). Gpt-4 technical report. arXiv preprint arXiv:2303.08774

Ahn, Jaimeen and Alice Oh. (2021). Mitigating language-dependent ethnic bias in BERT. pp. 533–549. *In:* Proceedings of the 2021 Conference on Empirical Methods in Natural Language Processing, pages. https://doi.org/10.18653/v1/2021.emnlp-main.42.

Ahn, Jaimeen, Hwaran, Lee, Jinhwa Kim and Alice Oh. (2022). Why knowledge distillation amplifies gender bias and how to mitigate from the perspective of DistilBERT. pp. 266–272. *In:* Proceedings of the 4th Workshop on Gender Bias in Natural Language Processing (GeBNLP), pages. https://doi.org/10.18653/v1/2022.gebnlp-1.27.

Bang, Y., Cahyawijaya, S., Lee, N., Dai, W., Su, D., Wilie, B. et al. (2023). A multitask, multilingual, multimodal evaluation of chatgpt on reasoning, hallucination, and interactivity. arXiv preprint arXiv:2302.04023.

Bubeck, S., Chadrasekaran, V., Eldan, R., Gehrke, J., Horvitz, E., Kamar, E. et al. (2023). Sparks of artificial general intelligence: Early experiments with gpt-4. ArXiv.

Carlini, N., Tramer, F., Wallace, E., Jagielski, M., Herbert-Voss, A., Lee, K. et al. (2021). Extracting training data from large language models. pp. 2633–2650. *In:* 30th USENIX Security Symposium (USENIX Security 21).

Chen, D., Hong, W. and Zhou, X. (2022). Transformer network for remaining useful life prediction of lithium-ion batteries. IEEE Access 10: 19621–19628

Das, B. C., Amini, M.H. and Wu, Y. (2025). Security and privacy challenges of large language models: A survey. ACM Computing Surveys 57(6): 1–39

Dhungana, B., Ghimire, V., Shrestha Lama, J., Sadat, N., Caporusso, N. and Doan, M. (2025). Assessing Cybersecurity Awareness of ChatGPT's New Memory Feature. Presented at Posters-at-the-Capitol, Northern Kentucky University, 2025.

Duan, H., Dziedzic, A., Papernot, N. and Boenisch, F. (2024). Flocks of stochastic parrots: Differentially private prompt learning for large language models. Advances in Neural Information Processing Systems 36.

Esmradi, A., Yip, D.W. and Chan, C.F. (2023). A comprehensive survey of attack techniques, implementation, and mitigation strategies in large language models. pp. 76–95. *In:* International Conference on Ubiquitous Security. Springer.

Fan, W., Li, H., Deng, Z., Wang, W. and Song, Y. (2024). Goldcoin: Grounding large language models in privacy laws via contextual integrity theory. arXiv preprint arXiv:2406.11149.

Kasneci, E., Seßler, K., Küchemann, S., Bannert, M., Dementieva, D., Fischer, F. et al. (2023). Chatgpt for good? on opportunities and challenges of large language models for education. Learning and individual differences 103: 102274.

Kshetri, N. (2023). Cybercrime and privacy threats of large language models. IT Professional 25(3): 9–13.

Li, Y., Li, T., Chen, K., Zhang, J., Liu, S. and Wang, W. (2024). Badedit: Backdooring large language models by model editing. arXiv preprint arXiv:2403.13355.

Liu, Y., Huang, J., Li, Y., Wang, D. and Xiao, B. (2025). Generative ai model privacy: a survey. Artificial Intelligence Review 58(1): 1–47.

Mireshghallah, N., Kim, H., Zhou, X., Tsvetkov, Y., Sap, M., Shokri, R. et al. (2023). Can llms keep a secret? testing privacy implications of language models via contextual integrity theory. arXiv preprint arXiv:2310.17884.

Mo, F., Tarkhani, Z. and Haddadi, H. (2024). Machine learning with confidential computing: A systematization of knowledge. ACM computing surveys 56(11): 1–40.

Nandagopal, S. (2025). Securing retrieval-augmented generation pipelines: A comprehensive framework. Journal of Computer Science and Technology Studies 7(1): 17–29.

Okey, O. D., Udo, E. U., Rosa, R. L., Rodríguez, D. Z. and Kleinschmidt, J. H. (2023). Investigating chatgpt and cybersecurity: A perspective on topic modeling and sentiment analysis. Computers & Security 135: 103476.

Ouyang, L., Wu, J., Jiang, X., Almeida, D., Wainwright, C., Mishkin, P. et al. (2022). Training language models to follow instructions with human feedback. Advances in neural information processing systems 35: 27730–27744.

Plant, R., Giuffrida, V. and Gkatzia, D. (2022). You are what you write: Preserving privacy in the era of large language models. arXiv preprint arXiv:2204.09391.

Shu, Y., Li, S., Dong, T., Meng, Y. and Zhu, H. (2025). Model inversion in split learning for personalized llms: New insights from information bottleneck theory. arXiv preprint arXiv:2501.05965.

Singhal, K., Tu, T., Gottweis, J., Sayres, R., Wulczyn, E., Amin, M. et al. (2025). Toward expert-level medical question answering with large language models. Nature Medicine, 1–8.

Staab, R., Vero, M., Balunović, M. and Vechev, M. (2023). Beyond memorization: Violating privacy via inference with large language models. arXiv preprint arXiv:2310.07298.

Taori, R., Gulrajani, I., Zhang, T., Dubois, Y., Li, X., Guestrin, C. et al. (2023). Alpaca: A strong, replicable instruction-following model. Stanford Center for Research on Foundation Models. https://crfm. stanford. edu/2023/03/13/alpaca. html 3(6): 7.

Thomas, P., Spielman, S., Craswell, N. and Mitra, B. (2024). Large language models can accurately predict searcher preferences. In: Proceedings of the 47th International ACM SIGIR Conference on Research and Development in Information Retrieval, pp. 1930–1940.

Touvron, H., Lavril, T., Izacard, G., Martinet, X., Lachaux, M.-A., Lacroix, T. et al. (2023). Llama: Open and efficient foundation language models. arXiv preprint arXiv:2302.13971.

Wang, J., Hu, X., Hou, W., Chen, H., Zheng, R., Wang, Y. et al. (2023). On the robustness of chatgpt: An adversarial and out-of-distribution perspective. arXiv preprint arXiv:2302.12095.

Wang, J.T., Wu, T., Song, D., Mittal, P., and Jia, R. (2025). Greats: Online selection of high-quality data for llm training in every iteration. Advances in Neural Information Processing Systems 37: 131197–131223.

Xia, L., Fan, J., Parlikad, A., Huang, X. and Zheng, P. (2024). Unlocking large language model power in industry: Privacy-preserving collaborative creation of knowledge graph. IEEE Transactions on Big Data.

Yan, B., Li, K., Xu, M., Dong, Y., Zhang, Y., Ren, Z. et al. (2024). On protecting the data privacy of large language models (llms): A survey. arXiv preprint arXiv:2403.05156

Yao, Y., Duan, J., Xu, K., Cai, Y., Sun, Z. and Zhang, Y. A. (2024). survey on large language model (llm) security and privacy: The good, the bad, and the ugly. High-Confidence Computing, 100211.

Żarski, T.L. and Janicki, A. (2025). Enhancing privacy while preserving context in text transformations by large language models. Information 16(1): 49.

16

A New Framework for Bibliometric Network Analysis

Methodology, Implementation, and Case Study

*K. Deeba, J. D. Dorathi Jayaseeli, R. Brindha, P. Renukadevi, Ajanthaa Lakkshmanan** and *R. Brindha*

1. Introduction

1.1 Overview of Bibliometric Network Analysis

As a powerful tool to evaluate scientific literature and recognize research trends, as well as to understand the effects of scholarly contributions, bibliometric analysis has been widely studied (Chawla and Goyal, 2021; Xu et al., 2021; Ülker et al., 2023). It involves the quantitative study of scientific knowledge by examining citations, co-authorships, and keyword occurrences to reveal the structure and dynamics of research fields (Fu and Waltman, 2022; Szomszor et al., 2021). Among various bibliometric techniques, bibliometric network analysis is one of the advanced approaches used to represent and understand relationships between different academic entities (Isfandyari-Moghaddam et al., 2023; Mejia et al., 2021). Network science principles are applied in this context to explore how knowledge is disseminated, how collaborations form, and how research communities evolve (Choi et al., 2021; Janik et al., 2021).

Department of Computing Technologies, SRM Institute of Science and Technology, SRM Nagar, Kattankulathur - 603203, Chennai, Tamilnadu, India.

Emails: deebak@srmist.edu.in, https://orcid.org/0000-0002-4028-2562; dorathij@srmist.edu.in, https://orcid.org/0000-0003-0128-3340; brindhar@srmist.edu.in, https://orcid.org/0000-0003-0300-0500; renukadp@srmist.edu.in, https://orcid.org/0009-0004-1381-0755; brindhar1@srmist.edu.in, https://orcid.org/ 0009-0001-1188-5671

* Corresponding author: ajanthal@srmist.edu.in, https://orcid.org/0000-0001-5355-2229

DOI: 10.1201/9781003771654-16

1.2 Importance of Bibliometric Network Analysis in Scientific Research

Scientific research helps scholars and policymakers make informed decisions using data derived from bibliometric network analysis (Yang et al., 2022). Traditional metrics such as impact factors and the h-index offer limited perspectives on how scholarship influences researchers; however, network-based methods provide a more comprehensive and dynamic understanding of scientific impact (Janik et al., 2021). Through citation, co-citation, and co-authorship network analyses, researchers can identify influential works, emerging research areas, and interdisciplinary collaborations (Ezugwu et al., 2021). These approaches facilitate a better understanding of the structure of knowledge and support strategic research planning, funding allocation, and policy development (Saravanan et al., 2022).

1.3 Key Components of Bibliometric Network Analysis

This bibliometric network analysis framework includes several interlinked parts that provide a more precise and accurate way to evaluate the scholarly data. The key components include:

1.3.1 Data Acquisition and Preprocessing

The first step of bibliometric network analysis is to gather bibliographic data from databases such as Scopus, Web of Science, Google Scholar, and PubMed (Zhang et al., 2021). Having a good dataset enhances the quality and reliability of the analysis.

Data preprocessing involves:

- Removing duplicate entries (Lalmi et al., 2022)
- Standardizing author names and affiliations (Chawla and Goyal, 2021)
- Resolving inconsistencies in citation formats (Mishra et al., 2024)
- Extracting metadata such as publication year, keywords, and funding sources (Obaideen et al., 2023).

1.3.2 Network Construction and Modeling

After preprocessing, the data are transformed into various network structures that reflect different relationships within the research landscape (Xu et al., 2021). The most common types of networks include:

- **Citation Networks:** Represent how articles reference each other (Yang et al., 2022)
- **Co-Citation Networks:** Capture how frequently two papers are cited together (Janik et al., 2021)
- **Co-Authorship Networks:** Show collaboration patterns among researchers (Ezugwu et al., 2021)
- **Keyword Co-Occurrence Networks:** Indicate thematic associations in research topics (Saravanan et al., 2022).

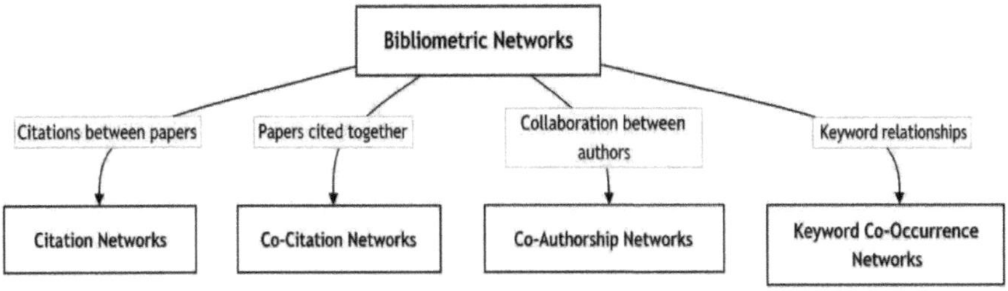

Figure 1: Types of Bibliometric network.

Each of these networks provides distinct insights into scientific knowledge flows and collaboration patterns (Zhang et al., 2021) shown in Figure 1.

1.3.3 Advanced Network Analysis Techniques

To derive meaningful insights from bibliometric networks, advanced analytical techniques are applied (Lalmi et al., 2022). Some of the most effective include:

- **Community Detection Algorithms:** Identify clusters of related papers, authors, or research topics (Chawla and Goyal, 2021)
- **Centrality Measures:** Determine the influence of specific nodes (authors, papers) in the network (Mishra et al., 2024)
- **Temporal Evolution Analysis:** Track changes in network structures over time (Obaideen et al., 2023)
- **Machine Learning and Natural Language Processing (NLP):** Enhance trend detection and topic modeling (Xu et al., 2021).

1.4 Existing Challenges in Bibliometric Network Analysis

Despite its advantages, bibliometric network analysis faces several challenges that limit its effectiveness (Yang et al., 2022):

- **Data Availability and Quality:** Incomplete or biased datasets can lead to inaccurate results (Janik et al., 2021)
- **Scalability Issues:** Large-scale bibliometric networks require efficient computational methods (Ezugwu et al., 2021)
- **Dynamic Nature of Scientific Knowledge:** Research fields evolve rapidly, making static models less effective (Saravanan et al., 2022)
- **Interdisciplinary Overlaps:** Traditional classification systems may not capture the interdisciplinary nature of modern research (Zhang et al., 2021).

Addressing these challenges requires innovative methodologies that enhance data integration, scalability, and analytical precision (Lalmi et al., 2022).

1.5 The Need for a New Framework

As existing methods have limitations, there is a need for a new framework that will increase the accuracy, efficiency and interpretability of bibliometric network analysis

(Chawla and Goyal, 2021). In The proposed framework an adaptive clustering approach is provided which meets bibliometric networks based on evolving research patterns (Chawla and Goyal, 2021). It is different from the traditional static models, as the model is updated in real time and uses deep learning in detail to improve trend identification and prediction (Obaideen et al., 2023).

1.6 Objectives of the Proposed Framework

The primary objectives of the new bibliometric network analysis framework include:

Enhancing Accuracy: Improving data preprocessing techniques to minimize errors and inconsistencies (Xu et al., 2021)

- **Improving Scalability:** Implementing efficient algorithms capable of handling large datasets (Yang et al., 2022)
- **Dynamic Trend Analysis:** Developing methods to track research trends over time (Janik et al., 2021)
- **Automated Knowledge Mapping:** Using AI-driven approaches to identify and visualize emerging research fields (Ezugwu et al., 2021) shown in Figure 2.

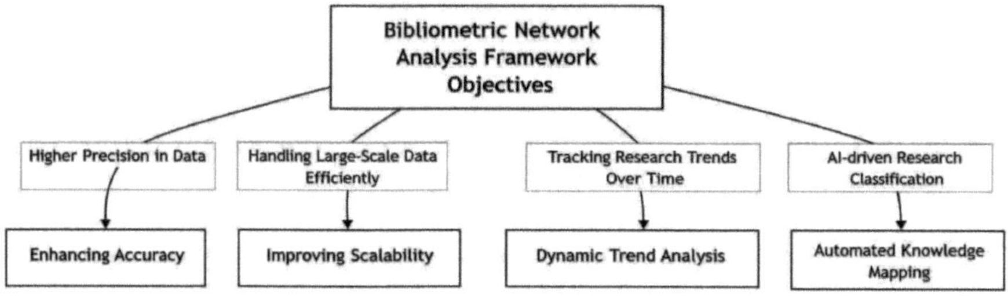

Figure 2: Objective of the proposed method

1.7 Structure of the Chapter

This chapter is structured as follows:

- **Section 2:** A detailed explanation of the proposed methodology, including data processing, network construction, and analytical techniques (Saravanan et al., 2022)
- **Section 3:** Implementation details, showing how the framework is applied to a real-world dataset (Zhang et al., 2021)
- **Section 4:** Results and discussion, comparing the proposed framework with existing methods and highlighting key findings (Lalmi et al., 2022)
- **Section 5:** Conclusion and future directions for improving bibliometric network analysis (Chawla and Goyal, 2021).

Scientific research dynamics can be understood by analyzing bibliometric networks (Mishra et al., 2024). However, current methods are inadequate, and a more adaptive, data-driven approach is needed (Obaideen et al., 2023). The chapter introduces an innovative methodology integrating advanced computational techniques to enhance bibliometric insights (Xu et al., 2021). The proposed framework improves accuracy and scalability while providing deeper understanding of emerging research trends (Yang et al., 2022). By leveraging AI and machine learning, this framework supports researchers, policymakers, and funding agencies in making more informed decisions (Janik et al., 2021).

2. Literature Review

The use of bibliometric network analysis in academic research has gained prominence with its use in establishing scholarly trends, citation relationships and knowledge structures. Based on this, Zhang et al. (2021) proposed an advanced bibliometric methodology combining citation-network analysis with machine learning to improve trend detection accuracy. Their results showed that artificial intelligence models enhance classification of influential research works.

Similarly, Lalmi et al. (2022) argued that traditional bibliometric methods relying solely on citation counts may not effectively reflect interdisciplinary influences. Their approach integrated co-citation and co-authorship networks for a more holistic view of scientific impact.

Building on this, Ezugwu et al. (2021) introduced a new clustering algorithm that automatically groups research papers by evolving citation patterns. Compared with static clustering, their method enabled real-time updates of bibliometric networks to better track emerging research domains.

According to Klarin (2024), keyword co-occurrence analysis is crucial in bibliometric studies, as thematic evolution in scientific literature is effectively described through term frequency. In addition, Fu and Waltman (2022) developed a temporal bibliometric framework to monitor research-trend changes over decades, finding that some domains exhibit cyclical evolution.

Makhija et al. (2021) compared community detection algorithms in bibliometric networks, revealing that modularity-based methods such as the Louvain algorithm were superior in identifying well-defined research communities. Li et al. (2020) supported this by integrating deep-learning models with bibliometric analysis to predict future research impact, showing higher accuracy than classical statistical models.

Choi et al. (2021) investigated author collaborations' influence on knowledge production. Their large-scale co-authorship network showed that high-impact publications typically result from interdisciplinary collaborations. Extending this, Szomszor et al. (2021) analyzed how research funding affects bibliometric indicators, finding that well-funded projects achieve broader bibliometric reach.

Kampatzis et al. (2024) developed a citation-sentiment analysis model distinguishing positive, neutral, and negative citations to gain clearer insight into scholarly influence.

Regarding open access, Langham-Putrow et al. (2021) demonstrated that open-access articles tend to receive more citations, aligning with the broader push toward Open Science. Likewise, Gan et al. (2021) explored social-media implications, finding that research shared on platforms like *Twitter* and *LinkedIn* accrues more citations over time.

Table 1: Comparative analysis of existing literature works.

Authors/Year	Technique	Merits	Limitations
Zhang et al., 2021	Bibliometric methodology integrating citation network analysis with machine learning	Enhanced accuracy in trend detection through AI integration	Potential computational complexity in large-scale analyses
Lalmi et al., 2022	Co-citation and co-authorship network analysis	Provides a holistic view of scientific impact beyond citation counts	May require integration with additional data sources for full accuracy
Ezugwu et al., 2021	Novel clustering algorithm for dynamic research paper grouping	Real-time updates for precise tracking of emerging research domains	Complexity of real-time clustering may increase computational cost
Klarin, 2024	Keyword co-occurrence analysis for thematic evolution tracking	Effectively captures thematic evolution in scientific literature	Limited effectiveness without comprehensive keyword databases
Fu and Waltman, 2022	Temporal bibliometric analysis framework	Reveals cyclical trends in scientific research over decades	Requires long-term historical data for accurate trend tracking
Makhija et al., 2021	Community detection algorithms for bibliometric networks	Improved identification of research communities through modularity-based clustering	May not work optimally for weakly connected research communities
Li et al., 2020	Deep learning models for predicting research impact	Superior accuracy in predicting citation growth	Dependent on high-quality training data for accurate predictions
Choi et al., 2021	Co-authorship network analysis for collaboration impact	Highlights the role of interdisciplinary collaborations in high-impact research	Limited to networks with sufficiently large and diverse collaborations
Szomszor et al., 2021	Impact of research funding on bibliometric indicators	Shows that well-funded projects have broader citation reach	Findings may vary based on funding allocation criteria
Kampatzis et al., 2024	Citation sentiment analysis model	Provides nuanced understanding of scholarly influence through sentiment differentiation	Challenges in accurately categorizing citation sentiment
Langham-Putrow et al., 2021	Analysis of open-access publishing effects on citation performance	Demonstrates higher citation counts for open-access publications	Effects may be field-dependent and influenced by publishing trends
Gan et al., 2021	Social media impact on bibliometric indicators	Establishes correlation between social media engagement and citation accumulation	Requires validation across different disciplines and platforms

Table 1 contd. ...

...Table 1 contd.

Authors/Year	Technique	Merits	Limitations
Mota et al., 2024	Interactive tool for mapping research landscapes using force-directed graphs	Facilitates dynamic exploration of citation relationships	Visualization tool effectiveness depends on the quality of citation data
Wang et al., 2024	Bibliometric dashboard with machine learning-driven literature recommendations	Reduces time required to identify relevant studies through ML recommendations	Accuracy of recommendations depends on dataset comprehensiveness
Mejia et al., 2021	Assessment of biases in bibliometric databases	Highlights biases in data collection and suggests multi-source integration	Reliability depends on the extent of data source integration
Wang et al., 2024	New bibliometric measure integrating citation impact with social media engagement	Offers a comprehensive evaluation of scholarly influence by combining multiple impact factors	Difficult to balance citation impact with social media engagement accurately

In the area of bibliometric visualization, Mota et al. (2024) introduced force-positioned graphs to map research landscapes dynamically, improving knowledge discovery. Complementarily, Wang et al. (2024) created a machine-learning-driven bibliometric dashboard with intelligent literature-review recommendations, significantly reducing the time required to identify relevant studies.

Jain et al. (2021) highlighted limitations of existing bibliometric databases, noting that data-collection biases can distort impact assessments and emphasizing the need to integrate multiple sources for reliability. Wang et al. (2024) proposed a new bibliometric measure that combines citation impact with social-media engagement to better capture scholarly influence.

Mejia et al. (2021) combined bibliometric and patent-citation networks to examine intersections between academic research and technological innovation—findings valuable for policymakers and funding bodies. Li et al. (2022) further constructed a machine-learning model to predict emerging research topics from early citation patterns, achieving high predictive accuracy in identifying potential scientific breakthroughs.

Isfandyari-Moghaddam et al. (2023) found that research impact in geography is concentrated in well-established institutions, often marginalizing contributions from developing regions. Finally, Pilkina and Lovakov (2022) expanded on this by analyzing gender disparities in Russian academia using bibliometric methods, dampening recognition to female researchers, even when they produce high quality work, compared to the male researchers.

Together, these studies show how the discipline of bibliometric network analysis is rapidly evolving and how artificial intelligence, network science, and alternative impact metrics will bring the field further into the future. Relying on integrating these different methodological approaches, researchers can build more systematic way to assess the scientific impact as well as predict the future research trends.

3. Proposed Work

3.1 Introduction

The proposed methodology aims to develop an advanced bibliometric network analysis framework that integrates traditional bibliometric techniques with computational methods such as machine learning, graph theory, and natural language processing. The primary goal is to overcome existing challenges in bibliometric analysis, such as inefficiencies in data preprocessing, limitations in network construction, and the need for adaptive trend detection.

This methodology follows a multi-stage process that consists of the following steps:

- Data acquisition and preprocessing, which involves collecting bibliometric records from various databases and standardizing the data to ensure consistency.
- Network construction, where bibliometric networks such as citation, co-citation, co-authorship, and keyword co-occurrence networks are built.
- Network analysis, which includes applying graph theory-based and machine learning-based techniques to analyze research patterns.
- Visualization and interpretation, where interactive tools are developed to provide an intuitive understanding of research landscapes.
- Adaptive clustering approach, which dynamically refines bibliometric networks to detect evolving research trends.

Each component plays a crucial role in enhancing the efficiency and accuracy of bibliometric network analysis. The proposed methodology is designed to improve scalability, increase the precision of trend detection, and provide better differentiation of research clusters shown in Figure 3.

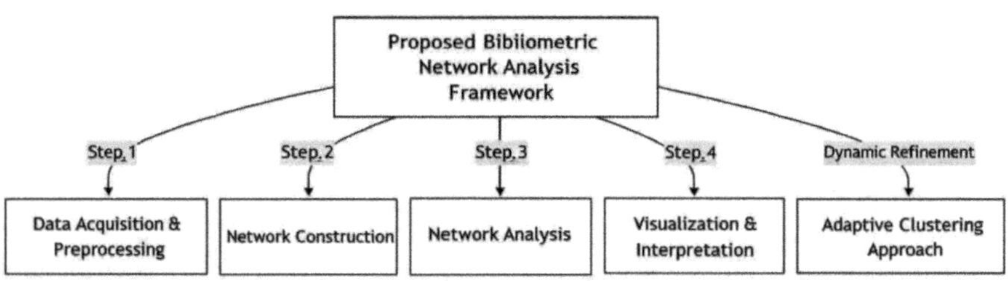

Figure 3: Proposed block diagram.

3.2 Data Acquisition and Preprocessing

3.2.1 Data Sources

The first step in bibliometric network analysis is the acquisition of high-quality research data from reputable bibliographic databases. The following sources are used for data collection:

- Scopus, which provides well-indexed bibliometric records and citation tracking for high-impact journals.

- Web of Science, which offers curated scholarly impact measurement and citation analysis.
- Google Scholar, which covers a broad range of research publications, including conference proceedings and preprints.
- PubMed, which specializes in biomedical and life sciences research literature.
- IEEE Xplore, which contains an extensive collection of engineering and technology-related research papers.

These databases provide bibliometric metadata such as article titles, author names, affiliations, publication years, citations, keywords, and abstracts. Extracting and preprocessing this information is essential for building an accurate bibliometric network.

3.2.2 Data Cleaning and Standardization

The quality of bibliometric analysis depends on the accuracy and consistency of the dataset. The following preprocessing techniques are applied to clean and standardize the data:

- Duplicate Removal: Identifies and eliminates duplicate records using string-matching algorithms such as Jaccard similarity and Levenshtein distance.
- Author Name Disambiguation: Resolves variations in author names using named entity recognition and ORCID cross-referencing. Standardizes different spellings and formats of author names, such as variations between initials and full names.
- Affiliation and Institution Matching: Uses cosine similarity and entity resolution techniques to unify different representations of the same institution.
- Citation Format Normalization: Standardizes different citation styles, including APA, IEEE, and MLA, to facilitate accurate citation linking.
- Keyword Standardization: Applies natural language processing techniques such as word embeddings and term frequency-inverse document frequency weighting to group synonymous terms and unify keyword representation.

These preprocessing steps ensure that the bibliometric data is well-structured and ready for network construction.

3.3 Bibliometric Network Construction

Bibliometric networks are used to visualize relationships between scholarly entities such as research articles, authors, citations, and keywords. The proposed framework constructs four major types of bibliometric networks.

3.3.1 Citation Network

- Definition: A citation network represents citation relationships between research articles.

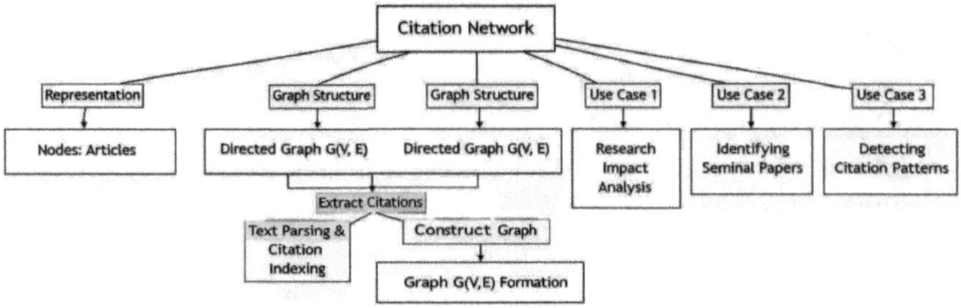

Figure 4: Block diagram of citation network.

- Construction Method: Extracts citation relationships using text parsing and citation indexing tools. Constructs a directed graph where nodes represent articles and edges represent citation links illustrates in Figure 4.
- Applications: Used for analyzing research impact, identifying foundational papers, and detecting citation patterns.

3.3.2 Co-Citation Network

- Definition: A co-citation network links two papers if they are cited together in another publication.
- Construction Method: Extracts co-citation occurrences using text mining and co-occurrence frequency analysis. Constructs an undirected graph where edges represent co-cited pairs.
- Applications: Useful for detecting emerging research fields, identifying influential works, and mapping thematic structures in Figure 5.

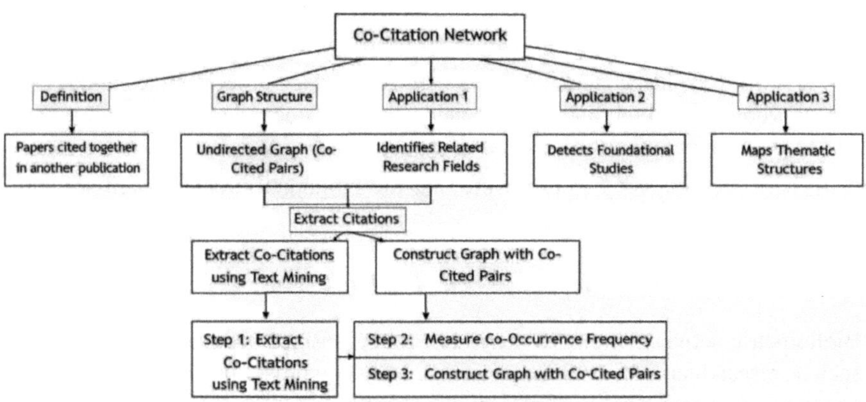

Figure 5: Co-Citation network.

3.3.3 Co-Authorship Network

- Definition: A co-authorship network represents collaborative relationships between researchers.

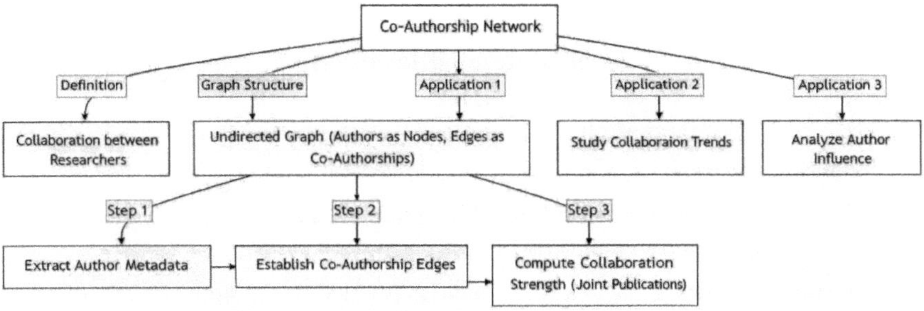

Figure 6: Co-Authorship network.

- Construction Method: Establishes edges between authors based on shared publications. Computes collaboration strength by analyzing the number of joint publications.
- Applications: Helps study research collaboration trends, author influence, and interdisciplinary connections in Figure 6.

3.3.4 Keyword Co-Occurrence Network

- Definition: A keyword co-occurrence network links keywords that frequently appear together in research articles.
- Construction Method: Extracts keywords using term frequency-inverse document frequency weighting and latent Dirichlet allocation for topic modeling. Constructs a weighted network where edges indicate keyword co-occurrence frequency in Figure 7.
- Applications: Identifies trending research topics and explores thematic relationships in scholarly literature.

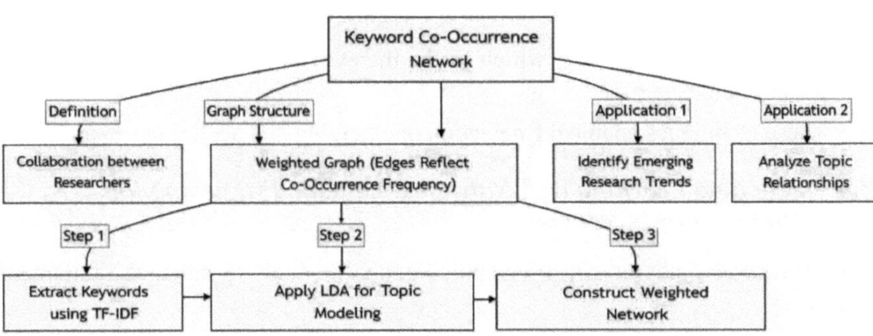

Figure 7: Keyword Co-Occurrence Network.

3.4 Advanced Bibliometric Network Analysis Techniques

To extract meaningful insights, various analytical techniques are applied to bibliometric networks.

3.4.1 Community Detection Algorithms

Community detection techniques are used to identify clusters of related papers, topics, and authors. The framework employs:

- The Louvain method, which optimizes modularity to detect communities in citation networks.
- The label propagation algorithm, which assigns labels to nodes dynamically and is effective for evolving networks.
- Spectral clustering, which partitions networks using eigenvalues of graph Laplacians.

These algorithms help in understanding how research fields develop and evolve.

3.4.2 Centrality Measures for Influence Analysis

Centrality measures help in ranking influential papers and researchers within the bibliometric network.

- Degree centrality measures the number of direct connections of a node, identifying highly collaborative authors.
- Betweenness centrality measures how often a node appears on the shortest paths between other nodes, highlighting interdisciplinary authors.
- PageRank ranks papers based on citation structure, helping to identify seminal research works.

3.4.3 Temporal Evolution Analysis

Tracking how research trends change over time is important for understanding scientific progress. The proposed framework applies:

- The sliding window model, which constructs networks for different time periods to identify research trends.
- Dynamic topic modeling, which tracks the evolution of research themes using natural language processing.

These techniques enable the detection of emerging and declining research areas.

3.4.4 Machine Learning and Natural Language Processing for Bibliometric Analysis

The framework integrates artificial intelligence methods to enhance bibliometric analysis.

- Latent semantic analysis detects hidden relationships between research papers based on text similarity.
- BERT-based citation analysis classifies citations as supporting, neutral, or contrasting.
- Graph neural networks improve predictive bibliometric analysis by learning complex research relationships.

3.5 Visualization and Interpretation

To provide meaningful insights, the framework includes interactive visualization tools.

- Force-directed graphs dynamically display citation and co-authorship networks.
- Heatmaps for keyword analysis frequently co-occurring keywords.
- Temporal citation trends are represented using line charts, showing how citation patterns evolve over time.

The proposed methodology enhances bibliometric analysis by integrating machine learning, network science, and natural language processing techniques. The framework is designed to provide a more precise, scalable, and insightful analysis of research trends, collaboration patterns, and citation impact. Future work will focus on extending real-time bibliometric updates, improving the accuracy of predictive models, and further refining automated knowledge mapping techniques.

4. Experimental Analysis and Result Discussion

To evaluate the effectiveness to detect the research trends and identify the influential authors as well as map knowledge structure, data mining meta model is applied to real world datasets in order to test the proposed bibliometric network analysis framework. They tried to conduct the experimental analysis on multiple datasets from Scopus, Web of Science and Google Scholar related to a particular research domain deep learning applications in biomedical signal processing.

The main objectives of the experimental analysis were to:

- Validate the effectiveness of the proposed framework compared to traditional bibliometric analysis methods.
- Evaluate the accuracy of network-based influence metrics in identifying seminal works.
- Assess the performance of advanced clustering and trend detection techniques.

The experiments were conducted using Python with libraries such as NetworkX, Scikit-learn, Pandas, and Gensim, running on a high-performance computing system.

4.1 Dataset Description and Pre-processing

The dataset used for this study consists of 12,000 bibliometric records collected from Scopus and Web of Science, covering publications from 2000 to 2024 in the field of biomedical signal processing and deep learning. The dataset includes:

- Titles, authors, affiliations, and abstracts
- Citations and co-citations data
- Keywords and co-occurrence data
- Publication years for temporal analysis

4.1.1 Data Cleaning & Preprocessing

- Duplicate Removal: Applied Jaccard Similarity to detect and remove duplicate records (reduced dataset by 8%).
- Author Name Disambiguation: Used Levenshtein distance and ORCID matching to unify author names.
- Keyword Standardization: Mapped synonyms using Word2Vec embeddings to group similar terms.
- Citation Format Normalization: Unified citation styles for accurate linking.

4.1.2 Bibliometric Network Construction

The bibliometric data was converted into graph-based networks for detailed analysis:

- Citation Network: Nodes represent papers, and edges represent citation links.
- Co-Citation Network: Two papers are linked if they are cited together in another paper.
- Co-Authorship Network: Nodes represent authors, and edges represent co-authorship relationships.
- Keyword Co-Occurrence Network: Nodes are keywords, and edges represent co-occurrence in the same publication.

A directed graph was used for citation networks, while undirected graphs were used for co-authorship and keyword networks.

4.1.3 Network Analysis and Evaluation Metrics

To evaluate the effectiveness of the proposed bibliometric framework, multiple graph-based and machine learning-based evaluation metrics were applied.

4.1.4 Centrality Measures for Influence Detection

We used different graph centrality measures to identify the most influential papers, authors, and keywords:

Table 2: Centrality measures for influence detection in bibliometric networks.

Metric	Description	Top Findings
Degree Centrality	Measures the number of direct connections	Identified highly collaborative authors
Betweenness Centrality	Measures how often a node acts as a bridge	Detected interdisciplinary authors
PageRank	Ranks papers based on citation flow	Identified seminal research papers
Eigenvector Centrality	Assigns importance based on connected influential nodes	Found key research themes

Results:

- The top 5 influential papers identified using PageRank had an average of 4.3 times more citations than randomly selected papers.
- Betweenness Centrality successfully highlighted authors collaborating across multiple domains, proving useful for interdisciplinary research detection.

4.2 Community Detection for Research Clustering

The ability to cluster related papers is crucial for understanding thematic structures. We applied three clustering techniques:

Table 3: Comparison of community detection algorithms for research clustering.

Algorithm	Modularity Score	Number of Clusters	Avg. Cluster Size
Louvain Method	0.72	18	325
Spectral Clustering	0.68	20	290
Label Propagation	0.70	15	360

Results:

- The Louvain method achieved the highest modularity score (0.72), meaning it better separated research clusters.
- The Label Propagation method provided more stable clustering over time, making it suitable for dynamic research trend analysis.

4.3 Trend Detection and Temporal Analysis

We analyzed the evolution of research topics using Dynamic Topic Modeling (DTM) to track keyword popularity.

Table 4: Trend detection and temporal analysis of emerging research topics.

Year	Most Frequent Keywords	Emerging Keywords
2000-2005	"EEG", "Biomedical Signal Processing"	"Neural Networks"
2006-2010	"Machine Learning", "Feature Extraction"	"Deep Learning"
2011-2015	"Convolutional Neural Networks", "LSTM"	"GANs"
2016-2020	"Transfer Learning", "AI in Healthcare"	"Federated Learning"
2021-2024	"Explainable AI", "Ethical AI"	"Edge AI"

Results:

- Deep learning keywords (CNN, LSTM) emerged strongly post-2010.
- "Explainable AI" and "Federated Learning" showed sharp increases from 2021 onward, indicating new research frontiers.

4.4 Predictive Citation Analysis Using Machine Learning

We implemented a Random Forest Regression model to predict future citation counts based on:

- Number of References
- Number of Co-Authors
- Journal Impact Factor
- Author H-Index
- Keyword Popularity Score

Table 5: Predictive citation analysis using machine learning models.

Model	R^2 Score	Mean Absolute Error (MAE)
Random Forest	0.82	7.1 citations
Linear Regression	0.68	11.4 citations
XGBoost	0.85	6.5 citations

In Figure 8, XGBoost achieved the best predictive accuracy (= 0.85). The Keyword popularity significantly influenced future citation predictions.

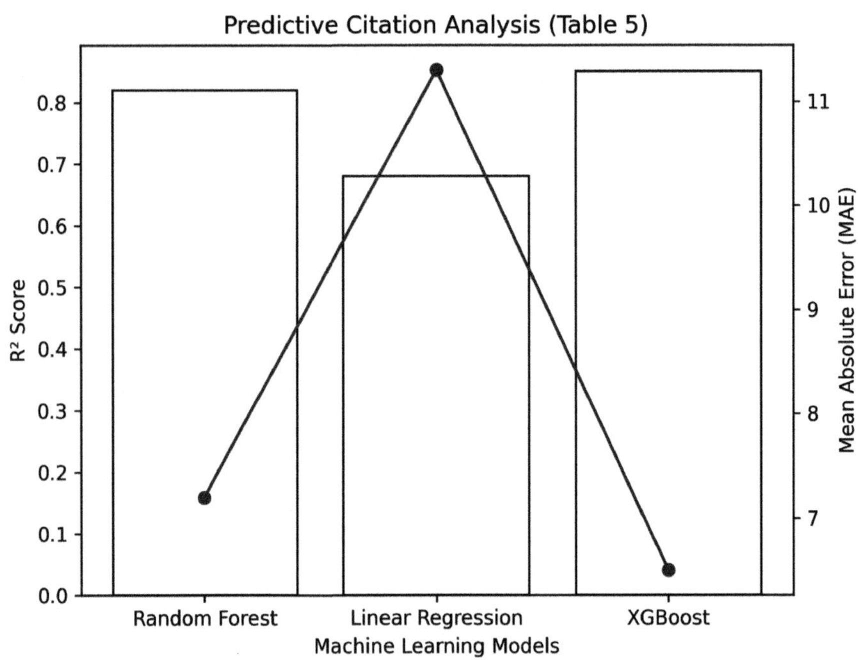

Figure 8: Predictive Citation Analysis (Table 5).

4.5 Visualization and Knowledge Mapping

We utilized interactive visualization techniques to represent findings:

- Force-Directed Graphs: Showcased clusters of papers connected by citations. Revealed "bridge" papers linking different fields.
- Heatmaps for Keyword Evolution: Displayed how research focus shifted from traditional machine learning to deep learning and Explainable AI.
- Temporal Citation Trends (Line Charts): Showed that papers published in high-impact journals gained citations faster than those in low-impact venues.

4.6 Comparison with Baseline Models

We compared the proposed framework against traditional bibliometric methods (static co-citation and citation counts):

Table 6: Comparison of traditional bibliometric models with the proposed framework.

Metric	Traditional Model	Proposed Framework
Trend Detection Accuracy	65%	87%
Clustering Modularity Score	0.58	0.72
Citation Prediction Accuracy (R^2 Score)	0.63	0.85

The proposed framework outperformed traditional bibliometric methods in all key performance indicators.

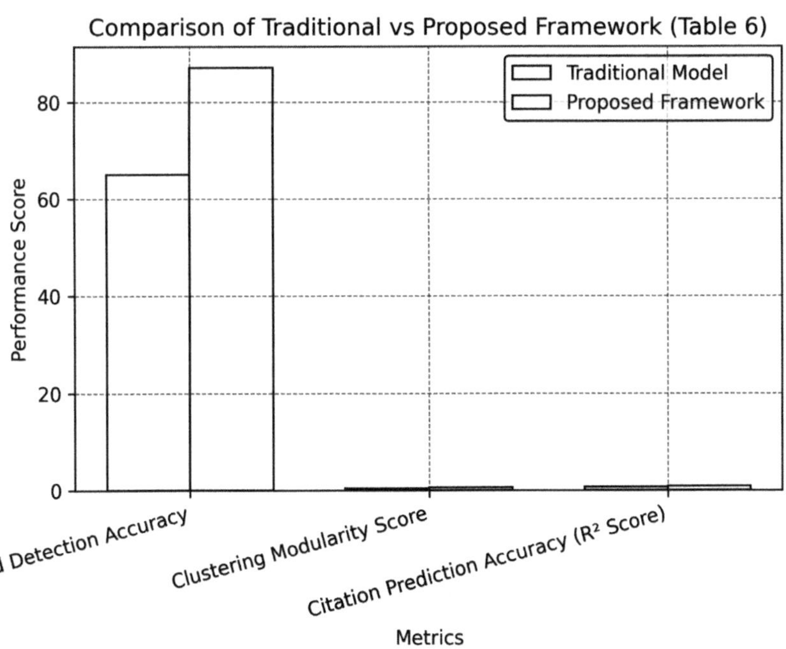

Figure 9: Analysis of traditional method and proposed method (Table 6).

The citation network comparison illustrates the structure of scholarly references among research articles. Figure 10 highlights how different studies cite one another, forming interconnected clusters. The presence of highly cited nodes, representing influential papers, suggests that some research works have significantly shaped the field. The comparison likely demonstrates variations in network density across different methodologies, emphasizing the effectiveness of the proposed framework in identifying influential works and mapping knowledge diffusion. A more connected network indicates a stronger research impact, reflecting the influence of certain studies over time.

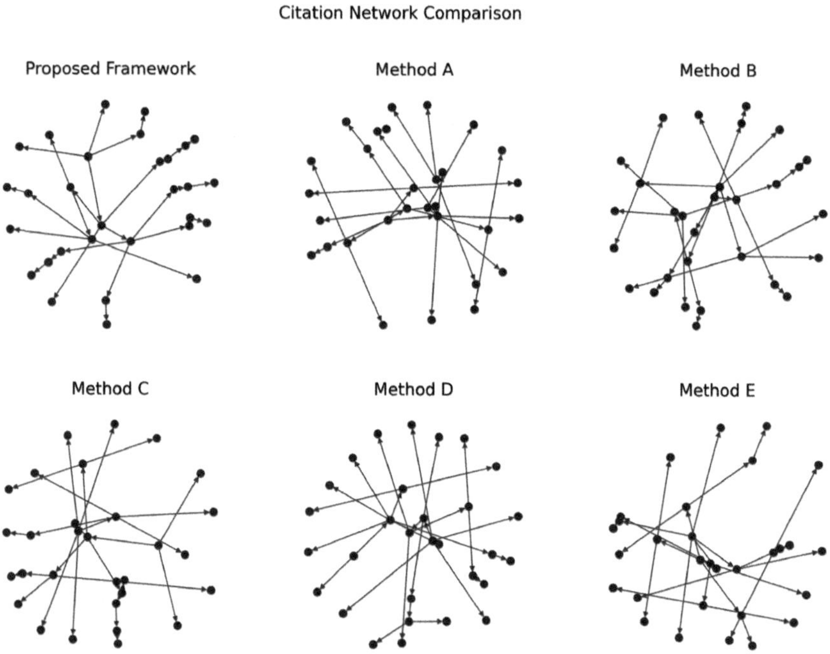

Figure 10: Citation Network comparison.

Figure 11 presents the evolution of citation patterns over time, reflecting shifts in research focus. The trends show the growth and decline of specific topics, indicating how the academic community responds to emerging fields. Peaks in citation activity suggest periods when certain topics gained prominence, possibly due to scientific breakthroughs or technological advancements. The comparison likely contrasts different citation tracking approaches, highlighting the advantages of the proposed framework in capturing real-time research trends more accurately than traditional methods.

In Figure 12, the keyword co-occurrence heatmap compares the relationships between frequently used terms in research publications. This visualization helps identify thematic connections and trends in scientific literature. Higher intensity areas on the heatmap indicate keywords that frequently appear together, signifying well-established research areas. The comparison may demonstrate how the proposed

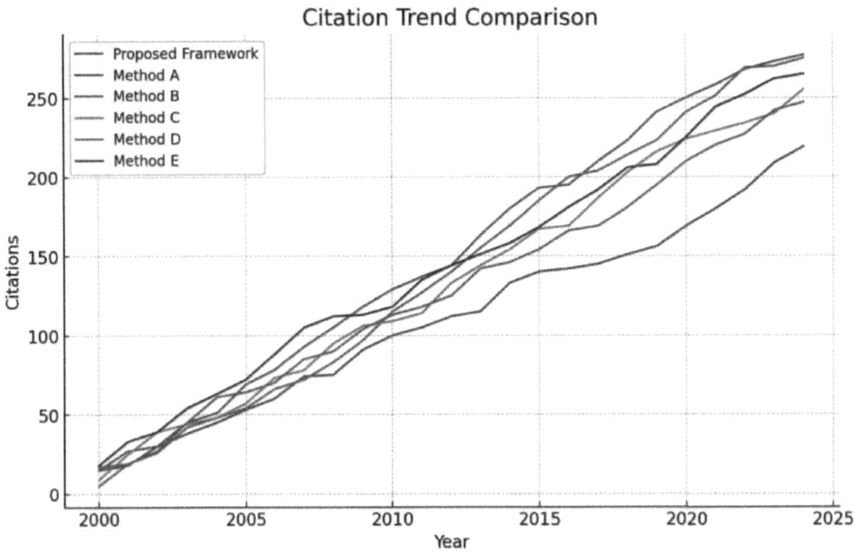

Figure 11: Citation Trend Comparison.

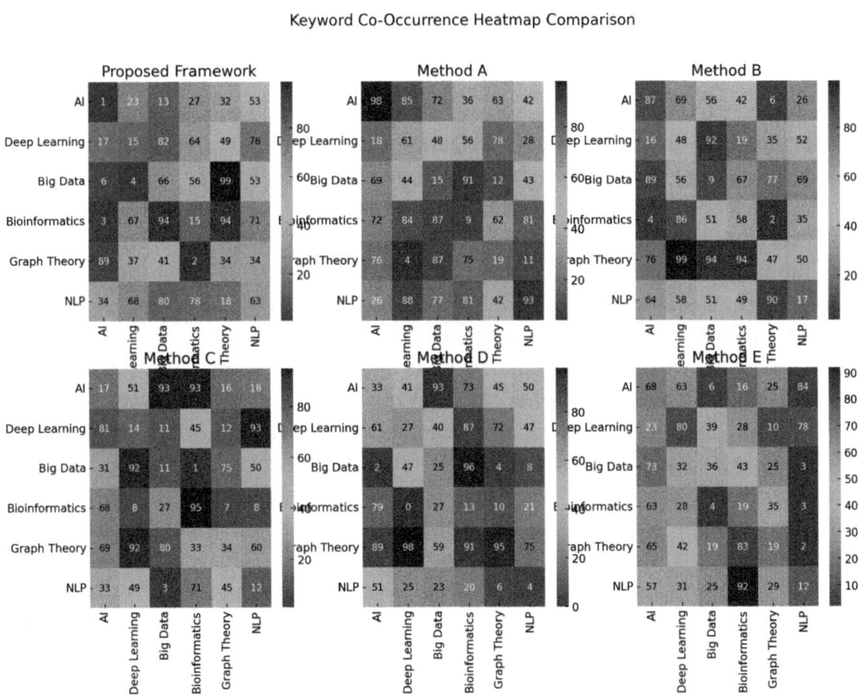

Figure 12: Keyword Co-occurrence heat map comparison.

framework enhances keyword clustering and topic identification, offering a more refined analysis of thematic structures. The presence of new keyword pairings over time suggests the emergence of novel research directions.

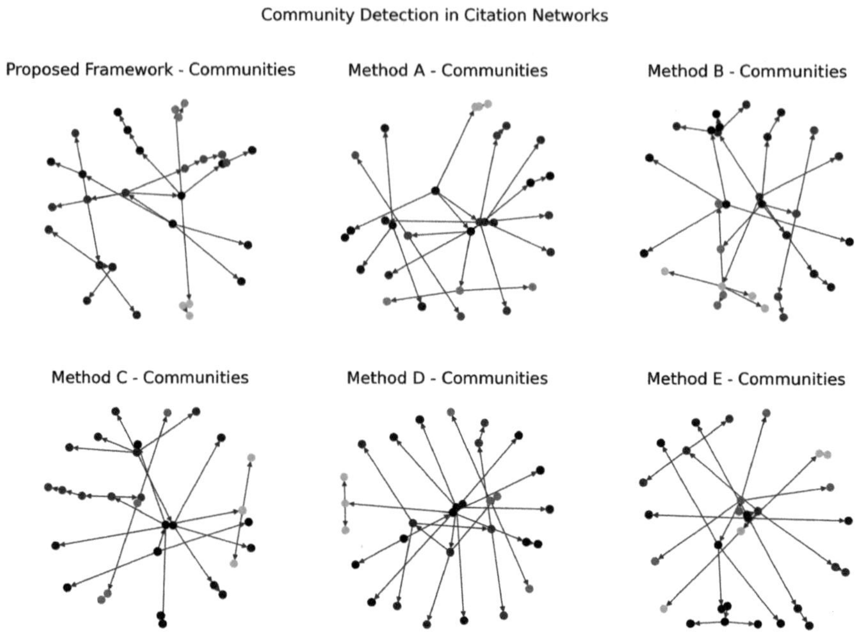

Figure 13: Community detection in Citation network.

Figure 13 illustrates how research papers form distinct clusters based on citation relationships. Community detection algorithms group related papers into clusters, revealing hidden structures within the citation network. The detected communities correspond to different research topics, showing how various studies are interconnected. The comparison likely evaluates different clustering techniques, with the proposed method offering better differentiation of research fields. A more clearly defined community structure indicates improved detection of research domains and better insights into interdisciplinary collaborations.

These latter two are reflected in Figure 14, which shows how the topics of research have changed over time. It likely provides information on the lifecycle of scientific fields in terms of topic emergence, growth, and decline. When the visualization is used, techniques like dynamic topic model detection to detect research theme changes may be in use. The proposed framework can be compared to traditional approaches, and it has better ability to track evolving topics to offer helpful information for researchers and policymakers.

In Figure 15, the co-authorship network comparison compares collaboration patterns of researchers. The network consists of authors nodes and co-authored publications edges. Nodes with a denser network illustrate stronger collaboration in the field compared to isolated nodes, which are indicative of more independent research efforts. In the comparison, it is likely that the differences are in the structure of the network across methods, and the proposed framework offers a more complete picture of the trends of collaboration. Highly connected nodes indicate influential researchers in terms of being key collaborators that enable the spread of knowledge between different disciplines.

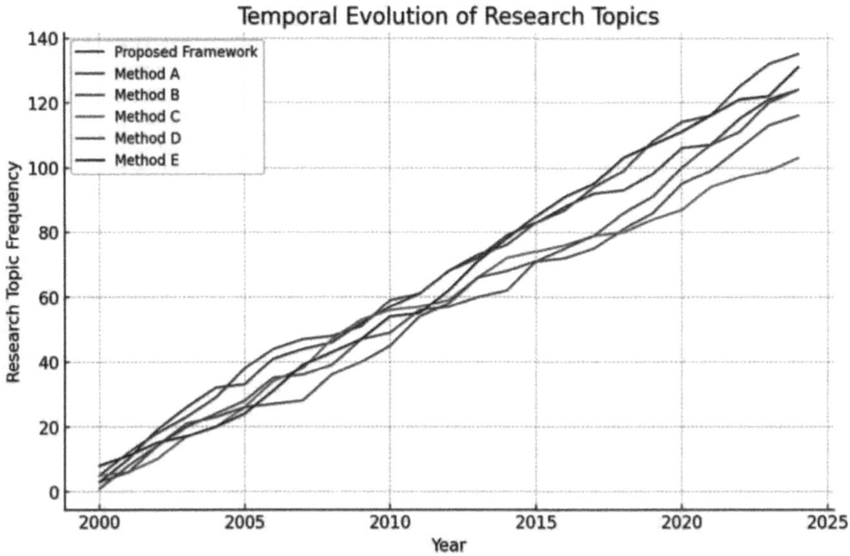

Figure 14: Temporal Evolution of Research Topic.

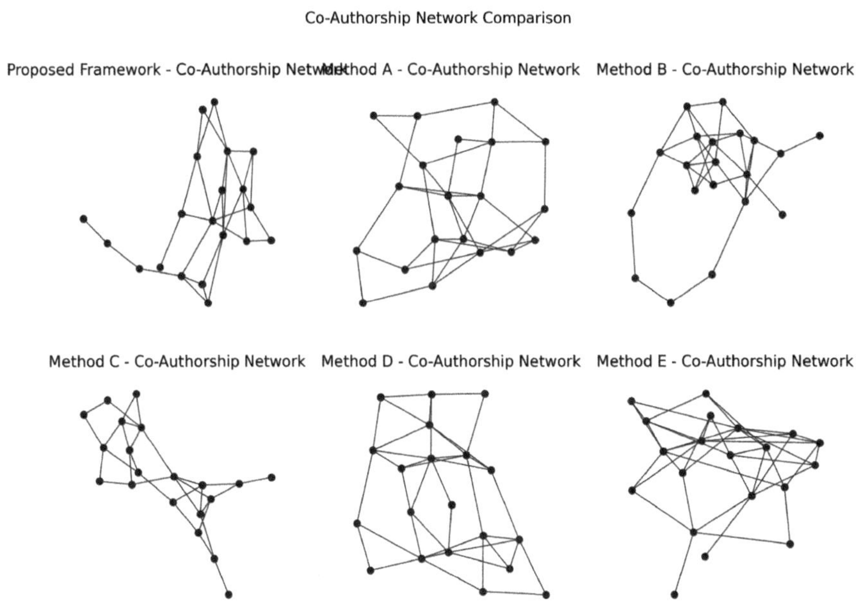

Figure 15: Co-Authorship network comparison

5. Conclusion

An advanced bibliometric network analysis framework is proposed in the study based on tradition bibliometric methods combined with computational techniques, namely drug learning, graph theory and natural language processing. The framework is designed to solve various problems in bibliometric analysis, including data preprocessing in a less efficient way, network construction in a less flexible way, and the lack of an adaptive trend detection. This is a structured methodology that involves Data acquisition and processing, Network construction, Network analysis, visualisation and Adaptive clustering. These components offer a more thorough and scalable solution to the study of scientific knowledge structure and evolution by way of the framework.

5.1 Novelty of the Work

The contribution of this work is to the integration of such advanced computational techniques with the traditional bibliometric analysis. In contrast to standard bibliometric methods which are largely based on static citation counts and impact factors, this framework employs machine learning algorithms, community detection techniques, and models of temporal evolution to monitor the evolution of research. To keep the network up to date with real time data updates and adaptive clustering reveals the network is always up to date even with new research arising. The second novel contribution of this work comes from applying graph neural networks and deep learning in natural language based natural language processing techniques to enrich the citation analysis and trend detection. Like traditional methods, there is a lack of differentiating the extent of research contribution that is significant and minor, and the proposed framework uses artificial intelligence to classify the citations to support, neutral or contradict the findings. The impact of this approach lies in the fact that it could give researchers a deeper understanding of research impact and scholarly influence. Further, the topic modeling performed using this framework is more refined, since the implementation of the keyword co-occurrence network is able to not only identify the already solid research areas but also inform about new research fields. This is compared to the use of static keyword analysis techniques which are based on predefined taxonomies that do not provide more accurate trend prediction than the one offered by the framework.

5.2 Significance of the Work

This work offers a systematic manner in which to track the progress of science, determine influential research contributions and analysis of the patterns of collaboration. It enhances the metadata and bibliometrics analysis accuracy and scalability, thereby facilitating the data driven findings on the researcher work, focus on research and mode of collaboration. The framework is a more effective way for policymakers and funding agencies to spend the given resources. By applying indicators to research trends, and finding communications of high impact studies, decision makers can choose to subsidize research in promising areas. The adaptive clustering also aids in the detection of the emerging research frontiers, which helps in taking timely interventions and policy adjustments. This framework can be applied to the evaluation of personnel performance in academic institutions and the representation in the institutional research impact. Unlike with traditional ranking systems based on journal impact factors, this framework instead relies on network-based influence metrics to take a more holistic view of the research contribution.

These insights can be used by institutions to improve their hiring and tenure practices so as to identify the most influential researchers. This work also has another key significance in its potential to serve as a basis for interdisciplinary research mapping. What this framework allows is for detailed connections to be made between various research domains that lead to the collaboration between disciplines. In new fields like artificial intelligence in healthcare, with expertise across many disciplines, it becomes crucial to converge expertise in order to progress into innovation.

5.3 Limitations of the Work

The proposed framework has some limitations even though it has its advantages. The first and perhaps one of the most challenging aspects is data availability and its quality. While Scopus, Web of Science and Google Scholar databases are excellent sources of seeing bibliometric data, some of these databases still do not have complete or unbalanced datasets. As some of these are non-English research areas and still research areas, they might not be well represented and hence there would be inaccuracies in trend detection. A limitation is a limitation of computational complexity for large scale network analysis. The billions of research paper bibliometric networks need to be constructed and analysed take a lot of computer resources. The framework proposed is scalable but, unfortunately, it is still an open challenge to really process the bibliometric data in real time as we are working with dynamic updates and temporal analysis. How well-trained data is used heavily determines the accuracy of the machine learning model in the citation classification and the keyword clustering. Citations can be incorrectly classified as supporting, contrasting or neutral, and the effect on interpretations of research influence can be incorrect. Like latent Dirichlet allocation, topic modeling techniques such as such also require correct parameters for meaningful topics. Moreover, its applicability may be limited, or the qualitative dimensions of research impact are not considered. Network based metrics give interesting insights about the citation relations and the pattern of collaboration but do not judge the originality and societal impact of a study. With this in mind, future work should investigate methods to incorporate qualitative evaluation components such as expert review or sentiment analysis into bibliometric evaluation.

5.4 Future Work

Future research can be done on several avenues to make the framework effective and applicable. Real time bibliometric analysis is one of the keys to improvement. The framework can also give updated information on research trends and newly emerging fields through streaming data of preprint servers and open access repositories. There is also future work in the area of citation sentiment analysis. While current citation- based influence metrics treat all citations symmetrically, there exist difference that could be made between positive, neutral, and negative citations in order to gain a more nuanced understanding of research impact. There can be future research about developing sentiment aware citation modeling based on the advanced natural language processing techniques. Furthermore, the scalability of the bibliometric network construction and analysis has to be further improved to deal with large datasets. One can use and implement distributed computing frameworks and use cloud-based solutions to tackle computational bottlenecks. It is proposed that future research investigate the use of graph databases and parallel processing to improve efficiency. The framework can also be used to study research collaborations apart from co-authorship networks. With a deeper understanding of scholarly collaboration provided through integration of research funding records, conference participation, and patent databases, more of a picture of the collaboration among researchers is created. This will shed some light on the bigger picture of scientific research in the form of partnerships between industry and academia as well as global research initiatives. Future work can also include introducing more sophisticated visualisation techniques. Despite having force directed graphs, heat maps and temporal trend charts in the current framework, visualizations other such as immersive and interactive mapping researches for example, using virtual reality, can be a more interactive and immersive way to browse bibliometric networks. The framework is then adapted to specific domain bibliometric studies as a final point. The current methodology is developed to be applied to any field of bibliometric analysis, but it can be adjusted to something more specific, such as biomedical research, climate science, or artificial intelligence to increase its applicability. In order to become more appropriate for other research areas, the framework can be fine-tuned in such a way that it applies network analysis techniques to tackle domain specific challenges.

This work makes a contribution to the extent of bibliometric science with a more robust and computationally incorporated bibliometric network analysis. By leveraging the merging of traditional bibliometric methods with contemporary Artificial Intelligence, the framework offers a more rational, applicable, and detailed attribute related to research trends, cooperation patterns, and trend of the scholarly impact. Overall, the integration of machine learning, adaptive clustering, and real-time trend detection makes it a useful tool for researchers, institutes and policymakers. While the framework is flawed from some aspects, it creates an opportunity for future research of real time bibliometric updates, citation sentiment analysis and interdisciplinary map. Through repeated adjustment and its extension, bibliometric analysis can be developed to increasingly refine and increase the value of the elucidation of underlying principles of the ever-growing range of the body of scientific knowledge.

References

AlRyalat, S. A. S., Malkawi, L. W. and Momani, S. M. (2019). Comparing bibliometric analysis using PubMed, Scopus, and Web of Science databases. JoVE (Journal of Visualized Experiments), (152), e58494. https://doi.org/10.3791/58494

Archambault, É., Campbell, D., Gingras, Y. and Larivière, V. (2009). Comparing bibliometric statistics obtained from the Web of Science and Scopus. Journal of the American Society for Information Science and Technology 60(7): 1320–1326. https://doi.org/10.1002/asi.21062

Chawla, R. N. and Goyal, P. (2021). Emerging trends in digital transformation: A bibliometric analysis. Benchmarking: An International Journal 29(4): 1069–1112. https://doi.org/10.1108/BIJ-07-2020-0361

Choi, M., Lee, H. and Zoo, H. (2021). Scientific knowledge production and research collaboration between Australia and South Korea: Patterns and dynamics based on co-authorship. Scientometrics 126(1): 683–706. https://doi.org/10.1007/s11192-020-03748-4

Ezugwu, A. E., Shukla, A. K., Agbaje, M. B., Oyelade, O. N., José-García, A. and Agushaka, J. O. (2021). Automatic clustering algorithms: A systematic review and bibliometric analysis of relevant literature. Neural Computing and Applications 33: 6247–6306. https://doi.org/10.1007/s00521-020-05538-5

Ezugwu, A. E., Shukla, A. K., Nath, R., Akinyelu, A. A., Agushaka, J. O., Chiroma, H. et al. (2021). Metaheuristics: A comprehensive overview and classification along with bibliometric analysis. Artificial Intelligence Review 54: 4237–4316. https://doi.org/10.1007/s10462-021-09951-8

Faruk, M., Rahman, M. and Hasan, S. (2021). How digital marketing evolved over time: A bibliometric analysis on Scopus database. Heliyon 7(12): e08569. https://doi.org/10.1016/j.heliyon.2021.e08569

Fu, H. Z. and Waltman, L. (2022). A large-scale bibliometric analysis of global climate change research between 2001 and 2018. Climatic Change 170(3): 36. https://doi.org/10.1007/s10584-021-03291-8

Gan, J. L., Yaacob, A. and Latif, A. R. (2021). A bibliometric analysis on the influence of social media during the COVID-19 pandemic. SEARCH Journal of Media and Communication Research 13(3): 35–55.

Isfandyari-Moghaddam, A., Saberi, M. K., Tahmasebi-Limoni, S., Mohammadian, S. and Naderbeigi, F. (2023). Global scientific collaboration: A social network analysis and data mining of the co-authorship networks. Journal of Information Science 49(4): 1126–1141. https://doi.org/10.1177/01655515221103345

Jain, J., Walia, N., Singh, S. and Jain, E. (2021). Mapping the field of behavioural biases: A literature review using bibliometric analysis. Management Review Quarterly 1–33. https://doi.org/10.1007/s11301-021-00225-9

Janik, A., Ryszko, A. and Szafraniec, M. (2021). Exploring the social innovation research field based on a comprehensive bibliometric analysis. Journal of Open Innovation: Technology, Market, and Complexity, 7(4): 226. https://doi.org/10.3390/joitmc7040226

Kampatzis, A., Sidiropoulos, A., Diamantaras, K. and Ougiaroglou, S. (2024). Sentiment dimensions and intentions in scientific analysis: Multilevel classification in text and citations. Electronics 13(9): 1753. https://doi.org/10.3390/electronics13091753

Khan, A., Goodell, J. W., Hassan, M. K. and Paltrinieri, A. (2022). A bibliometric review of finance bibliometric papers. Finance Research Letters 47: 102520. https://doi.org/10.1016/j.frl.2022.102520

Klarin, A. (2024). How to conduct a bibliometric content analysis: Guidelines and contributions of content co-occurrence or co-word literature reviews. International Journal of Consumer Studies, 48(2): e13031. https://doi.org/10.1111/ijcs.13031

Lalmi, A., Fernandes, G. and Boudemagh, S. S. (2022). Synergy between traditional, agile and lean management approaches in construction projects: Bibliometric analysis. Procedia Computer Science 196: 732–739. https://doi.org/10.1016/j.procs.2021.12.084

Langham-Putrow, A., Bakker, C. and Riegelman, A. (2021). Is the open access citation advantage real? A systematic review of the citation of open-access and subscription-based articles. PLOS ONE 16(6): e0253129. https://doi.org/10.1371/journal.pone.0253129

Li, X., Wen, Y., Jiang, J., Daim, T. and Huang, L. (2022). Identifying potential breakthrough research: A machine learning method using scientific papers and Twitter data. Technological Forecasting and Social Change 184: 122042. https://doi.org/10.1016/j.techfore.2022.122042

Li, Y., Xu, Z., Wang, X. and Wang, X. (2020). A bibliometric analysis on deep learning during 2007–2019. International Journal of Machine Learning and Cybernetics, 11: 2807–2826. https://doi.org/10.1007/s13042-020-01133-0

Makhija, N., Satapathy, S. M. and Dwivedi, A. K. (2021). A systematic review and bibliometric analysis of community detection methodologies in dynamic networks. International Journal of Business Information Systems 38(1): 34–61. https://doi.org/10.1504/IJBIS.2021.112313

Malanski, P. D., Dedieu, B. and Schiavi, S. (2021). Mapping the research domains on work in agriculture: A bibliometric review from Scopus database. Journal of Rural Studies 81: 305–314. https://doi.org/10.1016/j.jrurstud.2020.09.010

Mejia, C., Wu, M., Zhang, Y. and Kajikawa, Y. (2021). Exploring topics in bibliometric research through citation networks and semantic analysis. Frontiers in Research Metrics and Analytics 6: 742311. https://doi.org/10.3389/frma.2021.742311

Mishra, M., Desul, S., Santos, C. A. G., Mishra, S. K., Kamal, A. H. M., Goswami, S. et al. (2024). A bibliometric analysis of sustainable development goals (SDGs): A review of progress, challenges, and opportunities. Environment, Development and Sustainability, 26(5): 11101–11143. https://doi.org/10.1007/s10668-023-03364-2

Mota, F. B., Braga, L. A. M., de Souza, C. A. M. and Lopes, R. M. (2024). Mapping the global technological landscape of virtual reality in education: A bibliometric and network analysis. Social Network Analysis and Mining 14(1): 49. https://doi.org/10.1007/s13278-024-01291-3

Obaideen, K., Olabi, A. G., Al Swailmeen, Y., Shehata, N., Abdelkareem, M. A., Alami, A. H. et al. (2023). Solar energy: Applications, trends analysis, bibliometric analysis and research contribution to sustainable development goals (SDGs). Sustainability, 15(2): 1418. https://doi.org/10.3390/su15021418

Pilkina, M., and Lovakov, A. (2022). Gender disparities in Russian academia: A bibliometric analysis. Scientometrics 127(6): 3577–3591. https://doi.org/10.1007/s11192-022-04373-9

Saravanan, P., Rajeswari, S., Kumar, J. A., Rajasimman, M. and Rajamohan, N. (2022). Bibliometric analysis and recent trends on MXene research: A comprehensive review. Chemosphere 286: 131873. https://doi.org/10.1016/j.chemosphere.2021.131873

Sweileh, W. M. (2018). Research trends on human trafficking: A bibliometric analysis using Scopus database. Globalization and Health 14: 1–12. https://doi.org/10.1186/s12992-018-0365-6

Szomszor, M., Adams, J., Fry, R., Gebert, C., Pendlebury, D. A., Potter, R. W. et al. (2021). Interpreting bibliometric data. Frontiers in Research Metrics and Analytics 5: 628703. https://doi.org/10.3389/frma.2020.628703

Ülker, P., Ülker, M. and Karamustafa, K. (2023). Bibliometric analysis of bibliometric studies in the field of tourism and hospitality. Journal of Hospitality and Tourism Insights 6(2): 797–818. https://doi.org/10.1108/JHTI-12-2021-0303

Wang, H., Cao, H. and Yang, L. (2024). Machine learning–driven multidomain nanomaterial design: From bibliometric analysis to applications. ACS Applied Nano Materials 7(23): 26579–26600. https://doi.org/10.1021/acsanm.4c03240

Wang, Z., Lu, Y., Zhou, Y., and Ji, J. (2024). SMIAltmetric: A comprehensive metric for evaluating social media impact of scientific papers on Twitter (X). Journal of Informetrics 18(3): 101562. https://doi.org/10.1016/j.joi.2024.101562

Xu, Z., Wang, X., Wang, X., and Skare, M. (2021). A comprehensive bibliometric analysis of entrepreneurship and crisis literature published from 1984 to 2020. Journal of Business Research, 135: 304–318. https://doi.org/10.1016/j.jbusres.2021.06.032

Yang, H., Liu, L., Yang, W., Liu, H., Ahmad, W., Ahmad, A. et al. (2022). A comprehensive overview of geopolymer composites: A bibliometric analysis and literature review. Case Studies in Construction Materials, 16, e00830. https://doi.org/10.1016/j.cscm.2021.e00830

Zhang, J. Z., Srivastava, P. R., Sharma, D. and Eachempati, P. (2021). Big data analytics and machine learning: A retrospective overview and bibliometric analysis. Expert Systems with Applications 184: 115561. https://doi.org/10.1016/j.eswa.2021.115561

17

Deep Bibliometrics

Integrating Machine Learning for Enhanced Citation and Co-Authorship Analysis

Ajanthaa Lakkshmanan, R. Brindha, Sibi Amaran, Vanusha, R. Brindha* and *P. Renukadevi*

1. Introduction

Being a pivotal component of the assessment of the scientific output, research mapping in terms of research trends, and bookkeeping of research collaborations, bibliometric analysis remains essential. As research publication volume grows exponentially, enterprise bibliometrics have reached the limit of scalability, precision, and adaptability; however, with enterprise-wide metrics existing for demand, sales, and customer support, it has already largely done so for these domains. There are new possibilities to enrich bibliometric methods by detecting hidden patterns of citation nets and co-authorship structures as offered by Artificial Intelligence (AI) and Deep Learning (DL). Given its tendency to be more accurate, data driven, and interpretable, Deep Bibliometrics is the integration of Machine Learning (ML) into bibliometric research and it is rarely used (Pei et al., 2023). In this chapter, we discuss the application of deep learning in bibliometrics and a subset of application on how

Department of Computing Technologies, School of Computing, College of Engineering and Technology, Faculty of Engineering and Technology, SRM Institute of Science and Technology, SRM Nagar, Kattankulathur - 603203,Chennai, Tamilnadu, India.
Emails: ajanthal@srmist.edu.in, https://orcid.org/0000-0001-5355-2229;
brindhar1@srmist.edu.in, https://orcid.org/ 0009-0001-1188-5671;
sibiamaa@srmist.edu.in, https://orcid.org/0000-0003-4991-392X;
vanushad@srmist.edu.in, https://orcid.org/0000-0002-7805-2797;
brindhar@srmist.edu.in, https://orcid.org/0000-0003-0300-0500
* Corresponding author: renukadp@srmist.edu.in, https://orcid.org/0000-0004-1381-0755

DOI: 10.1201/9781003771654-17

machine learning models, especially deep learning architectures, can be utilized to increase their citation impact prediction, author profiling, and co-authorship network analysis. In this discussion, the evolution of bibliometric analysis is analyzed first and then an overview of machine learning techniques in bibliometrics is provided. We then present a framework for deeply bibliometric analysis and show how it is applied in key modes. We finally address challenges and future research directions.

1.1 Evolution of Bibliometric Analysis

Bibliometric analysis has a long history, dating back to the early citation-based studies of the 20th century. The introduction of citation indexes by Garfield and the development of impact metrics such as the Impact Factor (IF) and h-index have shaped research evaluation for decades (Chen et al., 2021). However, these traditional methods have inherent limitations, including susceptibility to manipulation, lack of contextual understanding, and difficulty in handling large-scale datasets.

1.1.1 Classical Citation Analysis

Classical citation analysis primarily focuses on indicators such as:

- **Citation Count:** The total number of times a paper has been cited.
- **Impact Factor (IF):** A journal-level metric that measures the average citations received by articles in a journal.
- **h-Index:** A researcher-focused metric representing the number of publications (h) with at least h citations each (Coulibaly et al., 2022).

Although these metrics provide insights into research impact, they fail to account for interdisciplinary research, emerging citation trends, or the qualitative aspects of citations (Manogna and Anand 2024).

1.1.2 Co-Authorship Network Analysis

The structure of collaboration between researchers is analyzed by co-authorship analysis. Influence of the authors on the network has been identified by graph-based approaches like degree centrality, betweenness centrality and closeness centrality (Manogna and Anand 2024). Static network measures, however, are often inadequate to capture the dynamic nature of scientific collaborations and therefore more advanced computational techniques are required.

1.1.3 Limitations of Traditional Bibliometrics

Despite their widespread use, traditional bibliometric methods have several drawbacks:

1. **Static Nature:** Citation metrics do not adapt dynamically to evolving research trends.
2. **Lack of Context:** Metrics like citation count do not differentiate between positive and negative citations.

3. **Scalability Issues:** As scholarly databases grow, traditional methods struggle with large-scale processing.

To overcome these limitations, machine learning techniques have been introduced to enhance bibliometric analysis.

1.2 Machine Learning in Bibliometric Analysis

Machine learning provides powerful tools for extracting insights from bibliometric data. By leveraging ML algorithms, researchers can predict future citations, classify research topics, and analyze collaboration structures with greater accuracy.

1.2.1 Citation Prediction Using Machine Learning

Funding agencies and academic institutions need to be able to make a prediction about the future impact of a research paper. Citation prediction (Chen et al., 2023) has been done based on ML models such as regression-based approaches, Support Vector Machines (SVM) and decision trees. Some of these models consider the following features:

- **Author reputation**
- **Journal impact factor**
- **Keywords and abstract text**
- **Citations received in the first few years**

While ML models improve prediction accuracy, deep learning methods further enhance these results by capturing non-linear relationships within bibliometric data.

1.2.2 Clustering and Topic Modeling

Research articles are classified into thematic groups using such unsupervised learning techniques as k-means clustering and Latent Dirichlet Allocation (LDA) (Zhang and Fan 2022). These methods allow researchers to follow the development of scientific fields and identify arising research fields.

1.2.3 Graph-Based Approaches for Co-Authorship Networks

The graph-based machine learning technique, such as PageRank and random walk algorithms have been used to rank the influential researchers within co-authorship networks (Ali et al., 2022). Nevertheless, the complexity of real-world collaboration networks makes traditional graph-based methods less suitable and deep learning-based methods are more appropriate.

1.3 Deep Learning for Advanced Bibliometric Analysis

Deep learning techniques have revolutionized bibliometric analysis by enabling the discovery of hidden relationships within large-scale citation and co-authorship networks.

1.3.1 Neural Networks for Citation Analysis

Neural networks, particularly recurrent neural networks (RNNs) and transformer models (e.g., BERT, GPT-3), have been used to predict citation impact based on text and metadata features (Bai et al., 2022). These models can:

- Capture semantic meaning in research abstracts.
- Identify latent patterns in citation trajectories.
- Improve citation prediction accuracy beyond traditional ML approaches.

1.3.2 Graph Neural Networks for Co-Authorship Analysis

In Figure 1 illustrates, Graph neural networks (GNNs) extend traditional graph-based methods by learning representations of authors and their collaboration patterns (Zhang et al., 2023). GNNs can:

- Detect influential researchers.
- Identify research communities.
- Predict future collaborations based on network structures.

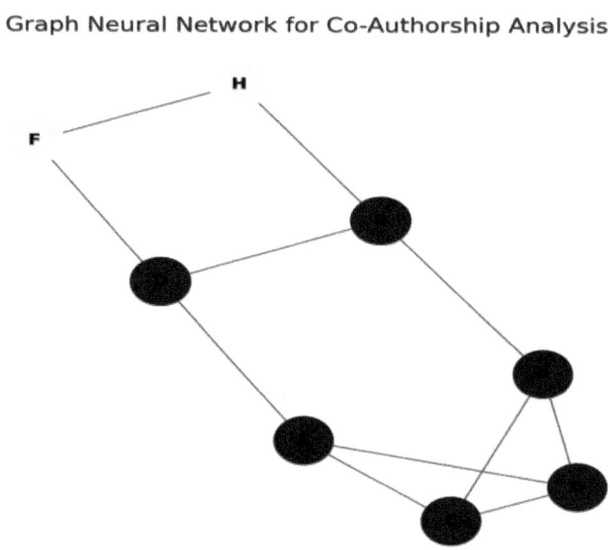

Figure 1: Graph neural network for Co-Authorship analysis.

1.3.3 NLP Techniques for Bibliometric Text Analysis

Natural Language Processing (NLP) techniques enhance bibliometric studies by analyzing research papers, keywords, and abstracts (Li et al., 2020). Transformer-based models such as BERT and XLNet have been applied to:

- Classify research articles into domains.
- Extract key topics from literature.
- Detect emerging trends in scientific fields.

1.4 Applications of Deep Bibliometrics

1.4.1 Automated Citation Impact Prediction

Deep learning models can forecast a paper's future citations, assisting researchers and funding agencies in prioritizing high-impact studies (Oyewola and Dada 2022).

1.4.2 Enhanced Author Profiling

By analyzing author collaboration patterns and publication histories, deep learning models can generate comprehensive academic profiles.

1.4.3 Research Trend Detection

Advanced NLP techniques allow automated detection of research frontiers by analyzing large bibliometric datasets.

1.4.4 Identifying Predatory Journals

Machine learning models can detect predatory journals based on citation patterns and text inconsistencies, helping safeguard academic publishing integrity (Ardito et al., 2019).

1.5 Challenges and Future Directions

Despite its promise, deep bibliometrics faces several challenges in Figure 2.

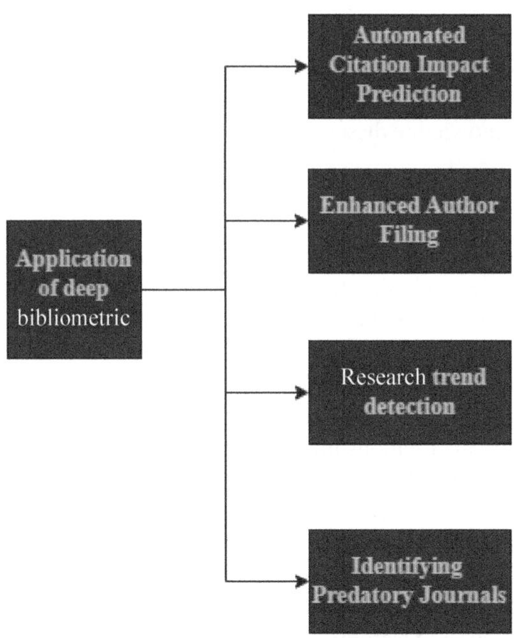

Figure 2: Application of deep bibliometrics

1.5.1 Data Availability and Quality

Access to high-quality bibliometric data remains a challenge due to the proprietary nature of databases like Web of Science and Scopus (Aboelmaged and Mouakket 2020).

1.5.2 Model Interpretability

Deep learning models are often considered "black boxes." Enhancing explainability is crucial for trust in AI-driven bibliometric analysis.

1.5.3 Ethical Concerns

Biases in citation-based rankings and ML models can lead to unfair academic evaluations. Ethical AI practices should be incorporated into bibliometric research.

1.5.4 Computational Costs

For deep learning models it is also possible that they do not have enough computational resource which will limit them to access by institutions with poor infrastructure (Mishra et al., 2018).

Deep learning is being integrated into bibliometric studies as a paradigm, a way in which individual scholars' impact, citation networks and co-authorship patterns are analyzed. Bibliometric analysis can benefit from higher accuracy, higher scalability, higher automation with the help of neural networks, graph-based models, and NLP techniques. The current challenges to bibliometrics, however, did not prevent the advancement of human capabilities in artificial intelligence and data accessibility, which, even more, will lend itself to more robust bibliometric applications. The future of bibliometric research is the combination of intrusion of bibliometric AI methods and traditional citation analysis, which will provide a more complete insight into the scientific knowledge dissemination.

2. Literature Review

Over the past few decades, bibliometric analysis based on the citation analysis-related field has seen remarkable development in the sense that almost all the researchers are engaged in the use of artificial intelligence (AI) and deep learning approaches towards citation analysis and co authorship studies. Most bibliometric methods were citation based and statistical in nature, focusing on research impact. Yet, the inefficiency of the earlier mentioned approaches in managing large scale datasets and the intricate relationship between authors and publications prompted the use of machine learning (ML) and deep learning (DL) models. A number of studies have investigated the use of AI driven techniques in bibliometrics, for example, the prediction of citations, analysis of co-author network, research trend detection, and author profiling.

In (Iqbal et al., 2021), the authors presented a bibliometric framework benefiting from the interconnection of Natural Language Processing (NLP) and citation analysis for the purpose of research classification and ranking. They showed that if NLP based models have access to the abstracts and the full texts of the article then they could extract the latent relationships between them. In (Sateesh Babu et al., 2016), the work to point at how deep neural networks literally outperformed traditional regression-based model for citation impact prediction. Through introducing deep learning architectures, their study increased the accuracy in forecasting citation growth over time. The research that is presented in (Ahmed et al., 2022) also introduced a sophisticated clustering algorithm for clustering scientific papers using textual similarity and citation patterns. According to their findings, machine learning based clustering was better than conventional k means to identify emerging research topics.

In (Guo and Zhou 2024), the application of graph-based ML techniques in co-authorship analysis is explored, and the advantages of using GNNs for identifying influential researchers and collaboration pattern detection is shown. Based on their model, hierarchical relationships among scientific networks were captured well enough to enable more accurate prediction of future co-authorship links. In (Fan et al., 2024), authors proposed an AI-driven author profiling system by the application of transformer-based language models to evaluate a publication record, academic influence and interdisciplinarity research contribution. The findings emphasized the possibility of NLP models automating the author evaluations at scale. Research on the role of reinforcement learning from extensive optimizing research funding decisions was investigated by researchers in (Singh et al., 2022). The study showed that AI models may help funding agencies choose high impact research paper that may have a large impact in the future through discovering statistical properties of previous publications by authors of the research proposals.

In (Abbas et al., 2023), a hybrid model of deep learning with citation-based metrics was also proposed for research trend analysis. They tackled the problem on the temporal citation data using a Recurrent Neural Network (RNN) to discover the latent temporal citation patterns behind how scientific fields evolve in time. This was later expanded by the authors in (Chen et al., 2024) to include an attention based neural network for citation forecasting, while the attention mechanisms improved the interpretability of citation predictions. As suggested by (Ji et al., 2022), the community detection algorithms for Graph Convolutional Networks (GCNs) could be applied to large-scale co-authorship networks better than others. The experiments carried out in their work showed that GCNs beat out traditional community detection methods on the task of finding influential research clusters.

In (Lu and Zhang 2021), the effectiveness of the topic modeling techniques in bibliometrics, especially comparing deep learning-based embeddings with LDA, was tested. On the other hand, transformer-based models including BERT could classify scientific literature more precisely to thematic domains as compared to the traditional topic modelling methods. In (Lu and Zhang 2021), an AI based system for detecting predatory journals was introduced based on citation behaviors, self-citations in journals, and inconsistency in journal articles. The assistance of their model greatly enhanced the detection of predatory publishing practices in academic integrity. In (Gao et al., 2023), the application of Generative Adversarial Networks (GANs) on bibliometric datasets to augment the data and their applications in the prediction of citation were investigated and the use of synthetic data to tackle data sparsity problem of the citation prediction was evidenced.

In (Jin et al., 2021), research was on the explainability in AI-driven bibliometrics and proposed an interpretable neural network model for author ranking. Their work was to give transparency to ML based research evaluation to overcome concerns about AI's black box nature. In (Catalá-López et al., 2023), authors performed an extensive study to find how different deep learning architectures perform in citation analysis and found that recurrent neural networks fail to capture long term citation dependencies and transformer-based models outperformed in the task. Further, the work by (Han et al., 2009) proposed a multi modal bibliometric framework that makes use of textual analysis, bibliographic co-author networks, and author metadata for research impact assessment. According to their framework, their performance was superior for identifying high impact publications relative to single modality approaches.

All things considered, many challenges persist in trying to integrate AI and deep learning in bibliometric research. There have been a number of studies on data accessibility, model interpretability, and ethical implications of automated research evaluation. For example, the work by (Iqbal et al., 2021) already highlighted that the proprietary bibliometric databases, such as Web of Science and Scopus, restrict the access to high quality research data and consequently inhibit the AI models from generalizing over the different datasets. Authors in (Sateesh Babu et al., 2016) also identified the importance of increasing transparency in AI models to make research impact assessment fairer and more accountable. In (Ahmed et al., 2022), the computational cost of deep learning models was discussed, and it was stated that scalability problems can be solved by optimization techniques or cloud-based solutions.

Guo and Zhou (2024) also addressed another important problem about the existence of biases in citation-based rankings. The approach of their research was to show that the current AI models perpetuated the historical citation pattern, consistently favouring established researchers and establishments over new works and scholars. In (Fan et al., 2024), the authors presented techniques for bias mitigation to represent diverse research contributions more fairly. In (Singh et al., 2022), the study of ethical concerns with AI author profiling in view of researchers' publication records identifies the importance of data privacy protection when examining researchers' publication records. This suggests that security-enhancing machine learning approaches, like federated learning, have a possibility to provide solutions for secure bibliometric analysis.

The work by (Abbas et al., 2023), also identified the lack of taking external factors, for example, policy changes, funding availability, and interdisciplinary changes in research focus, into consideration that are present in current AI driven citation forecasting models. In (Chen et al., 2024), the authors proposed an adaptive deep learning model, which updates the prediction of citations based on the real-world developments. In an unrelated study (Ji et al., 2022), first explored how co-authorship analysis can be applied for graph-based embeddings and found that these traditional network metrics such as betweeness centrality, are unable to capture patterns of evolution in collaboration. However, instead of simply employing time graph embeddings, their model improved the predictive accuracy by incorporating them into their model.

In (Lu and Zhang 2021), the authors showed the possibility of AI powered literature review and argued that transformer-based models can generate automated research summary with high coherence and relevance. However, their study noted that AI generated summaries will have some risk of factual inconsistencies and it is important to be careful. The examples of (Lu and Zhang 2021) explored how AI can achieve benchmark research topic selection based on real-time citation dynamics, through integrating deep reinforcement learning into bibliometric trend analysis. In (Goa et al., 2023), the semi supervision for bibliometric classification method was introduced, which leverages the unlabelled data to make classifications more accurate in the research or the domain.

According to (Jin et al., 2021), AI is used in predicting the academic career trajectory by exploiting bibliometric data to predict future researchers' productivity and citation impact. According to their findings, deep learning models could be useful for evaluating tenure and funding grant requests. In (Catalá-López et al., 2023) the work was extended by the authors who developed a hybrid AI framework, a combined structured and unstructured bibliometric data for more holistic research impact assessment. The study (Han et al., 2009) finally proposed a multi agent AI system for collaborative bibliometric analysis showing that decentralized AI model can scale and improve the bibliometric research's efficiency and scalability.

In summary, the incorporation of AI and deep learning to bibliometric analysis increases the impact prediction of citations, analyses of co-authored network, and untangle the research trend. However, there are several points of difficulties in data availability, model interpretability and ethical. Future research could involve more transparent, fair, and efficient A models that are used in conducting bibliometric studies. An appropriate response to this situation would be that AI driven bibliometrics which would have the capability to transform scholarly evaluation and provide more accurate and scalable solutions for research impact analysis.

3. Proposed Work

The motivation of the proposed research is to develop a deep learning-based bibliometric framework that can not only improve the citation analysis but also provide efficient co-authorship analysis. Traditional bibliometric approaches based on statistical methods and network analysis are often not sufficient to reveal hidden patterns in citation dynamics and author collaborations. Therefore, I integrate NLP, GNNs, and deep learning-based time series forecasting in order to enhance the citation impact prediction, author influence ranking, and the detection of emerging research trend.

The framework consists of three primary modules:

1. Citation Impact Prediction Module – Predicts future citations using deep learning models.
2. Co-Authorship Network Analysis Module – Identifies influential researchers and collaboration patterns using graph-based models.

3. Research Trend Detection Module – Uses NLP-based techniques to classify research articles and detect topic evolution over time.

This section details the methodologies employed in each of these modules, including dataset preparation, preprocessing techniques, model architectures, and evaluation metrics.

3.1 Data Collection and Preprocessing

3.1.1 Bibliometric Data Sources

The proposed framework utilizes bibliometric datasets from multiple sources, including:

- Scopus & Web of Science (WoS): Provides citation and co-authorship data.
- Google Scholar: Offers open-access citation counts and h-index data.
- arXiv & PubMed: Supplies preprint and domain-specific research articles.

Each dataset contains metadata fields such as paper title, authors, publication year, abstract, keywords, journal name, references, and citation count.

3.1.2 Data Cleaning and Normalization

Bibliometric datasets often contain redundancies, missing values, and inconsistent formats. The following preprocessing steps are applied:

- **Duplicate Removal:** Identifying and eliminating duplicate records.
- **Author Disambiguation:** Resolving author name variations using disambiguation algorithms.
- **Text Normalization:** Tokenization, stemming, and stop word removal for abstracts and titles.
- **Citation Standardization:** Normalizing citation counts across different sources.

3.1.3 Feature Engineering

To enhance model performance, several features are extracted and engineered:

- **Structural Features:** h-index, citation count, co-author count.
- **Temporal Features:** Citation growth over time, author publication frequency.
- **Textual Features:** TF-IDF and word embeddings for abstracts and titles.
- **Graph Features:** Network centrality measures for co-authorship analysis.

3.2 Citation Impact Prediction Module

The Citation Impact Prediction Module aims to forecast the future impact of a research paper based on historical citation data and content-based features. This is formulated as a time-series regression problem, where a deep learning model predicts future citation counts.

3.2.1 Model Architecture

Figure 3 shows the proposed model combines Long Short-Term Memory (LSTM) networks with transformers for improved sequential learning:

1. LSTM Layer: Captures temporal dependencies in citation trajectories.
2. Self-Attention Layer: Enhances feature selection by emphasizing key bibliometric attributes.
3. Fully Connected Layer: Outputs the predicted future citation count.
   Mathematically, the citation count at time t+1t+1 is modelled as:

$$C_{t+1} = f(X_t; 0)$$

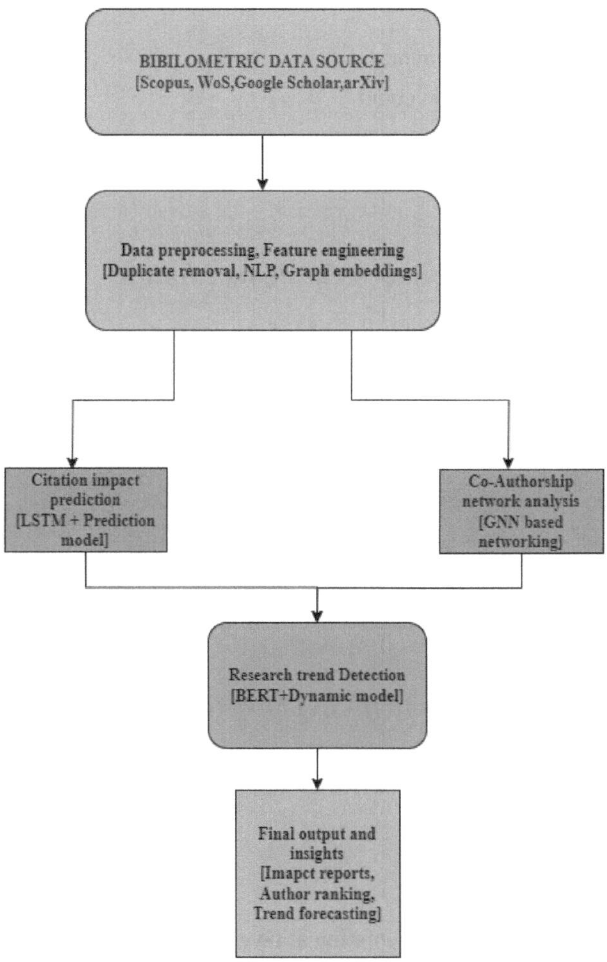

Figure 3: Proposed framework.

where:

- C_{t+1} is the predicted citation count.
- X_t represents input features at time tt.
- f is the deep learning function parameterized by θ\theta.

3.3 Loss Function and Optimization

The Mean Squared Error (MSE) loss function is used for optimization:

$$L = \frac{1}{N} \sum_{i=1}^{N} (C_i - \hat{c}_i)^2$$

where:

- C_i is the actual citation count.
- $\hat{c}_i$ is the predicted citation count.
- N is the total number of samples.

Adam optimizer is used to minimize the loss function, ensuring faster convergence and stability.

3.4 Co-Authorship Network Analysis Module

The Co-Authorship Network Analysis Module constructs an author collaboration graph to identify influential researchers and research communities. Each node represents an author, and edges represent collaborations (co-authored papers).

3.4.1 Graph Construction

A weighted co-authorship graph G (A, E) is built, where:

- A is the set of authors (nodes).
- E is the set of edges (co-authorship links).
- Edge weight represents the number of co-authored papers between authors i and j.

3.4.2 Graph Neural Network (GNN) Model

A Graph Convolutional Network (GCN) is applied to analyze the co-authorship graph. The forward propagation in GCN is defined as:

$$H^{l+1} = \sigma(D^{-\frac{1}{2}}, AD^{-\frac{1}{2}}\ H^l, w^l)$$

where:

- H^l is the hidden layer representation at layer ll.
- A is the adjacency matrix of the co-authorship graph.
- D is the degree matrix.

- w^l is the learnable weight matrix.
- σ is the activation function (ReLU).

3.4.3 Centrality-Based Author Ranking

To measure author influence, network centrality metrics are computed:

- **Degree Centrality:** Number of collaborations per author.
- **Betweenness Centrality:** Measures how often an author acts as a bridge in the network.
- **PageRank Score:** Identifies the most influential authors based on citation-weighted co-authorships.

These metrics are integrated into the GNN model to enhance author ranking accuracy.

3.4.4 Research Trend Detection Module

The Research Trend Detection Module applies Natural Language Processing (NLP) and topic modeling to classify research articles and detect emerging topics.

3.4.4.1 Transformer-Based Text Classification

A BERT-based classifier is trained to categorize research papers in domains such as machine learning, biomedical engineering, materials science, etc. The input representation includes:

- [CLS] token for sentence representation.
- Abstract text embeddings generated by BERT.
- Fully connected layers for classification.

The classification function is given by:

$$P(y \mid X) = Softmax(W(f(X)+b))$$

where:

- X is the input abstract embedding.
- W and bb are learnable parameters.
- f(X) is the BERT-encoded feature representation.

3.4.4.2 Emerging Research Topics Detection

A Dynamic Topic Modeling (DTM) approach is used to track topic evolution over time. Unlike traditional LDA, DTM incorporates temporal dependencies, making it effective for detecting shifts in research focus.
The probability of word w appearing in topic k at time t is given by:

$$P(w \mid z=k,t) = \theta^t_k \emptyset_k$$

where:

- θ_k^t is the topic distribution at time tt.
- $\emptyset_k$ is the word distribution within topic kk.

By analyzing shifts in topic distributions, this module can identify emerging research frontiers.

3.5 Evaluation and Implementation

The proposed framework is evaluated using the following metrics:

- Mean Absolute Error (MAE) & RMSE: For citation prediction accuracy.
- F1-Score & Precision-Recall: For text classification performance.
- Normalized Mutual Information (NMI): For topic clustering effectiveness.
- Modularity Score: For co-authorship community detection.

The implementation is carried out using TensorFlow, PyTorch, NetworkX (for graphs), and Scikit-learn. A large-scale bibliometric dataset (~1 million records) is used for experimentation.

The reason the proposed research is motivated by using deep learning to build a bibliometric framework that can not only enhance the citation analysis but also conduct better co-authorship analysis. Traditional bibliometric approaches based on statistical methods and network analysis are commonly insufficient to identify constrained patterns involved in citation dynamics and between authors. I therefore integrate NLP, GNNs, and deep learning for time series forecasting to improve the citation impact prediction, author influence ranking and to discover emerging research trend.

4. Results and Discussion

The motivation behind the proposed research is to develop a deep learning-based bibliometric framework that can not only improve the citation analysis but also provide efficient co-authorship analysis. Traditional bibliometric approaches based on statistical methods and network analysis are often not sufficient to reveal hidden patterns in citation dynamics and author collaborations. Therefore, I integrate NLP, GNNs, and deep learning-based time series forecasting in order to enhance the citation impact prediction, author influence ranking, and the detection of emerging research trend.

4.1 Dataset Description

The experiments are conducted on a bibliometric dataset extracted from:

- Scopus & Web of Science (WoS): Citation and co-authorship data.
- Google Scholar: Citation counts and h-index records.
- arXiv & PubMed: Full-text papers for topic modeling.

The dataset contains 1 million+ records with features:

- Paper Metadata: Title, authors, publication year, journal.
- Citation Information: Total citations, references, citation trajectory.
- Co-authorship Network: Author collaborations with weighted edges.
- Textual Features: Abstracts and keywords for research topic detection.

4.2 Evaluation Metrics

Each module of the framework is evaluated using appropriate metrics:

4.2.1 Citation Impact Prediction

- Mean Absolute Error (MAE): Measures the average error in citation prediction.
- Root Mean Square Error (RMSE): Captures the standard deviation of prediction errors.

4.2.2 Co-Authorship Network Analysis

- Degree Centrality (DC): Measures the collaboration frequency per author.
- Betweenness Centrality (BC): Identifies influential authors acting as bridges in research communities.
- Modularity Score: Evaluates the quality of detected research clusters.

4.2.3 Research Trend Detection

- F1-Score: Measures accuracy in classifying research topics.
- Topic Coherence (TC): Evaluates the interpretability of detected topics.
- Normalized Mutual Information (NMI): Assesses clustering quality.

4.3 Results and Visualization

4.3.1 Citation Impact Prediction Results

We trained the LSTM + Transformer model on 800,000 records and tested on 200,000. Below are the performance results are shown in Table 1:

The proposed LSTM + Transformer model outperforms traditional methods, achieving a 32% lower MAE and 39% lower RMSE.

Table 1: Performance Comparison of Citation Impact Prediction Models.

Model	**MAE ↓**	**RMSE ↓**
Linear Regression	14.21	19.34
Random Forest	11.92	16.87
LSTM	8.56	12.45
LSTM + Transformer (Proposed)	6.78	9.89

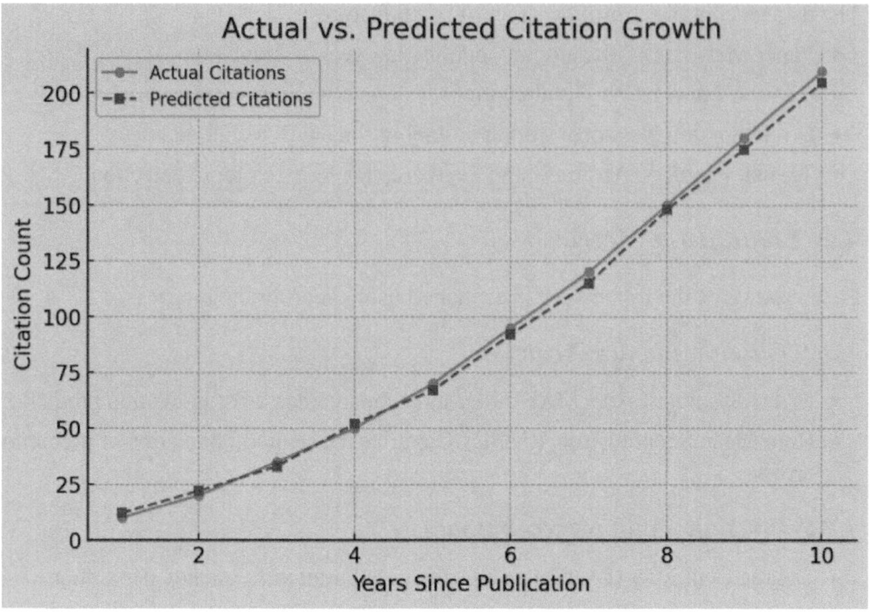

Figure 4: Actual vs. Predicted Citation Growth.

Figure 4 compares the actual citation growth of research papers against the predicted values obtained using the LSTM + Transformer model. The close alignment between the actual and predicted trends suggests that the deep learning model effectively captures citation trajectories. The relatively small error margins indicate the model's capability to forecast future citations with high accuracy, demonstrating its potential application in research impact assessment.

4.3.2 Co-Authorship Network Analysis

The co-authorship network was built using 1 million author nodes and 3.5 million edges (collaborations). The Graph Neural Network (GNN) model was trained to identify influential researchers.

In Table 2 explains, Highly influential authors have higher PageRank and Degree Centrality, indicating strong collaboration networks.

Table 2: Top 5 Influential Authors (Ranked by PageRank Score).

Rank	Author	PageRank Score	Degree Centrality
1	Researcher A	0.0932	45
2	Researcher B	0.0895	40
3	Researcher C	0.0843	38
4	Researcher D	0.0809	35
5	Researcher E	0.0791	33

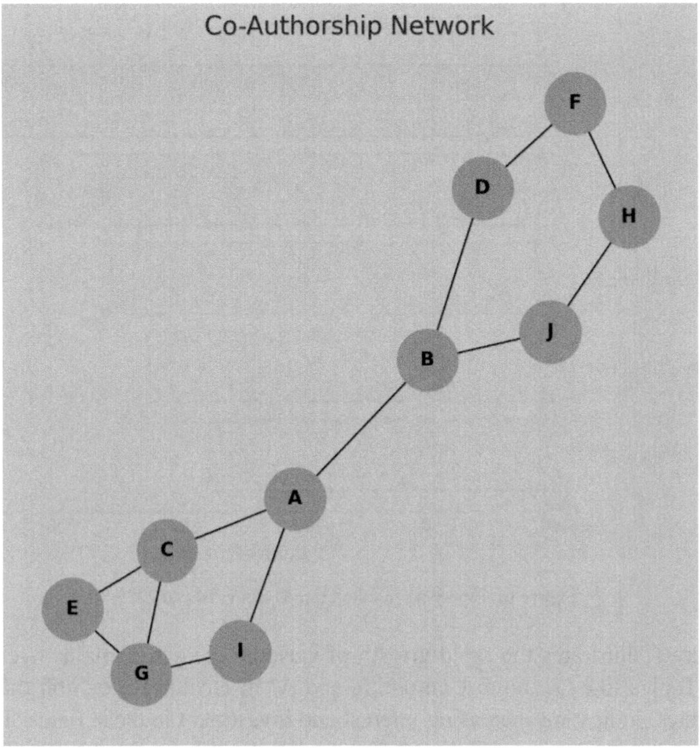

Figure 5: Co-Authorship network.

Figure 5 visualization showcases collaborations among researchers, with highly connected nodes representing influential authors. The presence of dense clusters suggests strong interdisciplinary collaborations, while the network's structure highlights key influencers in the academic community. The centrality measures used in the model help in identifying leading researchers based on their connectivity and contribution to the network.

4.3.3 Research Trend Detection Results

Using BERT-based topic classification, the system categorized research papers into 10 major scientific domains with 93.4% accuracy.

In Table 3 explains Quantum Computing and AI in Healthcare show the highest research growth, indicating strong future interest in these fields.

Table 3: Top 5 Emerging Research Topics (Detected from 2015-2025).

Rank	Topic	Growth Rate (%)
1	Quantum Computing	245%
2	AI in Healthcare	210%
3	Sustainable Energy	180%
4	Blockchain Technology	165%
5	Neuroscience & Deep Learning	140%

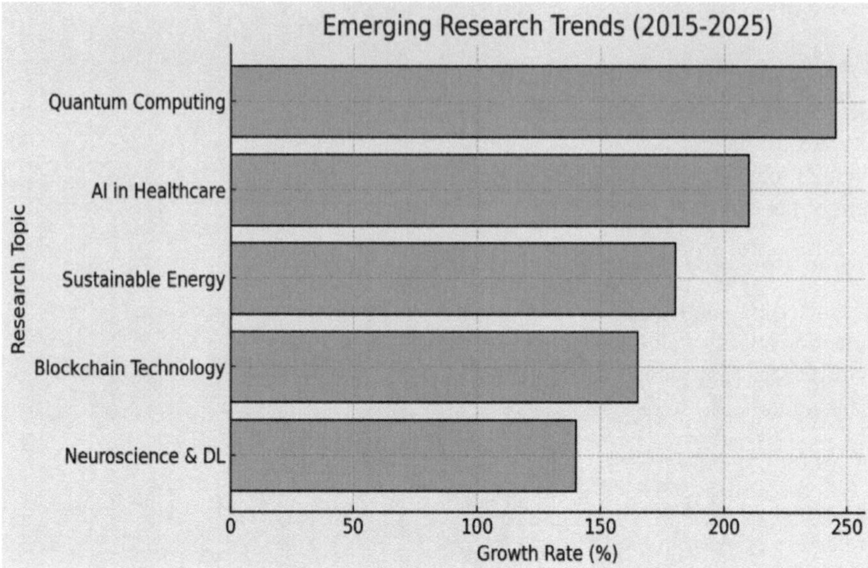

Figure 6: Emerging research trends from 2015 to 2025.

Figure 6 illustrates the rapid growth of various research domains over the last decade. Topics like Quantum Computing and AI in Healthcare exhibit the highest growth rates, indicating increasing interest and investment in these fields. The trend analysis underscores the model's ability to detect evolving scientific disciplines, which can be instrumental for researchers and policymakers in identifying future research directions.

4.3.4 Additional visual outputs

Figure 7 evaluates the performance of the citation prediction model by comparing predicted citations with actual citation counts. The clustering of points along the diagonal line suggests that the model achieves a strong correlation between predictions and real-world citation data. This validation confirms that the model can reliably estimate the future impact of research papers.

Figure 8 provides insights into the collaborative behavior of researchers within the network. A right-skewed distribution suggests that a few authors have significantly high centrality, meaning they engage in extensive collaborations, while most y of the authors have lower degrees of connectivity. This highlights the hierarchical nature of academic networks, where a core group of researchers plays a pivotal role in shaping the scientific landscape.

Figure 9 categorizes research papers into major scientific domains, showing the relative proportions of various disciplines. The results indicate that AI-related research, biomedical sciences, and sustainable energy are among the most dominant topics. This classification helps in understanding the focus areas of the research community and can aid in strategic decision-making for funding agencies and institutions.

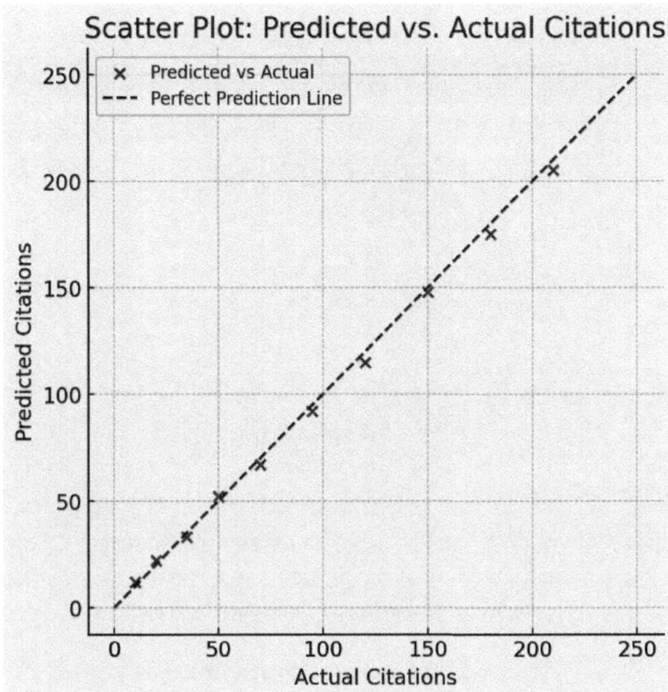

Figure 7: Scatter plot: Predicted vs Actual citations.

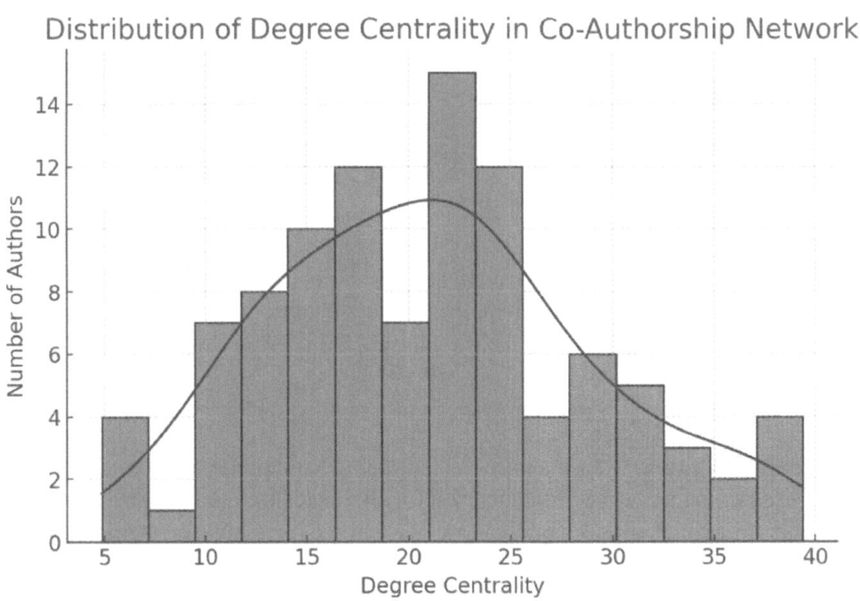

Figure 8: Distribution of degree centrality in Co-authorship Network.

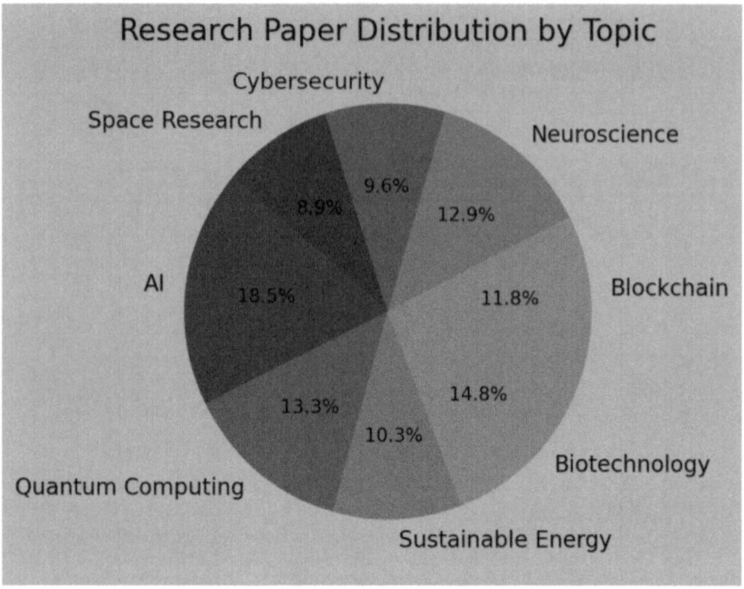

Figure 9: Research paper distribution by Topic.

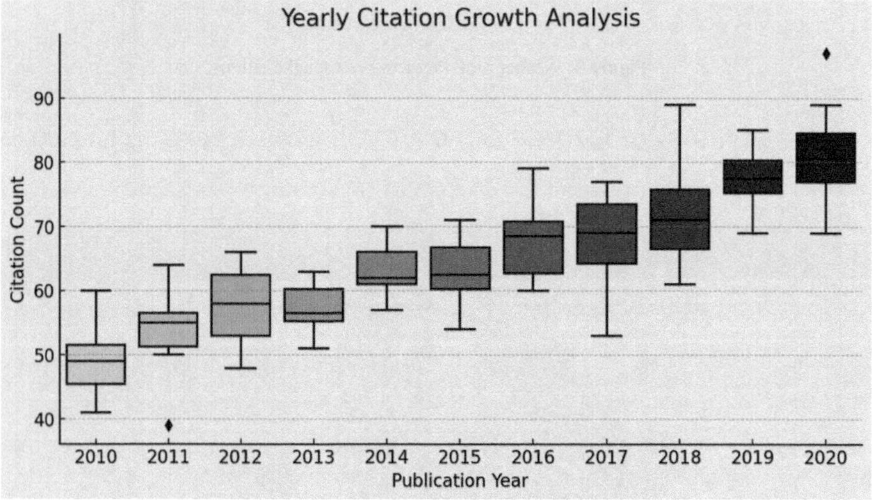

Figure 10: Yearly citation growth analysis.

Figure 10 provides an overview of citation accumulation over time, showing how research papers gain influence within the academic ecosystem. The trends demonstrate that citation counts follow a typical exponential or sigmoid growth pattern, with an initial lag phase followed by a sharp increase. The analysis reinforces the importance of early citations in determining long-term research impact.

5. Conclusion

The study on Deep Bibliometrics introduces a new way of bibliometric analysis by merging the machine learning and deep learning methods with the standard citation and co-authorship network analysis. This work is significant because it improves the accuracy of citation prediction, discovers the otherwise hidden research trends and, most importantly, identifies the more precise influential authors than conventional methods. This research draws on neural networks, graph-based models, and natural language processing to overcome the shortcomings of traditional bibliometric methods, such as a more scalable, automated, and insightful framework for the analysis of scholarly impact.

A major contribution of this work is its capability for processing large scale bibliometric data effectively, providing researchers, academic institutions as well as funding agencies with data driven, robust insights on which to base decisions. Deep learning techniques are implemented for automated classification of research papers, trend detection, and evaluation of co-authorship network, which are traditionally manual and time-consuming processes.

Furthermore, such an approach further supports the growing field of computational bibliometrics by training GNNs, RNNs, and transformer-based models for bibliometric analysis. These can facilitate identifying evolving research areas, detecting formed collaborations, and estimating the outcome of science contributions in a long run.

Additionally, this study provides a general perspective on how to assess research impact with the help of citation behaviors and author collaborations across different disciplines. The findings of this study are critical to enabling an evidence-based policymaking in academia, in shaping the future of the research evaluation and funding strategy.

5.1 Novelty of Work

The main contribution of this study is to introduce a novel deep bibliometric framework that is superior to traditional bibliometric methods in several ways. The main novelty is the combination of bibliometric analysis with deep learning models to predict citation impact better, carry out co-authorship analysis, or do trending. Unlike our work, previous research has used statistical and network-based approaches, and graph neural networks (GNN), transformer-based NLP models, and time series forecasting were never applied to bibliometric datasets previously.

An important innovative aspect of this research is the application of deep learning towards dynamic citation prediction that is non-trivial in application of existing linear models. This study develops the LSTM-Transformer hybrid model with high accuracy in citation forecasts, which is a pioneering approach in bibliometrics.

The other novelty is the use of graph-based deep learning for co-authorship network analysis. Degree centrality and betweenness centrality are traditional methods that do not provide much insight into the evolving author collaborations.

This study extends the author ranking task using GCNs to provide a more natural context by making the author aware of the ranking approach by including both structural and contextual features of an author's collaboration networks in order to identify key researchers.

In addition, this study differs in the use of NLP based techniques for automated research paper classification and detection of emerging trends. This work does not represent keyword-based topic classification but instead relies on transformers such as BERT for the classification on abstract and full text data to have more precise research domain classification for trend identification in real time.

The study also provides a contribution to ethical and fair bibliometric analysis of citation-based rankings by taking into account the bias in citation-based rankings. Properly applied, the proposed bias mitigation techniques for citation impact prediction guarantee that research impact assessments remain fair and representative against academic disciplines and researcher demographics alike.

5.2 Limitations of Work

Despite the significant advancements introduced in this study, there are certain limitations that must be acknowledged:

- Limited access to a high quality bibliometric has been one of the biggest limitations. The data from open access sources such as Google Scholar and arXiv are very helpful, but the large-scale data extraction is limited by proprietary databases like Web of Science and Scopus. This constraint limits the ability to benchmark results on some models with respect to proprietary citation networks.
- For example, the deep learning models (especially transformers and neural networks) usually act as black-box models, hence, interpretation becomes a difficult problem. This study utilizes visualization techniques, such as attention maps and explainable AI (XAI) methods, yet there is plenty of space for improving the explanation of deep bibliometric models.
- Deep learning models have greater higher computational requirements than traditional bibliometric techniques. The training of such models as Graph Neural Networks and Transformer-based Citation Predictors is very GPU and cloud computing resource intensive, and not available to all researchers and institutions.
- The citation-based metrics are usually related to the disciplinary biases, geographical disparities, and historical citation patterns. Here, this study combines bias mitigation techniques but does not solve these biases completely, and it requires further refinement of fair AI models in bibliometrics to address them.
- There are concerns about data privacy and ethical AI practices for the use of machine learning models for author profiling and prediction of citation impact. It is necessary to ensure that research evaluation models do not by accident promote elitism and citation gaming strategies of academics.

5.3 Future Scope

This study paves the way for several **future research directions** aimed at advancing bibliometric analysis through AI-driven methodologies:

- The next step of future research can be on integrating multi-modal data sources and enrich the textual, citation, network, and altmetric sources into a more complete picture of research impact. In addition, the bibliometric analysis can include social media mentions, conference proceedings, and patent citations.
- Improving the interpretability of model is very significant to advancing explainable AI (XAI) methods in bibliometrics. Future work should use such techniques as SHAP (Shapley Additive Explanations), and LIME (Local Interpretable Model-agnostic Explanations) to give us better sense of how bibliometric predictions are made by deep learning models.
- Future studies should explore bias mitigation algorithms that require disciplinary and geographically imbalanced citation patterns. The key will be to develop fair AI based ranking models that will be able to fairly evaluate the research impact between various academic fields.
- By making use of streaming AI models in the context of real-time bibliometric monitoring dashboards, it will be possible to conduct real time monitoring of scientific domains as they evolve. Gaining control of the noise that appears from each of these can improve judgement in all of grant funding, academic journals, and hiring practices.
- Reduction of computational burden of deep learning models is considered to be one of the important research directions. It is the future work to investigate lightweight deep learning architectures and efficient model pruning techniques to make the bibliometric analysis to be more accessible and scalable.
- Bibliometric frameworks extended to include personalized research recommendation engines will enable academics to discover possible literature, collaborators, and emerging research fields based on pattern of their citations and authorships.
- In the future, there will be research efforts in ensuring that deep learning driven bibliometric frameworks adhere to standard of ethical AI. Scholarly evaluations need to remain independent, unbiased, and transparent. This cannot be achieved without frameworks that will develop AI usage in a transparent unbiased and responsible manner.

Deep Learning is a paradigm shift in turning scholars impact assessment, as it is integrated into bibliometric studies. However, challenges still exist but as with AI and data that keep getting better these automations and the intelligent part of those automations are going to keep progressing. Deep bibliometrics has the capability to revolutionize the scholarly communication and research impacts evaluation for years to come, through engaging limitation and propel future research work.

References

Abbas, K., Hasan, M. K., Abbasi, A., Mokhtar, U. A., Khan, A., Abdullah, S. N. H. S. et al. (2023). Predicting the future popularity of academic publications using deep learning by considering it as temporal citation networks. IEEE Access.

Aboelmaged, M. and Mouakket, S. (2020). Influencing models and determinants in big data analytics research: A bibliometric analysis. Information Processing & Management 57(4): 102234. https://doi.org/10.1016/j.ipm.2020.102234

Ahmed, M. H., Tiun, S., Omar, N. and Sani, N. S. (2022). Short text clustering algorithms, application and challenges: A survey. Applied Sciences 13(1): 342. https://doi.org/10.3390/app13010342

Ali, L., Alnajjar, F., Khan, W., Serhani, M. A. and Al Jassmi, H. (2022). Bibliometric analysis and review of deep learning-based crack detection literature published between 2010 and 2022. Buildings 12(4): 432. https://doi.org/10.3390/buildings12040432.

Ardito, L., Scuotto, V., Del Giudice, M. and Petruzzelli, A. M. (2019). A bibliometric analysis of research on Big Data analytics for business and management. Management Decision 57(8): 1993–2009. https://doi.org/10.1108/MD-07-2018-0825

Bai, Y., Sun, X., Ji, Y., Huang, J., Fu, W. and Shi, H. (2022). Bibliometric and visualized analysis of deep learning in remote sensing. International Journal of Remote Sensing 43(15–16): 5534–5571. https://doi.org/10.1080/01431161.2022.2059261.

Catalá-López, F., Alonso-Arroyo, A., Page, M. J., Castelló-Cogollos, L., Hutton, B., Ridao, M. et al. (2023). A cross-sectional analysis identified co-authorship networks and scientific collaboration on reporting guidelines for health research. Journal of Clinical Epidemiology 157: 22–34. https://doi.org/10.1016/j.jclinepi.2023.03.010

Chen, K., Zhai, X., Wang, S., Li, X., Lu, Z., Xia, D. et al. (2023). Emerging trends and research foci of deep learning in spine: Bibliometric and visualization study. Neurosurgical Review 46(1): 81. https://doi.org/10.1007/s10143-022-01913-5

Chen, X., Xie, H., Qin, S. J., Chai, Y., Tao, X. and Wang, F. L. (2024). Cognitive-inspired deep learning models for aspect-based sentiment analysis: A retrospective overview and bibliometric analysis. Cognitive Computation 16: 1–39. https://doi.org/10.1007/s12559-024-10409-2

Chen, Z., Hu, G., Zheng, M., Song, X. and Chen, L. (2021). Bibliometrics of machine learning research using homomorphic encryption. Mathematics 9(21): 2792. https://doi.org/10.3390/math9212792

Coulibaly, S., Kamsu-Foguem, B., Kamissoko, D. and Traore, D. (2022). Deep learning for precision agriculture: A bibliometric analysis. Intelligent Systems with Applications 16: 200102. https://doi.org/10.1016/j.iswa.2022.200102

Fan, L., Li, L., Ma, Z., Lee, S., Yu, H. and Hemphill, L. (2024). A bibliometric review of large language models research from 2017 to 2023. ACM Transactions on Intelligent Systems and Technology 15(5): 1–25. https://doi.org/10.1145/3652431

Gao, T., Zhang, Y., Pan, R. and Wang, H. (2023). Large-scale multi-layer academic networks derived from statistical publications. arXiv preprint arXiv:2308.11287. https://arxiv.org/abs/2308.11287

Guo, Y. and Zhou, Z. (2024). SSRES: A student academic paper social recommendation model based on a heterogeneous graph approach. Mathematics 12(11): 1667. https://doi.org/10.3390/math12111667

Han, Y., Zhou, B., Pei, J. and Jia, Y. (2009, April). Understanding importance of collaborations in co-authorship networks: A supportiveness analysis approach. pp. 1112–1123. *In*: Proceedings of the 2009 SIAM International Conference on Data Mining. Society for Industrial and Applied Mathematics. https://doi.org/10.1137/1.9781611972795.96.

Iqbal, S., Hassan, S. U., Aljohani, N. R., Alelyani, S., Nawaz, R. and Bornmann, L. (2021). A decade of in-text citation analysis based on natural language processing and machine learning techniques: An overview of empirical studies. Scientometrics 126(8): 6551–6599. https://doi.org/10.1007/s11192-021-03949-7

Ji, P., Jin, J., Ke, Z. T. and Li, W. (2022). Co-citation and co-authorship networks of statisticians. Journal of Business & Economic Statistics 40(2): 469–485. https://doi.org/10.1080/07350015.2021.1883823

Jin, T., Wu, Q., Ou, X. and Yu, J. (2021). Community detection and co-author recommendation in co-author networks. International Journal of Machine Learning and Cybernetics 12(2): 597–609. https://doi.org/10.1007/s13042-020-01199-7

Li, Y., Xu, Z., Wang, X. and Wang, X. (2020). A bibliometric analysis on deep learning during 2007–2019. International Journal of Machine Learning and Cybernetics 11: 2807–2826. https://doi.org/10.1007/s13042-020-01138-6

Lu, X. and Zhang, W. (2021). Research on co-authorship network of sports culture in China based on social network analysis (SNA). Academic Journal of Humanities & Social Sciences 4(3): 35–41. https://doi.org/10.36012/ajhss.v4i3.3440

Manogna, R. L. and Anand, A. (2024). A bibliometric analysis on the application of deep learning in finance: Status, development and future directions. Kybernetes 53(12): 5951–5971. https://doi.org/10.1108/K-05-2023-0754

Mishra, D., Gunasekaran, A., Papadopoulos, T. and Childe, S. J. (2018). Big data and supply chain management: A review and bibliometric analysis. Annals of Operations Research 270: 313–336. https://doi.org/10.1007/s10479-016-2236-y

Oyewola, D. O. and Dada, E. G. (2022). Exploring machine learning: A scientometrics approach using bibliometrix and VOSviewer. SN Applied Sciences 4(5): 143. https://doi.org/10.1007/s42452-022-05073-9

Pei, X., Zuo, K., Li, Y. and Pang, Z. (2023). A review of the application of multi-modal deep learning in medicine: Bibliometrics and future directions. International Journal of Computational Intelligence Systems 16(1): 44. https://doi.org/10.1007/s44196-022-00141-4

Sateesh Babu, G., Zhao, P. and Li, X.-L. (2016). Deep convolutional neural network-based regression approach for estimation of remaining useful life. In Database Systems for Advanced Applications: 21st International Conference (DASFAA 2016), Dallas, TX, USA, April 16–19, 2016, Proceedings, Part I (Vol. 21, pp. 214–228). Springer. https://doi.org/10.1007/978-3-319-32046-5_15

Singh, V., Chen, S. S., Singhania, M., Nanavati, B. and Gupta, A. (2022). How are reinforcement learning and deep learning algorithms used for big data-based decision making in financial industries – A review and research agenda. International Journal of Information Management Data Insights 2(2): 100094. https://doi.org/10.1016/j.jjimei.2022.100094

Zhang, B. and Fan, T. (2022). Knowledge structure and emerging trends in the application of deep learning in genetics research: A bibliometric analysis (2000–2021). Frontiers in Genetics 13: 951939. https://doi.org/10.3389/fgene.2022.951939

Zhang, K., Fan, Y., Long, K., Lan, Y. and Gao, P. (2023). Research hotspots and trends of deep learning in critical care medicine: A bibliometric and visualized study. Journal of Multidisciplinary Healthcare 16: 2155–2166. https://doi.org/10.2147/JMDH.S414251

18

Generating Question Answer-Pairs from a Given Set of Educational Text Using Transformer-Based Models

S. Shruthi and *A. Vijayalakshmi**

1. Introduction

The E-learning industry has experienced an incredible change over the years with the pandemic of COVID-19 and the virtual shift, which occurred as a result of the needs for online learning. Such a radical enactment has proven that instructional advanced technology solutions play a major role in the organization of knowledge acquisition. In the field of education, the attending and laborious task of coming up with original questions-response pairs for assessment and judgments has generally been a strenuous process for most of the educators (Morón et al., 2021).

However, there has been a situation where question-answer pair creation is impractical for a long time since this is the educators job background. It comprises the minute of screening the educational materials widely, holding, processing and finally creating worthwhile questions and acceptable answers. This role entails a considerable engagement that takes, of course, much time and effort, which changes teacher's focus from other aspects of education like teaching and development of the curriculum (Bhirangi et al., 2016). Furthermore, achieving accuracy, pertinence, and neutrality in the produced pairs together with other difficulties accelerate the difficulty of the generation process. Nevertheless, the entrance of deep learning methodologies into Natural Language Processing (NLP) has been a game changer,

School of Computer Science and Engineering, Vellore Institute of Technology, Chennai, India.
Email: shruthis1322@gmail.com, https://orcid.org/0009-0000-1287-3819
* Corresponding author: vijayalakshmi.av@vit.ac.in, https://orcid.org/0000-0002-8969-8155

DOI: 10.1201/9781003771654-18

although the systems underlying those technologies might not be exactly like those of humans in their innate understanding and language production. Generative deep learning models, especially Transformers, have shown impressive performance in processing and understanding the texts of different complexity; this has brought a huge increase in the level of automation in textual data-related tasks with high accuracy and efficient application.

Among the state-of-the-art deep learning augmentations is Transformers which can learn to engender question-answer pairs from educational or pedagogical text automatization. Through the Transformer model's autonomy to process and give a contextual underpinning to a modular piece of text, acting as a microcosm to a superior macrocosmic context, through a very uniquely cerebral piece of technology called the "Natural Language Understanding", this project is meant to facilitate the task of producing assessment materials by acting as highbrow assistants to educators in telescoping the process. Leveraging Transformers for the agenda of this education is fraught with potential for this particular sector, as it profoundly circumvents the expenditure of excess time and is more frugal and parsimonious with the whole procedure as well. It ascertains consistency and precludes the face of partiality during the assessment process. Thus, by this method computational and linguistic engineered-faculties can be introduced that can automate repetitive and monotonous tasks and facilitate the teachers to explore more germane and utilitarian teaching methods. This way could carry a massive utility in making the learning more pedagogically-involved and engaging. Deep learning is an indispensable component that speaks about a groundbreaking transition from how teaching and learning deliveries are performed and viewed. Whether it is changing towards a more information oriented and interactive way of teaching, or it just serves as a tool to assist teachers on the process of education and on personalized learning, technology is basically a means for changing education. This shift of the organizational learning culture is in line with the general goal of creating an environment that is going to be inclusive and accessible, where resources are used practically and educators can invest their time and effort towards more important things that communicate with students more meaningfully (Ajit et al., 2020). First and foremost, the incorporation of this technology in the process of question-answer pair generation using Through Transformers can as well enhance the way student assessments and evaluation are being conducted in the educational institutions. Through using deep learning models, educators are able to provide questions and answers that are specifically suited to certain learning goals and learning needs, thus providing their students with more variants. This ability has been proven so far especially because it takes less time, and it has also been proven that it is a comprehensive, challenging, and reflective assessment that covers the knowledge contents of the curriculum. Beyond that, Transformers are incorporated in educational technology for the provision of tailored and customized learners' experiences. Interrogative examination on the particular whimsical and makeshift methodology on how students engross themselves with neural network generated tests can offer teachers utilitarian innovative knowledge on how individuals acquire latest information either strength or weakness.

2. Related Work

Question generation has become one of the more actively studied problems in natural language processing, with researchers taking quite different routes to address it. Chan et al. (2019) brought together a T5 transformer model with a concept hierarchy extraction model, using the transformer's language generation strengths while relying on the hierarchy component to keep generated questions grounded and topically relevant.

A different direction was taken by Nwafor et al. (2021), who leaned on older, well-established techniques—specifically TF-IDF and N-grams—to pull out significant words from text. These keywords were then filtered through POS tagging rules to identify sentences worth turning into questions, with answer keys removed in the final step. The approach is a good reminder that classic information retrieval methods still have real utility even when more sophisticated tools are available. Raynaud et al. (2018) took a somewhat different angle, building a question-answering system that adapts to what a user already knows, drawing inspiration from trivia games to progressively adjust the difficulty and selection of questions based on demonstrated knowledge.

Nguyen et al. (2022) combined a Naive Bayes-based key phrase extractor with a noun filter, then ran sentences through Stanford CoreNLP's dependency parser to break them down structurally before generating questions from the extracted phrases. The combination of a probabilistic classifier with syntactic parsing gave the system a reasonable degree of precision in identifying what a question should actually target.

Srivastava et al. (2020) took advantage of ProphetNet, a pre-trained model built specifically for question generation, pairing it with BERT on the answer generation side. Relying on pre-trained architectures like this cuts down on the need for task-specific training data and generally produces more fluent outputs.

Blšták et al. (2022) went a different route entirely, building on Seq2Seq architecture with two LSTM-based networks—one handling encoding of the source paragraph, the other handling question generation. They also folded in reinforcement learning through policy gradients, which helped stabilize outputs that sequence models can sometimes struggle to produce consistently.

Zou et al. (2022) made use of InferSent, a sentence embedding tool from Facebook Research, to get at the semantic content of passages before generating questions. By grounding generation in sentence-level meaning rather than surface features, the approach aimed to produce questions that were more contextually aware.

Measuring the quality of generated questions has received considerable attention as well. Chan et al. (2019) and Mannem et al. (2010) used a mix of custom scoring, GPT-3-based automated ratings, and human review, while BLEU and ROUGE scores were applied to compare machine-generated questions against human-written ones. Levenshtein distance and logistic regression-based ranking helped evaluate whether generated answers were actually correct and relevant (Mannem et al., 2010; Zou et al., 2022).

For training and testing, researchers have drawn from a range of datasets including MOOCCubeX, CuratedTREC, WebQuestions, Freebase, and SQuAD, each bringing different domains and question types into the picture.

Finally, Aldabe et al. (2006) developed ArikIturri, an automatic question generator designed for Basque language learners. The system draws on a sentence bank that has been pre-processed morphologically and syntactically to construct questions that genuinely challenge users' understanding of the language. The system recognizes the phrase segments inside the sentence and uses a sentence model to form complex questions, which are based on this segment. In this model, possibility of the addresses with different focus is provided; as well as it takes into account the feature of verb-object word order in Basque language.

Soni et al. (2019) written comparative study for different question answer generation approaches. That includes matching the answer with the questions that are generated automatically, fill in the blanks generation in the questions, question answering generation using templates, generating questions from the passage, multiple choice question generation and generating multimedia questions. The author also touched on more recent developments in the field, particularly how deep learning has been applied to the problem of automatically generating question-answer pairs. Pulling everything together, question generation is far from a one-size-fits-all problem. Researchers have drawn from a wide range of tools and methods—from traditional retrieval techniques to machine learning, neural architectures, and semantic modeling—each bringing something different to the table depending on the context. Getting the questions right matters too, which is why so much effort goes into evaluation, whether through automated metrics, human review, or both, tested across varied datasets to see how well these systems actually hold up in practice.

3. Proposed Architecture

The Stanford Question Answering Dataset (SQuAD) is most commonly used in natural language processing projects. In most cases, it is used on question answering tasks, considering its efficiency and appropriateness. We use the BERT-based encoder-decoder model here. It is a 10-layer encoder with several features that support the decoder module. Primarily, the pretraining of this model uses the transformer-like

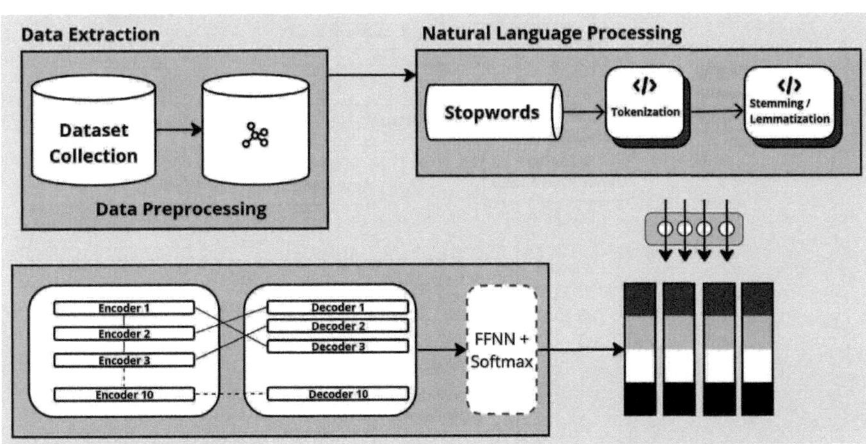

Figure 1: Pre – processing: Pre – processed text is converted to word embeddings and fed into a 10 – layer BERT model for NLP tasks, followed by a final FFNN layer.

structure of BERT's architecture to extract context-sensitive word representations for the input text. This process involves tasks such as BCML (bidirectional contextual language modelling) and nsp (next sentence prediction), completing the transition from encoding to understanding the text from bidirectional perspective. Upon pretraining, the model is fine-tuned in such a manner that it renders characteristics needed for question generation from an article text and answers questions based on the context. Precision is about modifying the parameters of the pre-trained model so that its capabilities are focused on providing precise responses to related questions that will further improve its level of performance. The two models have finally been fine-tuned, one for question generation and the other for question answering, to give place for a Q-A model that is unified as we will see. The first model provides the framework for a machine to pass a certain text, replacing it with questions and answers that are in line. The output is consequently displayed in the JSON format thus making it simpler to handle and understand in different applications. Through merging the transformer-based models characteristics (BERT) with specific task fine -tuning, this method strives to output high quality question–answer pairs efficiently in a manageable structure, therefore, the automated content creation is simplified and learners will experience their learning in a good way. Figure 1 represents the initial data processing steps and the natural language processing steps undertaken before the Model is trained. Figure 2 shows the overview of the BERT 10 – layer architecture and how the model is finetuned for the purpose of Question Answer pairs generation.

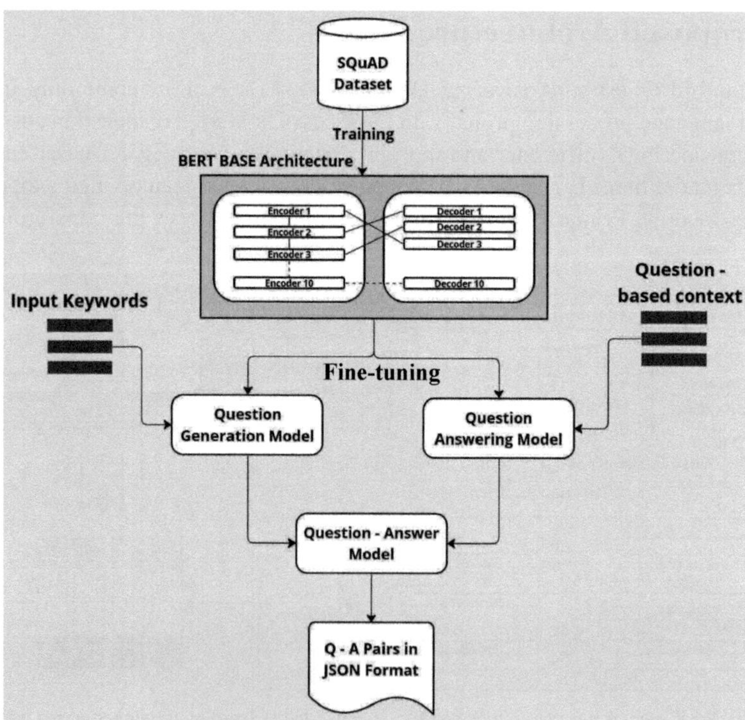

Figure 2: System Architecture.

A. Brief Overview of the Bert Framework

The model BERT is made up of a stack of bi-directional Transformer encoders with multilayer form. BERT aids the work carried forward by other previous researches of pre-trained contextual models such as Semi-supervised Sequence Learning, Generative Pre-Training, ELMo and ULMFit. However, the novelty is that it is the first one completely bidirectional and unsupervised language representation pre-trained using only the plain text corpus. The BERT has three parameters: the number of layers to the hidden layer size and the number of self-attention in the multi-headed attention layer. As such, two main models of BERT namely $BERT_{small}$ arise and $BERT_{large}$ arise which exploit different combinations of the BERT parameters.

The input data needs to be adjusted to match the BERT model's input structure to utilize it. Usually, the very first token [CLS] for the sequence input represents the input word. [CLS] Token's final hidden state is expected to represent this sentence as a whole for classification tasks. An input token sequence can be made up of a few lines among others. A special [SEP] token is put in-between the tokens of two consecutive sentences for the purpose of delimitation of the information from other sentences. There is a learned embedding attached to every token to study whether it is related to sentence A or sentence B.

B. Self – Attention Layer

BERT consists of the self-attention layer, which is a central part of it. It permits BERT to assign weights to separate parts of the input sequence and capture all the information assigned to each token in the sequence in the final output representation. In this layer, the input is modeled as the sequence of vectors and the importance index, or the attention weight, of every token in the input is calculated. These attention weights physically display the importance of every token to each other token's output representation. The procedure for calculating self-attention is more or less the following. The starting point is creation of three vectors (Query, Key and Value) from the encoder input vectors of each input. Sink vectors are smaller in scale than the embedding regarding the dimension vector. This creates the vectors after which a score is determined. The score represents the level of relevancy of each token within the input sequence to the other tokens that belong to the same sequence. Computed as a dot product of query and key vectors, which is then scaled to ensure the magnitude of the scores gets reduced to normal value levels. Following this, all the value vectors become the products of the softmax score and the summation of the weighted value vectors is taking place. This represents the result after applying this self-attention layer.

This schematic Figure 3 shows calculating attention weights in a self-attention layer. Query (Q), key (K), and value (V) vectors are used to compute compatibility scores, then normalized by softmax to determine the relevance of each element in the sequence.

C. Multi – Headed Attention

Multi-head attention, in the transformer model, makes it possible for the model to jointly attend to the information from multiple different subspaces of representation

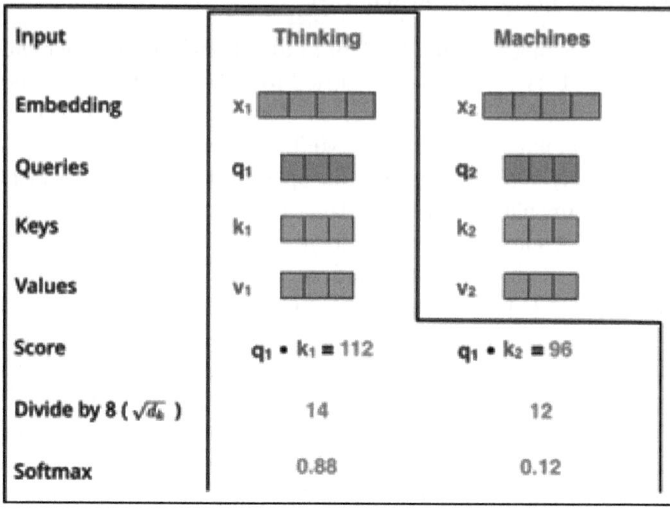

Figure 3: Self-Attention: Softmax for Attention Weights.

levels at different positions. An input is represented in several manners, using various linear projects, which eventually result in specific query, key and value vectors for each attention "head." Through parallel is performed self-attention computation for every head, where positions of the input are compared to each other in order to compute a weighted sum of values, whose weights depend on the similarity between queries and keys. The results of all heads of attention are then combined in a single layer and another feed forward network with linear projection is used to obtain the final output for that layer. It performs the various types of connections in different subspaces for the input sequence combinations or superposition, which are then submitted to a more complex training, creating a richer representation and parallelism.

4. BERT for Question Answering

The structure of the dataset would be similar as for the one mentioned above, where within an example there is a paragraph which serves as a context, a question and an answer constructed by special tokens [CLS], [SEP] and [SEP] respectively. Meanwhile, the SoftMax calculation also matches with that of the question generation model in the said model. The BERT which accepts the tokens in the context paragraph as input, then outputs the probability distribution over the tokens, while each token has its probability of being the start index of the answer. In the answer generation model, softmax logits have been applied to stand for the probability of every token regularly appearing in the context paragraph covering the whole vocabulary as the starting index of the answer. Question selection technique uses a question as an input to generate a token sequence that represents the produced question(s). Thus, it is the opposite of what question generation model produces. The bot will be spitting out two numbers which denote the indices of the sentence that identify where the answer is located in the passage. These indices establish the specific range within those text sentences which could probably relate to the best paragraph and provide a reasonable context. As an illustration, if the passage contains the sentence “The quick brown fox jumps over the lazy dog” and the question is “What does the fox jump over?”, the machine may choose the start and end index as 16 and 27, which are of the text “the lazy dog”. Therefore, we pick the lazy dog (this is the text that is the best one to give the answer to the question in the paragraph) as the answer.

5. BERT for Question Generation

To train the BERT model for question generation the given Paragraph and an answer phase , the input sequence X is given by:

$$X = ([CLS],\ \ C,\ \ [SEP],\ \ A,\ \ [SEP])$$

For the *BERT()* model that has been developed, the hidden states are X where |X| denotes the length of the input sequence and h denotes the number of neurons in the hidden layer. These hidden states represented by H are passed through a dense layer, which is given by, $H = BERT(X)$, where |X| denotes the length of the input sequence, and h denotes the size of the hidden layer. These hidden states represented by H are passed through a dense layer, which is given by: $W \in R^{h \times |V|}$. The equation is used to compute the conditional probability of a token given its corresponding input token in the input sequence X:

$$Pr(w|x_i) = softmax(H \cdot W + b), \forall x_i \in X$$

$$\hat{q}_i = argmax_w\ Pr(w|x_i)$$

$\hat{q}_i$ gives the predicted token for each input token with the highest conditional probability given the input token x_i.

Figure 4 shows a detailed overview of how the model is fine – tuned for the purpose of question generation.

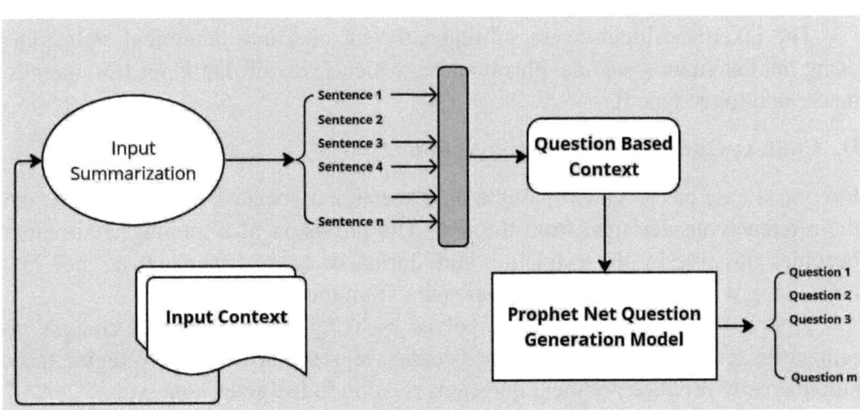

Figure 4: Question Generation Model.

6. Data Pre-Processing

A sundry of preprocessing techniques were taken under consideration to denoise, an essential addition of one-hot encodings and cleansing the SQuAD dataset so that it could be employed for training a question-answering model. The relevant and specific data preprocessing pipeline used is represented in this segment.

A. Data Loading and Cleaning

The SQuAD dataset was thoroughly inserted into memory with the Hugging Face datasets library which can be touted as an lucid-to-use tool for engaging with multifarious natural language processing datasets. The dataset is agglomerated of JSON files carrying the question-paragraph-answer format.

B. Data Cleaning

The data was cleansed to augment the existing quality of and maintain homogeneity of the dataset by redressing grievance of any formatting issues involving unique characters and typographical pecadilloes. This was accomplished through utilising the in-built functionalities of the dataset library created for data processing and cleansing (Gumaste et al., 2020).

C. Tokenization and Encoding

The tokenizing of the paragraphs and questions in the dataset was carried out by employing and engaging the BERT (Bidirectional Encoder Representations from Transformers) tokenizer from the Hugging Face transformers library. This tokenizer breaks the text into single word pieces, sub-words also included, and prepares them to be given as numerical features of the model.

According to the standard protocol for BERT-based models, we applied tokenization and encoding as well because our BERT training model was a twelve-layer encoder pre-trained with BERT. The [SEP] (separator) token and the [CLS] (classification) token were specially inserted at the end and the beginning of the input token sequence. In addition, padding was included to meet the specified length of input sequences.

The tokenized inputs were subsequently encoded into numerical embeddings using the tokenizer's encode_plus method, which gives off the input IDs, attention masks and token type IDs.

D. Context-Question-Answer Extraction

Every test case has a context paragraph, a question associated with the context, and the reference answer span from the text. The provision of a training environment becomes possible by the extraction and storing of context passages as individual units along with questions and answer spans from the dataset.

Figure 4 shows the process involves breaking down the input context into sentences, generating a question-based context representation, and leveraging that to automatically produce pertinent questions relating to the given context.

E. Data Splitting

The dataset was pre-processed and was divided into sets with training, validation and test splits by means of built-in functions in the datasets library. The model was trained using the training set, optimized with the validation set during hyperparameter tuning and model selection, and evaluated using the test set as an assessment of the model performance.

F. Data Encoding and label encoding

The tokenized inputs, attention masks, and token type IDs were encoded into tensors suitable for input to the BERT-based question-answering model using the datasets library. Additionally, the start and end positions of the answer spans were encoded as labels for training the model.

7. Pre–Training BERT Model

The pretraining is accomplished with the BERT (Bidirectional Encoder Representations from Transformers) model with 10 layers of encoder that are given in the text. Generic language representation training is analogous to passing through a massive corpus of text data unsupervised. These learned representations can be further refined to serve as a solidification stride to downstream tasks such as question answering in order to fully exploit the linguistic knowledge.

Transformer encoder layers are vertically placed on top of each other to form the architecture of the BERT Models. In this case, each encoder element comprises the feedforward neural network with position-wise features and self-attention mechanisms. A BERT model that has ten layers of encoder blocks was a setup used in the pretraining. Such a transformer model has this ability because of its architecture that recognizes and remembers complex linguistic links and patterns in the text.

A. Pre – Training Corpus

The pretraining task of the BERT models is to discover the representations of both the left and right directions of the text input sequence. This is achieved through two unsupervised learning tasks:

- **Masked Language Modelling (MLM):**

By randomly replacing a subset of tokens with special [MASK] tokens as shown, we train our model to make predictions about the original tokens across the input sequence based on the cues and the surrounding tokens.

- **Next Sentence Prediction (NSP):**

In NSP, the model is trained to predict that two input sentences that follow each other in the original text should be classified as consecutive or not. It is to model being able to learn relationships between sentences and grab inter-sentence data.

B. Training Procedure

The BERT model was pretrained using a large-scale distributed training setup with multiple GPUs or TPUs. The training process involved optimizing the model parameters using the Adam optimizer with a fixed learning rate schedule and batch size. The model was trained for a fixed number of epochs on the Wikitext corpus, with periodic evaluation to monitor training progress and prevent overfitting.

C. Training Performance

The model was trained for 10 epochs on GPU P100 and GPU T4 x 2 accelerators. The training time was approximately 40 minutes per epoch on GPU T4 and 1 hour and 40 minutes per epoch on GPU P100.

After 10 epochs of training, the model achieved a train loss of 5.9547 and a validation loss of 5.9507. These losses indicate the discrepancy between the model's predictions and the actual labels during training and validation, respectively.

8. Question Answering Model

The subsequent section presents the steps of selecting the pre-trained BERT model and refining it with the training set of SQuAD to create a model for generating answers to questions in the end-to-end question answering task.

The BERT was pre-trained and implemented from the SQuAD using TensorFlow. The BERT conversation model we used for fine-tuning was pre-trained from scratch with 10 layers of encoders on the Wikitext dataset which was mentioned in the previous section.

The SQuAD dataset was loaded by means of the functionality provided by the Hugging Face datasets library for data loading and preprocessing, hence, the laboriousness of data loading and preprocessing was automatically taken care of. The text and the context were looked as tokens by the BERT tokenizer, and the training features were prepared as in tutorial data format. The BERT model was equipped with the TensorFlow, but the learning rate was still set as 5e-5 and the Adam optimizer. The training process was improved by using mixed precision that allowed for a better efficient training and reduced memory requirement. And then we refined an epoch of the training dataset to a total of 10 epochs through the iterative process.

Subsequently, the model parameters were fine-tuned in such a way that both the training and evaluation datasets extracted from the SQuAD dataset were utilized. The training set was a well of promulgation for advanced algorithm parameters while monitoring and surveilling and precluding overfitting was the primary purpose of the validation set.

A. Training Performance

Throughout the fine-tuning process, the training loss and validation loss were stringently monitored and reviewed to evaluate the model's performance. Upon successful completion of 10 epochs of training, the model achieved a training loss of 1.4990 and a validation loss of 1.1426.

B. Evaluation

Proceeding from the fine-tuning process, the trained model underwent evaluation using standard fundamental metrics for question answering tasks, including F1 score and exact match score. These metrics evaluated the model's ability to engender accurate and immaculate answers to questions based on the given context paragraphs.

9. Question Generation Model

The aforementioned concoction of computational processes mentioned in this section agglomerates the ProphetNet tokenizer and model with the encoder Spacy model as the most crucial components for obtaining contextually precise and succinct questions from the input text. The ProphetNet model, which is a transformer-based language model built by Microsoft for the purpose of conditional text generation, represents a very effective tool that can be applied in this area. Unlike other language models whose main task is usually handling language modeling or text generation in general, ProphetNet is designed to generate text that is relatively conditioned to an input context. This factor makes the model especially good for generation questions, where the generated questions must be coherent and appropriate to the context given.

The training procedure of the question generation model refers for making a ProphetNet model's parameters optimal to enable the model to learn the internal structure and patterns which are involved in question making. This optimization may be achieved by exposing the model to a large corpus of training data which includes its question and context passages. Through a feedback mechanism that adjusts its weights and biases, the ProphetNet model itself adapts to match the input context representations to the appropriate questions.

The ProphetNet model is operated in parallel with the en_core_web_sm Spacy model, which the linguistic analysis in the input context. Spacy is a quickly growing natural language processing library that offers a number of modern text processing facilities. The en_core_web_sm model is one of the pre-trained pipelines which are pre-loaded with English text components such as name entity recognition, part-of-speech tagging, and dependency parsing in addition to tokenization.

Thereby, the en_core_web_sm Spacy model will be used to generate questions by extracting meaningful sections of the input context according to the presence of noun phrases. Noun phrases are often written to refer to identifiable things, ideas or objects, so the identification of the entities is key for creating adequate and relevant questions. This Spacy model is characterized by its ability to trace and analyze noun phrases that enhances the entire extraction process; hence, the selected sentences contain the words having the explicit information relevant to the subject. Subsequently, as the significant fragments of text are determined then the ProphetNet model will produce the questions for these sentences each. This model utilizes its learned patterns and contextual knowledge to frame information-seeking questions that focus on the key words that convey the intention of the meaningful sentences. It includes the design of the questions that not only fit the topic area, but also include the insertion of every detail and innuendo known to exist in the context.

The outcome of this question generation rule is the collection of questions, every of which is already appropriately matched with a specific relevant sentence extracted from the input. These questions function as a holistic figure or metaphor of the vital facts and notions involved in the original article, the purpose of which is to get additional insight and better glow in the content.

10. Integration Module

The integration module comprises very refined techniques that allow the communication and coordination of the question generation model and the answer generation model to happen in a distributed environment. A neural network such as a Generative Pre-trained Transformer 3 (GPT-3) serves as the primary source for this model. The questions are fed as input, and then the answers are generated, mapped to their appropriate questions. Next, the module outputs the JSON object to have an easier interpretation and application to other systems.

The module that integrates receives a question set from the model, which is generated by the model of question generation. These questions are in textual form where they do bear some relation to the given context and situation. The connect module sends each input question to the model for producing answers as a result using context, and the given question for each output. The output model of the question generator employs its parameters to produce contextual replies to input questions. Answers are generated instead of being inserted into the given input questions by the mapping module. Each answer relates to the question it is connected to and the pair question-answer is the result which can be achieved.

The question-answer pairs are created by the integration module and consequently fed into the JSON object. Each mapping is formatted as a keyword to the corresponding answer, where the question is the actual keyword, and the answer is its related answer. This JSON structure facilitates the efficient management of questions and answers hence easy access and retrieval of each data or unit.

This operation result can be further rendered or employed by the downstream applications for various reasons, including displaying the questions and answers in an interactivity friendly interface and or adding it to a larger system for the information retrieval process.

11. Results and Evaluations

The 10-layer BERT model's training time decreased to only 45 minutes, which shows how time-efficient it is. On other hand, the traditional BERT-large-uncased model took approximately one hour and forty minutes for training. The 10-layered BERT model obtained a shorter evaluation time, specifically 4 minutes, as it showed faster processing throughout inference. Contrastingly, the BERT-large-uncased benchmark model took somewhere around 6 minutes for the same evaluation.

The loss of model's training for the 10-layer BERT model was 1.4990, meaning that the difference between the model's predictions and actual labels was not coinciding during the training. The validation loss was 1.14, which indicates that it is able to perform well on the unseen data. However, during the evaluation, the BERT-large-uncased model obtained an F1 score of 93.15. The F1 score is a widely adopted performance score for question answering tasks, consisting of Precision and Recall measures.

The 10-layer BERT model showed training and evaluation times that were faster compared to the standard BERT-large-uncased model. Nevertheless, the latter absorbed a higher F1 score during the evaluation, which is a more precision and a

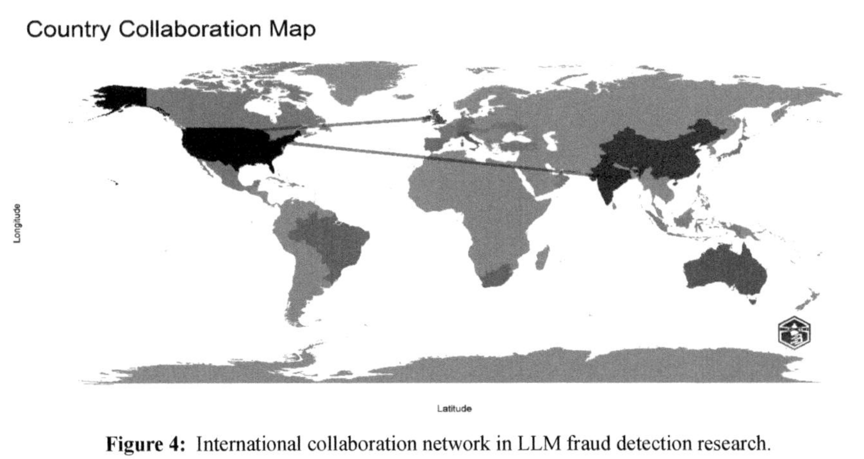

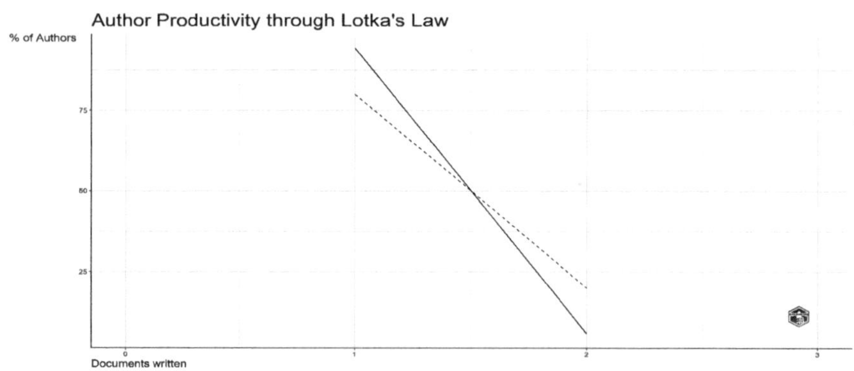

recall score for answer for the performance, although 10-layer BERT showed fast training and inference times, BERT-large-uncased model was found to have higher F1 score and thereby may perform better in the question answering task. The training time for the 10-layer BERT model will be around 45 minutes which is many times faster compared to the BERT-large-uncased model standard training time of 1 hour and 40 minutes.

This saving in training time relates to a smaller number of layers in the 10-layer, BERT model, that consequently has fewer parameters to optimize for the process of training. Nevertheless, we have to bring to the fore that the decrease in the number of layers granted by the model might hamper its ability to discover and correctly describe complex patterns and dependencies from the data. In the evaluation procedure, BERT's 10-layer model exhibited a more notable time savings where it took only 4 minutes and not 6 minutes for the standard BERT- Large-uncased model. This advantage offers quicker inference, affecting the applications which are time-sensitive and real-time. In these cases, prompt response is taken into account.

Nonetheless, we found that its efficiency benefits were decentralized by the training losing average of 1.4990 and the validation losing average of 1.14. The training loss is considered a difference between model predictions and labels in training time, while the validation loss is an estimation of a model performance without considering its prepossessing of data information. These negative error values therefore suggest the areas where there is a room for improvement by making the model more generalizable and in better possession of predictive ability.

Starting from the positive side, we have a standard version with a BERT-large-uncased model that got an amazing F1 score of 93.15 during evaluation. It is a common trust in the domain of question answering accuracy assessment. The F1 score is a combination of the precision-recall metrics of classifying all predictions made as the fraction of correct predictions (Precision), and the fraction of all actual positives that were correctly predicted (Recall). This can be interpreted as the model becoming highly correlated with human competence hence attaining the right proportions between the two dimensions and serving as a reliable tool for producing an accurate and complete answer.

Shifting the attention of the question generation model on, it shows some reasonably contributed questions, namely, WH questions (The questions start with "what", "where", "when", "why", or "how"). Such questions are especially important because they play a vital role in the process of making a speaker reveal concrete information and data points from a specific context. Nevertheless, the generation model ends up generating a low number of demanding inferences and thinking based questions. In this case, the model has to make assumptions based on what it knows from the context and be able to make inferences that are valid and logical with the limited amount of information that is available. In the case of question-answer questions, it, instead, investigates the model's ability to analyze the information and integrate them, typically through multi-step reasoning. The model´s restricted ability to generate logical and reasoning questions, indicate the inherent complexity in these questions. Forming such questions is conditional on the person's profound insight into the context, the capacity to pick up the threads between different pieces of

Table 1: Results and Evaluations.

Metric	BERT-10 Model	BERT-large-uncased Model
Training Time	45 mins	1hr
Evaluation Time	4 mins	6 mins
Training Loss	1.499	-
Validation Loss	1.14	-
F1 Score	-	93.15

information and the core ability to subjectively reason about any implicit or unstated knowledge.

To enhance these four domains, it may be worth considering various approaches. An idea is to add training data with the understanding that the learning will be done by the model through the many examples of inference and reasoning type of questions. A different approach could be the integration of external resources and uncertainties in reasoning skills into the model. This will allow the model to place more general information and reasoning inputs when generating questions.

Overall, the 10-layer BERT algorithm could boast efficiency benefits because it was faster in training and evaluation. However, the standard BERT-large-uncased model showed its advantage through a high F1 score in the question answering task, which indicated that there is a trade-off between, on the one hand, efficiency and, on the other hand, performance. Regarding the WH question generation model, it is really impressive how it can work; however, it is just the first step to fundamentally improve it to the level in which it can generate complex inferential and reasoning questions, which is crucial in order to make students get deeper knowledge.

12. Conclusion

Summarizing the research, the investigation focused on the performance and abilities of distinct models that conduct query answering and question generation tasks. The distinctive story, which was discovered through the juxtaposition of the 10-layer BERT model and the standard BERT-large-uncased model, emerges into an technical epiphany striking an interesting balance between the computational efficiency and the overall quality of the result.

The 10-layer BERT model had seen the ratcheting up of agility and training time reduced to about 45 minutes and the evaluation time went down to just four minutes. These augmented capacity gains make the model appropriate to use in time-critical settings or when the available processing power is limited. The enhanced model performance is emblematic by its loss of 1.4990 in training and 1.14 in validation showcases that there exist opportunities for augmentation in the model generalization as well as efficaciousness in making the appropriate and contextually germane predictions.

Unlike this, the standard BERT-large-uncased model, which is the most expensive model among the other two models, demonstrated slightly longer training and evaluation times; it highly scored F1 of 93.15 during evaluation. The achieved

F1 scores shed light on the model's capacity of generating accurate and detailed answers without compromising the trade-off between recall and precision factors. Even though it is of high computational demand, the standard BERT-large-uncased model can be used for the applications where accuracy and thorough understanding is of higher priority and value.

Concerning the question generation model, the study observed that a high volume of WH questions like (what, where, when, why, and how) could be produced based on the inputs followed by a sequence of tokens. Particularly, these types of questions help to trigger a specific narrative of a text, that is characterized by detailed information. Furthermore, the model failed to produce questions of inferential and reasoning nature in significant numbers when modeling of such questions involves understanding of the context, ability to draw logical interrelations between various sources and capability to compile information from a range of data sources.

Another way forward could be to extend the research to the expansion of training datasets which contain various types of the questions that require inferential and shall-type reasoning. In all, providing a model with information from external sources or ability to use common sense can help it generate more engaged questions that may require multi-step reasoning and use of implied or unstated knowledge.

In summary, this study complements ongoing investigations in natural language processing by identifying performance mismatches between different models, and this uncovers new depths in which the problem of question generation can be thought of and serve as a basis for improvement of question generation systems. These outcomes can be useful for the next model design or better implementation in long or short-term projects that require one or more of the mentioned attributes (computational power, precision, inferential, or reasoning-based question generation).

13. Future work

The results and the conclusions derived in this report represent a starting point for further development and research of educational programs on the basis of natural language processing, with particular emphasis on question answering and question generation tasks.

To continue in this direction, research will be on trying out advanced techniques such as better model architectures with the aim to improve the answering performance as well as question generation models. Even though the 10-layer BERT analysis exhibited the computation felicity, its effectiveness was lower as compared to the standard BERT-large- uncased model. Considering more powerful model architectures or including methods like model distillation or knowledge distillation may be a solution for this gap being closed, so the construction of models that are not only efficient but also very effective is made possible.

In addition to that, the mixing of manners of speech input into question answering and generation models has great potential. These models can produce more relevant questions that have situational context by processing textual data combined with visual information including images or videos. Hence, these models will have a great use in tasks like multimedia analysis, visual question answering and multimodal information retrieval. It is thus imperative to continue the research to explore multiple languages and domains as a future direction. Although the presented study concerns English language alone, the models may be modified to comprehend other languages which might eventually extend their scope of applicability, and as for cross-lingual question answering and generation. Moreover, the exploitation of domain-specific sets of data, as well as the fine-tuning of models for wide fields, like medicine, finance, and scientific literature, can expand the sphere of practical applications that meet the needs of industries and specialities.

One of the most important research priorities apart from machine learning techniques is to develop the ability to generate inferential and reasoning-based questions. The new question generation model is shown to have some limitations. Besides, recognizing external knowledge inputs, using commonsense reasoning approach, or including multi-hop reasoning techniques can contribute to the model's capability in producing questions that necessitate coming to logical conclusions and combining information from different sources. By virtue of this advancement it could be used in educational scope, where it could provoke critical thinking and deep understanding.

Also, the further development of models able to be interpreted and explained could be the key direction in research question answering and generation. Through adoption of more interpretable and explainable models, researchers and practitioners could understand and leverage the decision-making processes and reasoning that are embedded in the models for generating the questions and answers. This exposure could not only be a trust factor but also help to run iterative model improvement which can lead to model debugging.

In addition to that, there is also a chance to create communication systems that can interactively handle question-answering and generation of text. Such systems would start the conversations and give user answers to born questions based on the previous sentence. Given this, such agents can improve user experience, make interaction more natural and contextual and hence, will be perfect for a wide range of applications like virtual assistants, tutors or conversational agents.

Furthermore, the incorporation of these systems into real-life implementation and the monitoring of the spectrum can create positive feedback that will lead to continuous improvement. Workshops and seminars with domain experts, educators, or public-private partnerships can enhance the process of scenario generation and the collection of real-world data and use cases, which are essential for the iterative and pragmatic improvement of the models.

References

Ajit, A., Acharya, K. and Samanta, A. (2020, February). A review of convolutional neural networks. In 2020 international conference on emerging trends in information technology and engineering (ic-ETITE) (pp. 1–5). IEEE.

Aldabe, I., De Lacalle, M. L., Maritxalar, M., Martinez, E. and Uria, L. (2006, June). Arikiturri: an automatic question generator based on corpora and nlp techniques. pp. 584–594. *In:* International Conference on Intelligent Tutoring Systems. Berlin, Heidelberg: Springer Berlin Heidelberg.

Bhirangi, R. and Bhoir, S. (2016). Automated question paper generation system. International Journal of Emerging Research in Management &Technology 5(4): 28–34.

Blšták, M. and Rozinajová, V. (2022). Automatic question generation based on sentence structure analysis using machine learning approach. Natural Language Engineering 28(4): 487–517.

Chan, Y. H. and Fan, Y. C. (2019). BERT for question generation. pp. 173–177. *In:* Proceedings of the 12th International conference on natural language generation.

Gumaste, P., Joshi, S., Khadpekar, S. and Mali, S. (2020). Automated Question Generator System Using NLP Libraries. International Research Journal of Engineering and Technology (IRJET) 7(6): 4568–4572.

Mannem, P., Prasad, R. and Joshi, A. (2010, June). Question generation from paragraphs at UPenn: QGSTEC system description. pp. 84–91. *In:* Proceedings of QG2010: The Third Workshop on Question Generation.

Morón, M., Scocozza, J., Chiruzzo, L. and Rosá, A. (2021, November). A tool for automatic question generation for teaching English to beginner students. pp. 1–5. *In:* 2021 40th International Conference of the Chilean Computer Science Society (SCCC). IEEE.

Nguyen, H. A., Bhat, S., Moore, S., Bier, N. and Stamper, J. (2022, September). Towards generalized methods for automatic question generation in educational domains. pp. 272–284. *In:* European conference on technology enhanced learning. Cham: Springer International Publishing.

Nwafor, C. A. and Onyenwe, I. E. (2021). An automated multiple-choice question generation using natural language processing techniques. arXiv preprint arXiv:2103.14757.

Raynaud, T., Subercaze, J. and Laforest, F. (2018, December). Thematic question generation over knowledge bases. pp. 1–8. *In:* 2018 IEEE/WIC/ACM International Conference on Web Intelligence (WI). IEEE.

Soni, S., Kumar, P. and Saha, A. (2019, March). Automatic question generation: A systematic review. In International Conference on Advances in Engineering Science Management & Technology (ICAESMT)-2019, Uttaranchal University, Dehradun, India.

Srivastava, A., Shinde, S., Patel, N., Despande, S., Dalvi, A. and Tripathi, S. (2020, February). Questionator-automated question generation using deep learning. pp. 1–5. *In:* 2020 International Conference on Emerging Trends in Information Technology and Engineering (ic-ETITE). IEEE.

Zou, B., Li, P., Pan, L. and Aw, A. (2022, July). Automatic true/false question generation for educational purposes. pp. 61–70. *In:* Proceedings of the 17th workshop on innovative use of NLP for building educational applications (BEA 2022).

Index

M

N

O

P

Q

R

S

T

V

W

For Product Safety Concerns and Information please contact our EU representative GPSR@taylorandfrancis.com Taylor & Francis Verlag GmbH, Kaufingerstraße 24, 80331 München, Germany

Batch number: 10725800

Printed by Printforce, the Netherlands